Candida albicans: A Major Fungal Pathogen of Humans

Candida albicans: A Major Fungal Pathogen of Humans

Editor

Jonathan Richardson

MDPI • Basel • Beijing • Wuhan • Barcelona • Belgrade • Manchester • Tokyo • Cluj • Tianjin

Editor
Jonathan Richardson
King's College London
UK

Editorial Office
MDPI
St. Alban-Anlage 66
4052 Basel, Switzerland

This is a reprint of articles from the Special Issue published online in the open access journal *Pathogens* (ISSN 2076-0817) (available at: https://www.mdpi.com/journal/pathogens/special_issues/Pathogen_Candida_albicans).

For citation purposes, cite each article independently as indicated on the article page online and as indicated below:

LastName, A.A.; LastName, B.B.; LastName, C.C. Article Title. *Journal Name* **Year**, *Volume Number*, Page Range.

ISBN 978-3-0365-4555-4 (Hbk)
ISBN 978-3-0365-4556-1 (PDF)

Contents

Preface to *"Candida albicans*: A Major Fungal Pathogen of Humans"

Despite the tremendous advances being achieved in our understanding of the molecular mechanisms that underpin host–pathogen interactions, *Candida albicans* continues to be the most prevalent of life-threatening fungal pathogens. Exquisitely fine-tuned environmental sensing combined with genetic plasticity and the ability to rapidly adapt to the multiple niches of the human body make *C. albicans* a major adversary in medical mycology.

This book aims to highlight the increasing global threat posed by *C. albicans* and raise awareness of ongoing research which aims to combat *C. albicans* infections. Review articles and primary research are combined to showcase the current state of play regarding morphological switching, nutrient sensing and metabolism, and the impact of *C. albicans* infections on neonates. Current research investigating antifungal host responses and strategies toward the development of antifungal agents is also presented.

This Special Issue is aimed at those interested in *Candida albicans* biology, host–pathogen interactions, and medical mycology. I would like to thank all of the authors who contributed to this Special Issue. Your dedication and expertise are appreciated.

Jonathan Richardson
Editor

MDPI

Editorial

Candida albicans: A Major Fungal Pathogen of Humans

Jonathan P. Richardson

Centre for Host-Microbiome Interactions, Faculty of Dentistry, Oral & Craniofacial Sciences, King's College London, London SE1 1UL, UK; jonathan.richardson@kcl.ac.uk

Citation: Richardson, J.P. *Candida albicans*: A Major Fungal Pathogen of Humans. *Pathogens* **2022**, *11*, 459. https://doi.org/10.3390/pathogens11040459

Received: 7 March 2022
Accepted: 9 April 2022
Published: 11 April 2022

Publisher's Note: MDPI stays neutral with regard to jurisdictional claims in published maps and institutional affiliations.

Fungal infections kill ~1.6 million people every year [1]. The fungal pathogen *Candida albicans* causes > 150 million mucosal infections and ~200,000 deaths per annum due to invasive and disseminated disease in susceptible populations. Economically, yearly healthcare costs for *Candida* infections in the USA are ~\$2 billion [2], with similar per capita costs in the European Union. *C. albicans* accounts for ~75% of all *Candida* infections and is an enormous global health burden, the severity of which continues to escalate.

This Special Issue, "*Candida albicans*: A Major Fungal Pathogen of Humans", combines review articles [3–7] and original research [8–12] to explore recent advances in our understanding of *C. albicans* morphological switching, nutrient acquisition and metabolism, invasive infection in neonates, antimicrobial host responses and the development of potential therapeutics.

Morphological plasticity is considered a central tenet of *C. albicans* pathogenicity. While numerous forms of *C. albicans* are known to exist including chlamydospores, gray cells, GUT (gastrointestinal induced transition) cells, and white/opaque cells, the reversible yeast-to-hypha transition is widely regarded as a crucial weapon in the *C. albicans* arsenal. The intracellular signaling networks that activate morphological switching and the maintenance of sustained hyphal growth in response to diverse environmental cues are explored by Chow et al. [3]. The importance of chromatin-mediated epigenetic regulation in *C. albicans* morphological switching and biofilm formation is highlighted by Iracane et al. [4], in which an overview of *C. albicans* chromatin structure, histone modification, chromatin remodeling, and the influence of non-coding transcription and non-coding RNAs is presented.

C. albicans must assimilate nutrients acquired from a hostile host environment in order to thrive and persist; a process which occurs in the face of an often competitive microbiota. The versatility of amino acids as a nutrient source is addressed by Silao et al. [5], in which nutrient sensing, amino acid uptake and metabolism are explored with a particular emphasis on proline catabolism. Cellular metabolism is an important component of host–microbe interplay during commensal colonization and infection. In their review, Pellon et al. [6] discuss the metabolic flexibility of *C. albicans* in the context of commensalism and virulence, and the role of host metabolism in the control of innate immune responses to fungal infection.

Invasive fungal disease is a major cause of infection-related death among critically ill newborns. The risk factors associated with the development of invasive *Candida* infections following major surgery in neonates are investigated by De Rose et al. [7]. The authors also discuss fungal colonization of preterm infants, innate neonatal defence, and explore the epidemiology of fungal infection in neonatal intensive care units and the diagnosis and financial burden of invasive *Candida* infections together with prophylaxis.

The production of antimicrobial proteins is a key feature of host defence against active fungal infection. Research by Dishman et al. [8] provides an intriguing glimpse into the mechanistic action of "metamorphic proteins"; molecules that can reversibly switch between alternative structural conformations. One such protein is XCL1; a human chemokine capable of killing *Escherichia coli* and *C. albicans*. By locking XCL1 into distinct three-dimensional structures, Dishman et al. demonstrate that different conformations of

XCL1 kill *C. albicans* in vitro via two different mechanisms. Such intriguing findings may one day inform on the design of future therapeutics.

The unacceptably high mortality rate associated with invasive fungal infections is a reminder of the current shortcomings in antifungal therapy. A limited number of antifungal drugs combined with ever-increasing levels of antifungal resistance highlights the desperate need for new avenues of therapeutic intervention. In their article, Faria et al. [9] explore the antifungal activity and in vivo toxicity of a 1,3,4-oxadiazole derivative (LMM6). Application of LMM6 to *C. albicans* in vitro revealed encouraging antifungal and anti-biofilm activity, and reduced fungal burdens (kidney, spleen) in a murine model of systemic candidiasis.

The torsional stress experienced by DNA during replication and transcription is relieved by topoisomerase activity. Fungal DNA topoisomerases are highlighted as a potential antifungal target by Gabriel et al. [10], who demonstrate that capridine-β accumulates in *C. albicans* cells where it undergoes subsequent biotransformation into an inhibitor of fungal topoisomerase II activity in a strain-dependent manner.

Sphingolipids have recently garnered attention as potential targets for antifungal therapy due to their central role in fungal growth, morphogenesis, and virulence. Aureobasidin A and myriocin inhibit inositol phosphorylceramide synthase and glucosylceramide synthase, respectively, which are enzymes required for sphingolipid synthesis in fungi. In their article, Rollin-Pinheiro et al. [11] evaluate the antifungal activity of Aureobasidin A and myriocin against type strains and fluconazole-resistant clinical isolates of *C. albicans* and *C. glabrata*. Both compounds displayed encouraging antifungal activity in vitro and functioned synergistically with fluconazole, highlighting potential routes toward combinatorial therapy.

Attachment of *C. albicans* to epithelial cells is a prerequisite for commensal colonization and pathogenic infiltration of mucosal barriers [13]. Chemical strategies to reduce physical interaction between pathogenic fungi and host cells are yielding encouraging results. In their article, Martin et al. [12] describe the synthesis of a multivalent glycoconjugate in which an inhibitor of *C. albicans* adhesion is chemically coupled to a linear peptoid scaffold. Fungal adherence to buccal epithelial cells was reduced in vitro following treatment with the glycoconjugate formulation, and investigations to elucidate the precise mechanism of action are currently ongoing.

Despite significant advances in our understanding of *C. albicans* biology and immunopathology, selective pressures within the host environment continue to mould *C. albicans* into an ever-more formidable foe. Improvements in the breadth of the antifungal armoury are required if we are to overcome the challenges associated with increasing levels of resistance. Continued research aims to pave the way towards more positive patient outcomes.

I thank all of the authors that contributed to this Special Issue of *Pathogens*. Your time and expertise are appreciated.

Funding: This work was funded by an MSCA Individual Fellowship from the EC European Commission (VacCan: 101027512).

Conflicts of Interest: The authors declare no conflict of interest.

References

1. Brown, G.D.; Denning, D.W.; Gow, N.A.R.; Levitz, S.M.; Netea, M.G.; White, T.C. Hidden Killers: Human Fungal Infections. *Sci. Transl. Med.* **2012**, *4*, 165rv13. [CrossRef] [PubMed]
2. Benedict, K.; Jackson, B.R.; Chiller, T.; Beer, K.D. Estimation of Direct Healthcare Costs of Fungal Diseases in the United States. *Clin. Infect. Dis.* **2019**, *68*, 1791–1797. [CrossRef] [PubMed]
3. Chow, E.; Pang, L.; Wang, Y. From Jekyll to Hyde: The Yeast–Hyphal Transition of *Candida albicans*. *Pathogens* **2021**, *10*, 859. [CrossRef] [PubMed]
4. Iracane, E.; Vega-Estévez, S.; Buscaino, A. On and Off: Epigenetic Regulation of *C. albicans* Morphological Switches. *Pathogens* **2021**, *10*, 1463. [CrossRef] [PubMed]
5. Silao, F.G.S.; Ljungdahl, P.O. Amino Acid Sensing and Assimilation by the Fungal Pathogen *Candida albicans* in the Human Host. *Pathogens* **2021**, *11*, 5. [CrossRef] [PubMed]

6. Pellon, A.; Begum, N.; Nasab, S.D.S.; Harzandi, A.; Shoaie, S.; Moyes, D.L. Role of Cellular Metabolism during *Candida*-Host Interactions. *Pathogens* **2022**, *11*, 184. [CrossRef] [PubMed]
7. De Rose, D.; Santisi, A.; Ronchetti, M.; Martini, L.; Serafini, L.; Betta, P.; Maino, M.; Cavigioli, F.; Cocchi, I.; Pugni, L.; et al. Invasive *Candida* Infections in Neonates after Major Surgery: Current Evidence and New Directions. *Pathogens* **2021**, *10*, 319. [CrossRef] [PubMed]
8. Dishman, A.; He, J.; Volkman, B.; Huppler, A. Metamorphic Protein Folding Encodes Multiple Anti-*Candida* Mechanisms in XCL1. *Pathogens* **2021**, *10*, 762. [CrossRef] [PubMed]
9. Faria, D.; Melo, R.; Arita, G.; Sakita, K.; Rodrigues-Vendramini, F.; Capoci, I.; Becker, T.; Bonfim-Mendonça, P.; Felipe, M.; Svidzinski, T.; et al. Fungicidal Activity of a Safe 1,3,4-Oxadiazole Derivative Against *Candida albicans*. *Pathogens* **2021**, *10*, 314. [CrossRef] [PubMed]
10. Gabriel, I.; Rząd, K.; Paluszkiewicz, E.; Kozłowska-Tylingo, K. Antifungal Activity of Capridine β as a Consequence of Its Biotransformation into Metabolite Affecting Yeast Topoisomerase II Activity. *Pathogens* **2021**, *10*, 189. [CrossRef] [PubMed]
11. Rollin-Pinheiro, R.; Bayona-Pacheco, B.; Domingos, L.; Curvelo, J.d.R.; de Castro, G.; Barreto-Bergter, E.; Ferreira-Pereira, A. Sphingolipid Inhibitors as an Alternative to Treat Candidiasis Caused by Fluconazole-Resistant Strains. *Pathogens* **2021**, *10*, 856. [CrossRef] [PubMed]
12. Martin, H.; Masterson, H.; Kavanagh, K.; Velasco-Torrijos, T. The Synthesis and Evaluation of Multivalent Glycopeptoids as Inhibitors of the Adhesion of *Candida albicans*. *Pathogens* **2021**, *10*, 572. [CrossRef] [PubMed]
13. Nikou, S.-A.; Kichik, N.; Brown, R.; Ponde, N.O.; Ho, J.; Naglik, J.R.; Richardson, J.P. *Candida albicans* Interactions with Mucosal Surfaces during Health and Disease. *Pathogens* **2019**, *8*, 53. [CrossRef] [PubMed]

Review

From Jekyll to Hyde: The Yeast–Hyphal Transition of *Candida albicans*

Eve Wai Ling Chow [1], Li Mei Pang [2] and Yue Wang [1,3,*]

1   Institute of Molecular and Cell Biology (IMCB), Agency for Science, Technology and Research (A*STAR), 61 Biopolis Drive, Proteos, Singapore 138673, Singapore; wlechow@imcb.a-star.edu.sg
2   National Dental Centre Singapore, National Dental Research Institute Singapore (NDRIS), 5 Second Hospital Ave, Singapore 168938, Singapore; pang.li.mei@ndcs.com.sg
3   Department of Biochemistry, Yong Loo Lin School of Medicine, National University of Singapore, 10 Medical Drive, Singapore 117597, Singapore
*   Correspondence: mcbwangy@imcb.a-star.edu.sg

Abstract: *Candida albicans* is a major fungal pathogen of humans, accounting for 15% of nosocomial infections with an estimated attributable mortality of 47%. *C. albicans* is usually a benign member of the human microbiome in healthy people. Under constant exposure to highly dynamic environmental cues in diverse host niches, *C. albicans* has successfully evolved to adapt to both commensal and pathogenic lifestyles. The ability of *C. albicans* to undergo a reversible morphological transition from yeast to filamentous forms is a well-established virulent trait. Over the past few decades, a significant amount of research has been carried out to understand the underlying regulatory mechanisms, signaling pathways, and transcription factors that govern the *C. albicans* yeast-to-hyphal transition. This review will summarize our current understanding of well-elucidated signal transduction pathways that activate *C. albicans* hyphal morphogenesis in response to various environmental cues and the cell cycle machinery involved in the subsequent regulation and maintenance of hyphal morphogenesis.

Keywords: polymorphism; hyphal morphogenesis; hyphal activation; signal transduction pathways; cell cycle regulation

Citation: Chow, E.W.L.; Pang, L.M.; Wang, Y. From Jekyll to Hyde: The Yeast–Hyphal Transition of *Candida albicans*. *Pathogens* **2021**, *10*, 859. https://doi.org/10.3390/pathogens10070859

Academic Editor: Donato Gerin

Received: 14 June 2021
Accepted: 5 July 2021
Published: 7 July 2021

Publisher's Note: MDPI stays neutral with regard to jurisdictional claims in published maps and institutional affiliations.

1. Introduction

Candida albicans is a commensal fungus that is usually a benign member of the microflora in the gastrointestinal tract, genitourinary tract, mouth, and skin of most healthy individuals [1–4]. *C. albicans* is also an opportunistic fungal pathogen responsible for infections ranging from mild superficial infections to life-threatening candidemia [5]. The use of modern medical therapies such as broad-spectrum antibiotics, cancer chemotherapy, and solid organ transplant has led to an increase in the population vulnerable to invasive candidiasis [6,7]. *C. albicans* is a leading cause of hospital-acquired infections; in the intensive care unit (ICU), candidemia may represent up to 15% of nosocomial infections with an estimated attributable mortality of 47% [7–11].

C. albicans displays a wide range of virulence factors and fitness attributes, including its capacity for rapid evolution of resistance to commonly used antifungals (e.g., azoles, polyenes, and echinocandins) and its ability to form biofilms on medical devices, contributing to its success as a pathogen. One striking feature that allows *C. albicans* to cross the commensal-to-pathogen boundary is its ability to switch reversibly between two morphological forms, namely unicellular budding yeast, or filamentous form (hyphae and pseudohyphae), in response to various environmental cues that reflect the host environment [12–19].

Yeast, hyphal, and pseudohyphal forms of *C. albicans* are all present in tissues of human patients and animals with systemic invasive candidiasis [20,21]. Yeast cells exhibit a round-to-oval cell morphology that arises from budding and nuclear division [22]. In contrast, hyphae consist of tubular cells that remain firmly attached following cytokinesis without a constriction at the site of separation. Pseudohyphae share features resembling both yeasts and hyphae, which are branched chains of elongated yeast cells with constrictions at the septum. Both yeast and hyphal forms have crucial and complementary roles important for infection [23]. For instance, the yeast form is required for adhesion to endothelial cells and dissemination into the bloodstream, while the hyphal form is required for tissue penetration during the early stages of infection and yielding resistance towards phagocytosis [24–29]. Hyphae-specific virulence factors such as adhesins (Hwp1, Als3, Als10, Fav2, and Pga55), host tissue degrading proteases (Sap4, Sap5, and Sap6), and cytolytic peptide toxin (Ece1), aggrandize the host cell damage during infection [22,30].

Although *C. albicans* can undergo an array of morphological transitions such as the formation of chlamydospores, gray cells, GUT (gastrointestinally induced transition) phenotype, and white/opaque cells, the yeast-to-hyphae transition appears to be the most critical virulence trait [12,15]. Mutants locked in either the hyphal or yeast form have shown diminished virulence, suggesting that the ability to switch between the two morphological forms is essential for virulence [31,32]. Recent advances in mechanistic studies have provided insights into the morphological regulation, coordination, and interplay between environmental factors and genes associated with pathogenesis. This review provides an update on the signal transduction pathways involved in activating *C. albicans* hyphal morphogenesis and how the cell cycle progression and its machinery further aid the regulation and maintenance of sustained hyphal growth.

2. Environmental Cues Inducing the Yeast-to-Hyphae Transition

C. albicans has adapted to growth in the human host and can transit from yeast to hyphae under a diverse range of environmental cues, as shown in Figure 1 [16,17,19]. Depending upon the cues encountered, morphogenesis can be triggered via several pathways which activate different regulatory circuits.

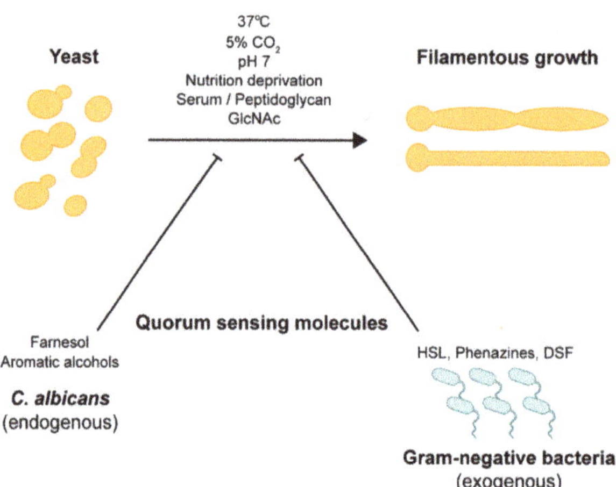

Figure 1. External hyphal-inducing signals. The yeast-to-hyphae transition in *C. albicans* can be triggered by various environmental cues such as high temperature (37 °C), high CO_2 concentration (~5%), pH 7, nutrition deprivation, serum, peptidoglycan, *N*-acetylglucosamine, and inhibited by quorum-sensing molecules from endogenous and exogenous sources.

2.1. Host Niches

The human host presents one of the most favorable environments for *C. albicans* morphogenesis due to the presence of multiple inducing factors such as elevated (body) temperature (37 °C), the presence of serum, elevated carbon dioxide (CO_2) levels (~5%), and low glucose content (0.1%) [33–36]. Hyphal initiation requires an increase in temperature to 37 °C and the release of quorum-sensing molecules (e.g., farnesol) for the temporary clearance of the major transcription repressor of hyphal morphogenesis Nrg1 [37]. Elevated temperature is also known to promote filamentation through the molecular chaperone Hsp90 and the transcription factor Sfl2 [38,39]. In combination with elevated temperature, the host serum is one of the strongest inducers of *C. albicans* hyphal morphogenesis [40]. Our group has previously demonstrated that bacterial peptidoglycans present in the host serum trigger the hyphal growth of *C. albicans* by directly activating the cyclic AMP (cAMP)-protein kinase A (PKA) signaling cascades through adenylyl cyclase Cyr1 [35,41]. Similarly, CO_2, another potent inducer of filamentous growth, is also known to activate the cAMP-PKA pathway by binding to Cyr1 [42]. In *C. albicans*, a carbonic anhydrase Nce3 is involved in CO_2 signaling and conversion of CO_2 to bicarbonate (HCO_3^-) [33]. Especially in host niches with limited CO_2 (e.g., on the skin), the CO_2/HCO_3^- equilibration controlled by Nce3 is crucial for the pathogenesis of *C. albicans*. The G-protein-coupled receptor Grp1 and the Gα protein Gpa2 act as the glucose-sensing network for *C. albicans* morphogenesis [43]. Low glucose concentration present in the bloodstream results in the maximal hyphal formation, while high glucose concentrations repress it [36]. The factors mentioned above have been shown to activate the fungal cAMP-PKA signaling pathway [33,44–46].

2.2. Hypoxia (Low Oxygen)

Hypoxia is a clinical characteristic of inflammatory conditions, representing zones of intense immune activity [47,48]. *C. albicans* can modulate the host response under hypoxia and anoxia (absence of oxygen) to evade immune responses [47]. As a commensal, *C. albicans* adapts to hypoxia condition by repressing the transcription factor of filamentous growth Efg1. Interestingly, Efg1 has a dual role in hyphal morphogenesis. Under hypoxia, it acts as a repressor at temperatures $\leq$ 35 °C, while under normoxia (normal oxygen level), Efg1 is a strong inducer of hyphal formation [48]. *efg1*Δ/Δ mutants displayed hyperfilamentous growth at temperatures $\leq$ 35 °C during hypoxic growth on agar surface or during embedment in agar but not during growth in liquid media [48,49]. In contrast, Ace2 is essential for hyphal morphogenesis under hypoxia while being dispensable under normoxia [50,51]. Efg1 and Ace2 share functional overlap; chromatin immunoprecipitation on microchips (ChIP) analyses revealed that hypoxic repressors (Efg1 and Bcr1) and hypoxic activators (Ace2 and Brg1) are connected in regulatory circuits in controlling hyphal morphogenesis under hypoxia conditions [48]. Additionally, Efg1 was implicated in the Cek1-mediated pathway under hypoxia at $\leq$35 °C; low Efg1 phosphorylation levels inhibit Cek1 and Cph1, preventing hyphal morphogenesis. The low Efg1 phosphorylation levels also inhibited hyphal morphogenesis through the cAMP-PKA pathway.

2.3. pH Conditions

C. albicans is constantly exposed to fluctuations in pH ranging from acidic to slightly alkaline in different human body niches such as the digestive tract, vagina, oral cavity, blood, and tissues [52]. pH sensing is mediated through Rim101, an important regulator of the yeast-to-hyphae morphological transition [53–55]. Upon activation, the transcription factor Rim101 enters the nucleus and mediates pH-dependent responses [56]. Remarkably, *C. albicans* is not only capable of sensing and adapting to environmental pH but can also modulate extracellular pH by alkalinizing its surrounding environment and auto-inducing hyphal formation [57]. Furthermore, alkalinization has been shown to counter the macrophage acidification during engulfment, promoting its survival in the macrophage [58].

2.4. N-Acetylglucosamine (GlcNAc)

GlcNAc is commonly found as a structural component of the mucosa of the gastrointestinal tract, bacterial cell wall peptidoglycan, and fungal cell wall chitin [59,60]. Given the ubiquitous nature of GlcNAc in host niches and microbial cells, it could potentially serve as a critical signaling molecule that regulates the switch between the commensalism and pathogenicity of *C. albicans* [61]. Figure 2 depicts an update on GlcNAc signaling pathways and their involvement in hyphal morphogenesis. Ngt1 was identified as a membrane transporter specific for GlcNAc, indicating the importance of GlcNAc in intracellular signaling [62]. However, metabolism or breakdown of GlcNAc intracellularly is not required in the *C. albicans* hyphal morphogenesis as triple deletion mutants that lack all three catabolic genes (*HXK1*, *NAG1*, and *DAC1*) can exhibit filamentous growth with the addition of exogenous GlcNAc [63]. Interestingly, genetic screens have revealed two novel transcription factors, *NGS1* and *RON1*, which play essential roles in both the GlcNAc catabolism and GlcNAc-induced filamentous growth [64]. *NGS1* encodes a protein similar to the GNAT family of histone acetyltransferase Gcn5, while *RON1* encodes a protein similar to the Ndt80-like DNA-binding domain [65]. Ngs1 was discovered as a novel GlcNAc signal sensor and transducer for GlcNAc-induced transcription in *C. albicans* [65]. Ngs1 targets the promoters of GlcNAc-inducible genes constitutively via the transcription factor Rep1 [65]. Ron1 was initially thought to act as both an activator and a repressor of hyphal morphogenesis. However, *ron1Δ/Δ* mutants constructed using the CRISPR/Cas9 method did not display observable GlcNAc-induced filamentous growth [64,66]. It is noteworthy that, upon the addition of GlcNAc, *ndt80 ron1* double deletion mutants could overcome the hyphal defects observed in *ndt80Δ/Δ* mutants. Collectively, it suggests that Ron1 functions as a repressor of filamentous growth in the absence of Ndt80 [66]. The GlcNAc signaling pathway was initially believed to be related to the cAMP-PKA pathway as *cyr1Δ/Δ* mutants cannot form hyphae under a broad range of conditions, including GlcNAc [67]. However, it was later discovered that the fast-growing *cyr1* pseudo revertant strains could undergo filamentous growth in a GlcNAc containing medium [68]. This indicates that GlcNAc can stimulate a signaling pathway independent of the cAMP-PKA pathway that has yet to be fully elucidated. An alternative pathway involved in the GlcNAc signaling is the pH-sensing Rim101 pathway. Production of excess ammonia during GlcNAc catabolism results in an increase in extracellular pH (>5), which indirectly stimulates the hyphal induction in *C. albicans* via the Rim101 pathway [60,69].

2.5. Amino Acids Sensing

C. albicans can utilize amino acids as alternative carbon sources during growth in glucose-poor, amino acid-rich conditions [57]. Amino acids that can be catabolized to arginine and proline are potent inducers of hyphal morphogenesis [70–72]. *C. albicans* detects extracellular amino acids via the plasma membrane-localized SPS (Ssy1-Ptr3-Ssy5) complex, which regulates two paralogous transcription factors, Stp1 and Stp2 (Figure 2) [73–75]. In the presence of extracellular amino acids, the amino acid sensor Ssy1 (Cys1) activates amino acid permease (AAP) genes [76], while the peripherally membrane-associated Ptr3 recruits casein kinase I (CKI), which activates the endoproteolytic activity of the endoprotease Ssy5 [76,77]. Ssy5 endoproteolytically cleaves the nuclear exclusion domain of Stp1 and Stp2, facilitating their translocation to the nucleus [74]. Processed Stp1 regulates the expression of *SAP2*, which encodes the major secreted aspartyl proteinase, and *OPT1*, which encodes an oligopeptide transporter. The active Stp2 activates the expression of a subset of AAP genes [74,75]. The endoplasmic reticulum (ER) chaperone protein Csh3 is required for the proper expression and plasma membrane localization of Ssy1 and AAPs [78]. *ssy1Δ/Δ*, *ptr3Δ/Δ*, *ssy5Δ/Δ*, *csh3Δ/Δ*, and *stp2Δ/Δ* mutants fail to respond to the presence of extracellular amino acids and display impaired filamentous growth [73–76,78]. Amino acid-induced morphogenesis has recently been shown to be dependent on proline catabolism, with a strict requirement for Ras1 activity [79]. Proline catabolism in the mitochondria leads to elevated cellular ATP levels, which exceed the critical threshold of ATP needed to induce cAMP synthesis, leading to hyphal morphogenesis [71,79,80].

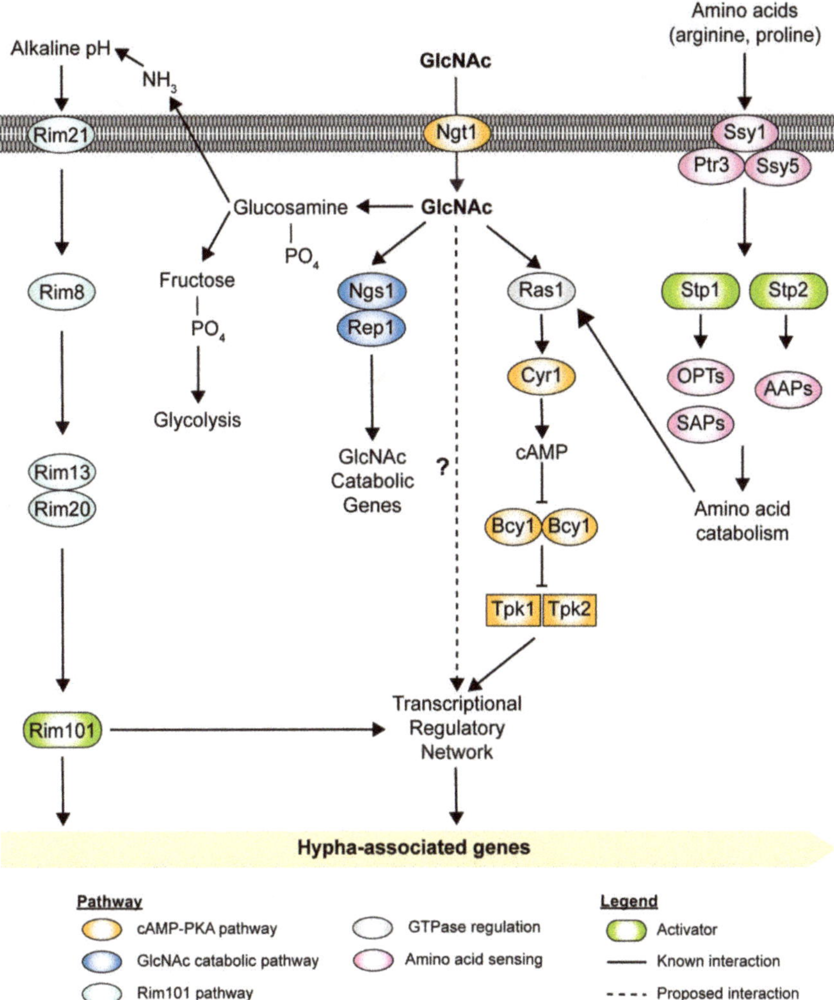

Figure 2. *N*-acetylglucosamine (GlcNAc) and amino acid-induced signal transduction pathways in *C. albicans*. Ngt1, localized in the plasma membrane, transports GlcNAc into the cell. However, when GlcNAc is present in high concentrations, it can enter the cell through diffusion. The main signal transduction pathway for GlcNAc-induced hyphal growth was initially thought to be the cAMP-PKA pathway. Recently, the transcription factors Ngs1 and Rep1, which are involved in GlcNAc catabolism, were found to stimulate hyphal growth via a cAMP-independent pathway. GlcNAc catabolism also increases the extracellular pH, which favors the hyphal growth via the alternate Rim101 pathway. Extracellular amino acids are detected via the plasma membrane-localized SPS (Ssy1-Ptr3-Ssy5) complex. The SPS-sensor activation leads to endoproteolytic processing at the nuclear exclusion domain of transcription factors Stp1 and Stp2. Processed Stp1 regulates the expression of secreted aspartyl proteinase (e.g., *SAP2*) and oligopeptide transporters (e.g., *OPT1* and *OPT3*), while processed Stp2 regulates the expression of amino acid permeases (APPs).

2.6. Quorum Sensing

In addition to host environmental cues, *C. albicans* morphogenesis is also regulated by several endogenous and exogenous quorum-sensing molecules (QSMs) [81–85]. Tyrosol and farnesol are well-known QSMs produced by *C. albicans*, which accelerate and inhibit the yeast-to-hyphae transition, respectively [81,86,87]. *C. albicans* also produces aliphatic alcohols (e.g., ethyl alcohol, isoamyl alcohol, 1-dodecanol, 2-dodecanol, and nerolidol) and aromatic alcohols (e.g., 2-phenylethyl alcohol and tryptophol) that inhibit filamentation and subsequent biofilm formation [88,89]. Farnesol and 1-dodecanol were implicated in the Ras1-cAMP signaling pathway, and hyphal defects can be restored upon the addition of dibutyryl-cAMP [89]. Farnesol also inhibits filamentous growth through the negative regulators Tup1 and Nrg1 [90]. The hyphal morphogenesis of *C. albicans* can also be regulated by interaction with other microorganisms found in the host environment [91,92]. For instance, the coexistence of *C. albicans* and Gram-negative bacteria, such as *Pseudomonas aeruginosa*, *Stenotrophomonas maltophilia*, and *Burkholderia cenocepacia*, is commonly found as mixed infections in the lungs of cystic fibrosis (CF) patients [93]. Exogenous QSMs, namely, 3-oxo-C12-homoserine lactone and phenazines (pyocyanin, phenazine methosulfate, and phenazine-1-carboxylate) secreted by *P. aeruginosa*, were found to inhibit the hyphal development of *C. albicans* [82,94]. Diffusible signal factor (DSF), representing a new class of widely conserved quorum-sensing signals from Gram-negative bacteria, has been implicated in inter-kingdom signaling between *C. albicans* and bacteria [95]. DSF (*cis*-11-methyl-2-dodecenoic acid) produced by *S. maltophilia*, and BDSF (*cis*-2-dodecenoic acid) produced by *B. cenocepacia* play a role in the yeast-to-hyphae transition [95]. DSF released by *S. maltophilia* has been reported to interfere with two key virulence factors of *C. albicans*: the yeast-to-hyphae transition and biofilm formation [96]. Recent microarray studies revealed the involvement of repressors (Ubi4 and Sfl1) and the activator (Sfl2) of filamentous growth in BDSF regulation of hyphal morphogenesis [97]. With the addition of BDSF, elevated levels of Ubi4 and Sfl1 and degradation of Sfl2 block the yeast-to-hyphae transition. *C. albicans* is also commonly found along with other microorganisms in inter-kingdom biofilms [98]. Many bacteria and fungi can secrete glucanases into the environment that digest glucan, the most abundant fungal cell wall component [99,100]. *C. albicans* itself secretes at least three glucanases (Xog1, Exg2, and Spr1) which are involved in cell wall remodeling during cell division and morphogenesis [101,102]. Interestingly, it has been found that β-1,3-glucanase, secreted by bacteria and fungi, can induce filamentous growth in *C. albicans* even at low temperatures (22 °C), in a cell density-dependent manner [103]. *cek1Δ/Δ* and *efg1Δ/Δ* mutants cannot form hyphae in response to β-1,3-glucanase, suggesting that the Cek1-mediated pathway is involved [103].

2.7. In Vitro Conditions

Hyphal growth can also be induced using synthetic growth media such as Lee's medium (pH 7), spider medium, and mammalian tissue culture M199 under laboratory conditions [104–106]. Nitrogen starvation-induced filamentation occurs in the low nitrogen SLAD medium via ammonium permease Mep2 sensing [107,108]. Methionine, an amino acid in Lee's medium, has been reported as the main inducer of yeast-to-hyphae transition via G-protein-coupled receptor Gpr1 sensing [36]. Recently, the methionine permease Mup1 and the S-adenosylmethionine decarboxylase Spe2 were discovered to be crucial for cAMP production in response to methionine [109]. Both nitrogen and amino acid catabolism activate hyphal morphogenesis via the cAMP-PKA pathway.

3. The cAMP-PKA Pathway

The cyclic adenosine monophosphate (cAMP)-protein kinase A (PKA) pathway is highly conserved in eukaryotes and regulates many cellular processes in *C. albicans* [44,110]. This pathway plays a critical role in morphogenesis, positively regulating filamentation [111–113]. One of the well-studied regulatory targets of the cAMP-PKA pathway is the transcription factor Efg1, which stimulates the expression of numerous hyphal-specific genes through the activation of the transcription factor Ume6 (Figure 3).

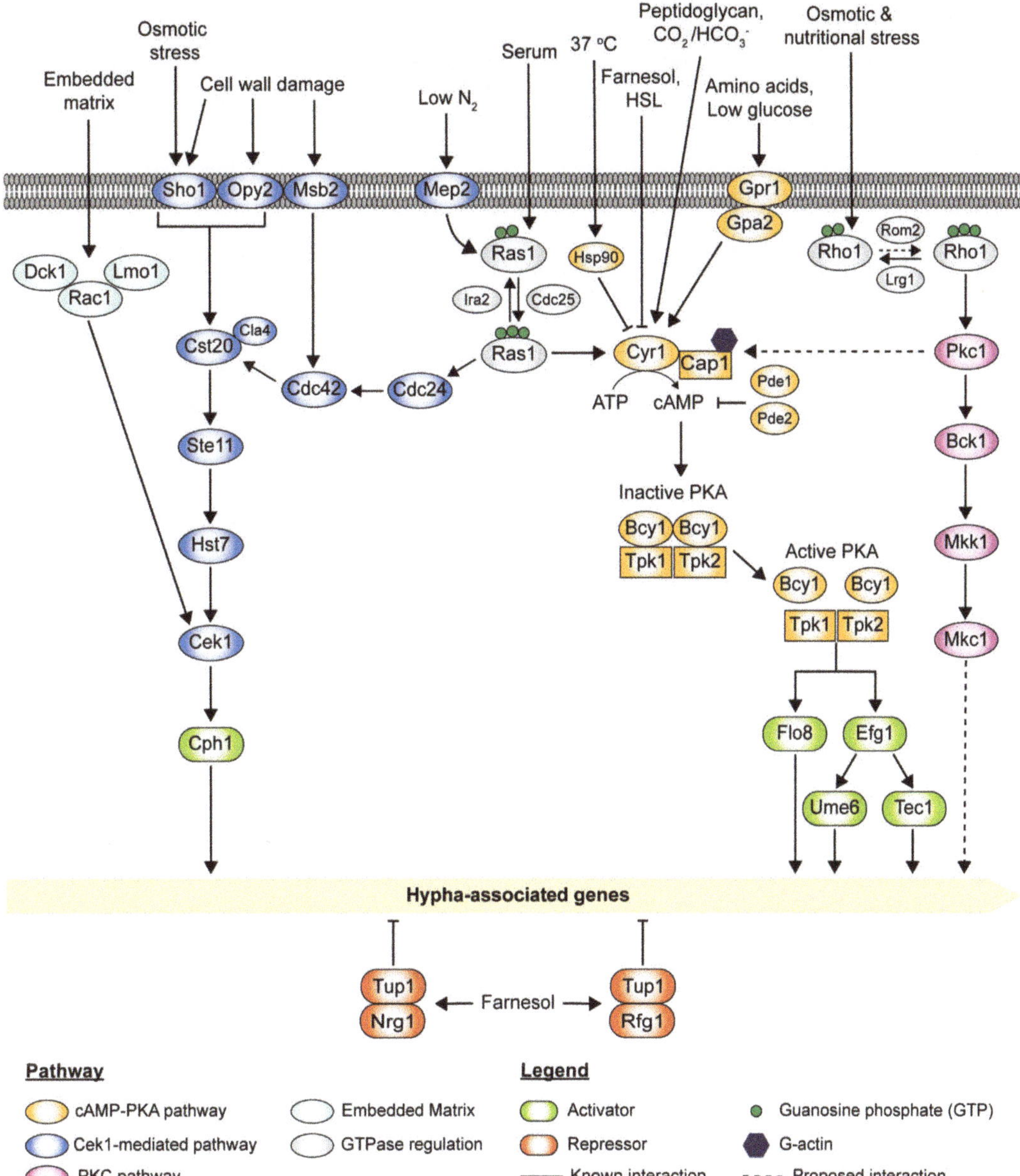

Figure 3. Signal transduction pathways that govern hyphal growth in *C. albicans*. Activation of filamentous growth in *C. albicans* by various environmental cues and signal transduction pathways; the cAMP-PKA pathway, the Cek1-mediated pathway, the PKC pathway, and the embedded matrix.

The cAMP-PKA pathway is triggered either directly through the adenylyl cyclase Cyr1 or via the small GTPase Ras1, which activates Cyr1, depending upon the stimuli encountered (Figure 3) [114]. Cyr1, in direct association with Cap1 (cyclase-associated protein), drives the conversion of ATP to cAMP [46]. PKA holoenzyme is activated upon cAMP binding to the homodimer regulatory subunit Bcy1, inducing a conformational change releasing the two catalytic subunits, Tpk1 and Tpk2, which then activate downstream target proteins or genes through phosphorylation or the binding of promoter regions to induce transcription (Figure 3) [110,115–117]. Tpk1 and Tpk2 are partially redundant. Tpk1 is required for hyphal formation on solid media, whereas Tpk2 is needed for hyphal formation in liquid media and invasive growth into solid media [116]. While the loss of either subunit does not block filamentation, loss of both Tpk1 and Tpk2 completely blocks filamentation [39,116]. cAMP levels are negatively regulated by Pde1 (low-affinity phosphodiesterase) and Pde2 (high-affinity cAMP phosphodiesterase), which increase the rate of cAMP degradation [118,119]. Loss-of-function mutations or deletion of *PDE2* increase cAMP levels, leading to constitutive activation of the pathway and hyperfilamentation [119,120]. The *pde2*Δ/Δ mutants exhibit reduced virulence due to reduced adhesion capability [121]. On the contrary, the *pde1*Δ/Δ mutants can still undergo filamentation [122]. Interestingly, *pde1 pde2* double deletion mutants exhibit attenuated virulence as compared to *pde2*Δ/Δ mutants [121].

The adenylyl cyclase Cyr1 is required for hyphal development and virulence but is not essential for basal growth in *C. albicans* [67]. Deletion of *CYR1* has a global impact on gene expression, resulting in many alterations in response to environmental cues [67,123]. Cyr1 contains several highly conserved functional domains, which include a Gα domain, a Ras-association (RA) domain, a leucine-rich repeat (LRR) domain, a cyclase catalytic (CYCc) domain, and a Cap1 (cyclase-associated protein 1) binding domain (CBD) [124,125].

The small GTPase Ras1, an upstream activator of Cyr1, transduces extracellular signals (serum in combination with elevated temperature or nitrogen starvation) to Cyr1 [44,107,114]. Ras1 usually exists in the cell in an inactive (GDP-bound) form, and its switch to the active form (GTP-bound) is regulated by the GTPase-activating protein (GAP) Ira2; the guanine nucleotide exchange factor (GEF) Cdc25 drives the GTP-Ras1-to-GDP-Ras1 switch (Figure 3) [126]. Active Ras1 directly interacts with Cyr1 through the RA domain, stimulating cAMP production [44,114]. Cyr1 activity depends upon the binding of Cap1 at the CBD domain and the binding of G-actin to Cap1 to form a tripartite complex, which serves to maintain the activation of the pathway [46,127]. Deletion of *CAP1* results in lowered cAMP levels and blocks in morphogenesis.

The presence of serum drives morphogenesis via Ras1 activation of the cAMP-PKA pathway. Deletion of *RAS1* impairs serum-induced filamentous growth, which can be overcome by supplementation with cAMP, and overexpression of cAMP signaling components rescues its defects [128,129]. The serum contains various active factors, such as glucose and bacterial peptidoglycan fragments, that can stimulate the pathway. The Gα domain of Cyr1 is the binding site for a heterotrimeric G-protein α subunit Gpa2, which is activated by the G-protein-coupled receptor Gpr1 in response to glucose and amino acids [34,130]. However, neither Gpr1 nor Gpa2 is required for serum-induced hyphal formation in liquid media [43]. Glucose-induced activation of cAMP synthesis appears to be mediated by Cdc25-Ras1 interaction and not Gpr1 binding of Cyr1 [36,43,45]. The LRR domain of Cyr1 recognizes and binds muramyl dipeptides (MDP), subunits of bacterial peptidoglycan present in serum [35,39,41]. Deletion or mutation of the LRR domain abolishes cAMP-PKA activation in the presence of MDPs [35]. CO_2 or HCO_3^- bind to the CYCc domain, stimulating the production of cAMP required for hyphal growth [33]. Both endogenous and exogenous QSMs farnesol and 3-oxo-C12-homoserine lactone (HSL) block the hyphal growth by binding to the CYCc domain and inhibiting the activity of Cyr1 [82].

Temperature-dependent morphogenesis via the cAMP-PKA pathway is governed by the heat shock chaperone protein, Hsp90, whose expression is regulated by the heat shock transcription factor, Hsf1 [131]. Under basal conditions, Hsp90 and its co-chaperone Sgt2 interact with Cyr1 and repress it [39,132]. Temperature elevation results in cellular stress leading to an increase in competing Hsp90 client proteins, thereby relieving Hsp90 repression of Cyr1. Inhibition of Hsp90 leads to filamentous growth under non-inducing conditions.

Cell cycle perturbation also induces morphogenesis via the Ras-cAMP-PKA signaling pathway. Disrupting cell cycle progression by treating with the DNA synthesis inhibitor hydroxyurea (HU) arrests cells in the S phase, while prolonged depletion of Cln3 arrests cells in the G1 phase [133,134]. Arrested cells switch to filamentous growth and re-enter the cell cycle via Ras1-activation of the cAMP-PKA pathway [133,134]. However, while cell cycle arrest in G1 and S phase induces morphogenesis, different mechanisms are involved. HU-induced filamentation does not require the downstream transcription factor Efg1, and the hyphal-specific transcription factor Ume6 and the G1 cyclin Hgc1 but involves other PKA target genes [133,135]. In contrast, filamentation due to Cln3 depletion requires Efg1, Ume6, and Hgc1 [133–135].

4. The MAPK Pathways

The mitogen-activated protein kinase (MAPK) signal transduction pathway consists of three components: the MAP kinase kinase kinase (MAPKKK), the MAP kinase kinase (MAPKK), and the MAP kinase (MAPK) (Figure 3). MAPK signaling is dependent on three phosphotransfer steps. Upon activation, MAPKKK becomes phosphorylated and triggers the phosphorylation of the MAPKK, which in turn phosphorylates the MAPK [136,137]. In *C. albicans*, the Cek1-mediated MAPK pathway and the PKC MAPK pathway are activated by different stimuli. They serve as patterns of cascades that are essential for its morphogenesis and virulence, as shown in Figure 3 [137,138]. Apart from the cAMP-PKA signaling pathway, Ras1 also signals through the MAPK signaling cascade (Cek1-mediated) to coordinate filamentation in response to nitrogen starvation conditions via the Mep2 sensor [107].

4.1. The Cek1-Mediated MAPK Pathway

C. albicans extracellular signal-regulated kinase (ERK)-like 1 (Cek1)-mediated MAPK pathway is involved in cell wall biogenesis and virulence [139,140]. The Cek1-mediated MAPK pathway also plays an important role in hyphal development through the activation of downstream transcription factor Cph1, a positive regulator of filamentation [106,141]. This pathway can be induced by several factors such as low nitrogen, cell wall damage, osmotic stress, and embedding matrix. Under nitrogen starvation conditions, this pathway is activated by the ammonium permease Mep2 via a Ras1-dependent manner [107]. Cdc42, an essential GTPase, and its GEF Cdc24 are required for filamentous growth and virulence [142,143]. Upon interactions with Cdc24 and Ras1, activated Cdc42 turns on downstream effectors, including p21-activated kinase (PAK) Cst20 and Cla4, which then triggers concerted phosphorylation of the Ste11 (MAPKKK), Hst7 (MAPKK), and Cek1 (MAPK) (Figure 3) [138]. Mutations in the Cek1-mediated cascade cause defects in hyphal development to a different degree under certain conditions and result in attenuated virulence in animal models [144,145]. The Cek1-mediated MAPK pathway can also be activated through its upstream transmembrane proteins via cell wall damage or osmotic stress. Sho1, Opy2, and Msb2 form a complex that interacts with Cdc42 and Cst20, triggering Cek1 phosphorylation [144]. *sho1Δ/Δ*, *opy2Δ/Δ*, and *msb2Δ/Δ* mutants display altered sensitivity to cell wall damaging agents such as Congo Red, zymoylase, and tunicamycin, suggesting their roles in cell wall biogenesis [146,147]. *sho1Δ/Δ* mutants are sensitive to osmotic stress (i.e., 1 M sodium chloride), suggesting its additional role in osmotic stress signaling. The Cek1-mediated MAPK pathway responds to embedded matrix conditions by initiating a signaling cascade that ultimately activates Cph1 via Cek1 [22]. Rac1, Lmo1, and its exchange factor Dck1 are essential for invasive filamentous growth in the embedding matrix [148,149]. In contrast to Cdc42, Rac1 is not required for serum-induced hyphal growth [150]. *rac1Δ/Δ*, *lmo1Δ/Δ*, and *dck1Δ/Δ* mutants were observed to exhibit filamentous growth defects on solid agar and increased sensitivity to cell wall damaging agents, such as Calcofluor White and Congo Red [148]. Intriguingly, the overexpression of the Cek1 MAP kinase in *rac1Δ/Δ*, *lmo1Δ/Δ*, and *dck1Δ/Δ* mutants restores invasive filamentous growth on solid media, suggesting that Rac1, Lmo1, and Dck1 function together upstream of the Cek1-mediated MAPK pathway during invasive filamentous growth [148]. The downstream transcription factor, Cph1, is essential for hyphal growth on solid agar but not in liquid media [106]. It was found that *efg1 cph1* double deletion mutants cannot form filaments under hypha-inducing conditions and are avirulent in animal models [31]. However, *efg1 cph1* double deletion mutants can form filamentation when embedded in the matrix, suggesting the involvement of other transcription factors for hyphal development under this condition [151].

4.2. The PKC MAPK Pathway

The protein kinase C (PKC) MAPK pathway is commonly known as the cell wall integrity pathway [137]. Pkc1 activation leads to a MAPK cascade activation of Bck1 (MAPKKK), Mkk1 (MAPKK), and Mkc1 (MAPK). Cellular morphogenesis in *C. albicans* is a highly dynamic process controlled by a master regulator, Rho1, in response to various stressors (Figure 3) [152]. Rho1, the master regulator of the cell wall integrity signaling cascade, is activated by the GEF Rom2 and inactivated by the GAP Lrg1 [153]. Recently, the PKC MAPK pathway was discovered to regulate *C. albicans* morphogenesis through the co-regulation of cAMP signaling [154]. Interestingly, Rho1 plays an important role in filamentation through Pkc1. Pkc1 was found to be a global regulator of *C. albicans* morphogenesis through the regulation of adenylyl cyclase Cyr1. A reduction of Cyr1 activity was observed in *pkc1Δ/Δ* mutants [154]. Lrg1 deactivates Rho1 by locking it in its inactive form, which suppresses the yeast-to-hyphae transition. *C. albicans* morphogenesis is independent of its canonical MAPK cascade. Deletion of *BCK1* or *MKC1* does not impair the filamentous growth in response to the Hsp90 inhibitor geldanamycin or serum [155]. Although the downstream transcription factors of Mkc1 have previously been proposed to be Efg1, Czf1, and Bcr1, to date, *C. albicans* morphogenesis through distinct effector(s) remains elusive [154,155].

5. Negative Regulators of Hyphal Morphogenesis

C. albicans morphogenesis is negatively regulated by the transcriptional repressors Tup1, Nrg1, and Rfg1 [156–158]. Tup1 is a global transcriptional repressor, and its inactivation leads to constitutive filamentous growth and derepression of hyphal-specific genes [130,156,159]. Nrg1 and Rfg1 are well characterized DNA-binding proteins, which regulate different subsets of hyphal-specific genes by recruiting co-repressor Tup1. A DNA microarray analysis revealed significant up-regulation of 61 genes in response to serum and 37 °C [160]. Approximately half of these genes are found to be repressed by the transcription factors Tup1, Nrg1, and Rfg1, suggesting their importance in repressing hyphal morphogenesis. *C. albicans* cells that lack these repressors develop into pseudohyphae with the expression of hyphal-specific genes [161]. Surprisingly, only *nrg1Δ/Δ* mutants form hyphae in response to serum. In addition, *nrg1Δ/Δ* mutants appear to display stronger hyphal phenotypes than *rfg1Δ/Δ* mutants, suggesting its predominant role in the negative regulation of hyphal growth [162].

5.1. The Farnesol-Mediated Inhibition Pathway

Though Tup1 is found to act independently of the cAMP-PKA and MAPK pathways to regulate morphogenesis, it seems to play a crucial role in the farnesol response pathway [90,159]. Farnesol, an endogenous QSM, is produced when the cell densities of *C. albicans* are high. While farnesol can block the yeast-to-hyphae transition, it cannot block the elongation of pre-existing filaments [163–165]. Morphological and transcriptional studies, which investigated the possible functional overlap between farnesol and hyphal transcriptional repressors, have demonstrated the direct involvement of Tup1 in the farnesol-mediated inhibition of filamentous growth [90]. *tup1Δ/Δ* and *nrg1Δ/Δ* mutants display elevated levels of farnesol and are constitutively filamentous even in the presence of exogenous farnesol. In the presence of farnesol, *TUP1* levels increase, but *NRG1* and *RFG1* levels are unaffected [90]. Further targeted studies on the farnesol-mediated inhibition pathway have unraveled its dedicated mechanistic control of filamentous growth [166]. Upon inoculation of cells, where farnesol inhibition is relieved, the transcriptional repressor Cup9 is constantly degraded by the N-end rule E3 ubiquitin ligase Ubr1, allowing the expression of kinase Sok1 and subsequent degradation of Nrg1. In contrast, the presence of farnesol inhibits the degradation of Cup9, thereby repressing Sok1 expression, which in turn blocks the degradation of Nrg1 and hyphal development [166].

5.2. The Roles of Negative Regulators Tup1 and Nrg1 in Hyphal Elongation

Critical regulators of hyphal initiation and the activation of hypha-associated genes, such as Efg1, Cph1, Czf1, and Flo8, are shown in Figure 4. Thereafter, a second regulatory network is required for the hyphal elongation process and long-term maintenance of hyphal growth through Hgc1, Eed1, and Ume6, which are negatively regulated by Tup1 and Nrg1 [144,167–169]. Eed1 was first identified in oral tissue infections from patients suffering from oral disease, and its associated regulatory network was explored through comprehensive transcriptomics analysis [167]. Eed1 is positively regulated by Efg1 as the overexpression of *EED1* partially rescues the hyphal defects in *efg1Δ/Δ* mutants. *EED1* expression is significantly up-regulated in the continuously filamentous *nrg1Δ/Δ* and *tup1Δ/Δ* mutants under non-hyphae-inducing conditions [167]. In contrast, under hyphae-inducing conditions, *EED1* levels were slightly decreased in *nrg1Δ/Δ* and *tup1Δ/Δ* mutants, but elevated 10-fold in wild-type cells. Collectively, this suggests that *EED1* is repressed by both Nrg1 and Tup1 in wild-type *C. albicans*. Ume6 acts downstream of Eed1 as the overexpression of *UME6* restored the hyphal elongation defect observed in *eed1Δ/Δ* mutants [167]. *UME6* expression levels were significantly down-regulation in *eed1Δ/Δ* mutants [167]. *HGC1* expression is detected within 5 min of hyphal induction, whereas *UME6* expression is only detected after 15 min upon induction [170]. This suggests that a Ume6-independent mechanism initially induces HGC1. Nrg1 and Tup1 negatively regulate both Ume6 and Hgc1 [161,169]. Ume6 could also be induced as a result of relief of transcriptional repression by the Nrg1-Tup1 complex.

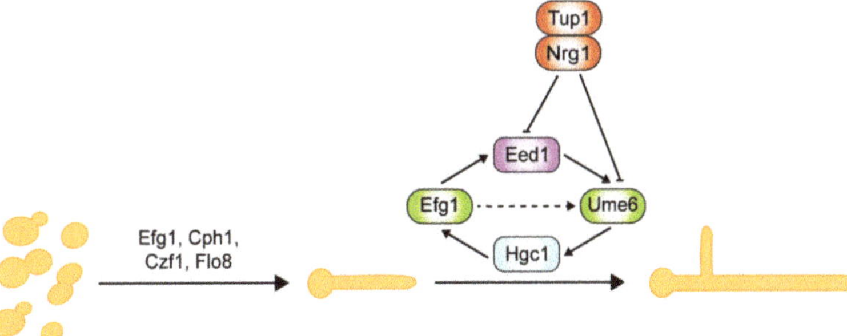

Figure 4. Regulation of hyphal elongation requires mechanisms for initiation and long-term maintenance. Initiation of hypha growth requires transcription factors such as Efg1, Cph1, Czf1, and Flo8. Subsequent elongation process and maintenance require the involvement of Hgc1, Eed1, and Ume6. Both Eed1 and Ume6 are negatively regulated by Tup1 and Nrg1.

5.3. O_2 and CO_2 Signaling Pathways for Sustained Hyphal Development

The stability of hyphae-specific transcription factor Ume6 is governed by two parallel pathways in response to O_2 and CO_2 concentrations [171,172]. Ofd1 negatively regulates the stability of Ume6 by E3 ubiquitin ligase Ubr1 under hypoxia conditions. *ofd1Δ/Δ* and *ubr1Δ/Δ* mutants can maintain hyphal elongation in atmospheric O_2 and 5% CO_2 [171,172]. However, deletion of *UBR1* does not block Ume6 degradation in atmospheric CO_2, suggesting the involvement of additional E3 ubiquitin ligase in response to CO_2 [172]. Recently, it was discovered that CO_2, an inducer of filamentous growth, also plays a critical role in the sustenance of hyphal growth in response to high CO_2 (5%) [172]. In the CO_2 signaling of sustained hyphal growth, a type 2C protein phosphatase (PP2C) Ptc2 and a cyclin-dependent kinase Ssn3 were identified to be the major positive and negative regulators, respectively [172]. High CO_2 induces Ptc2-mediated dephosphorylation of Ssn3. Consequently, the hypophosphorylated Ssn3 fails to phosphorylate Ume6 at the S437 residue. This prevents subsequent ubiquitination of Ume6 by the E3 ubiquitin ligase SCF^{Grr1}, resulting in stabilization of Ume6 for the sustenance of hyphal growth.

5.4. Negative Regulators as Potential Drug Targets

Recent discoveries have introduced novel compounds that inhibit *C. albicans* hyphal and biofilm formation through the up-regulation of negative regulators Tup1 and Nrg1 [173,174]. Treatment of *C. albicans* with novel synthetic SR analogues, 5-[3-substituted-4-(4-substituted benzyloxy)-benzylidene]-2-thioxo-thiazolidin-4-one derivatives, resulted in a 3 to 4-fold increase in the expression of *TUP1* and a 2-fold increase in the expression of *NRG1*, which effectively inhibits the hyphal morphogenesis [173]. Copper oxide nanoparticle (Cu_2O-NP) was found to inhibit the yeast-to-hyphae transition through the down-regulation of *RAS1* and up-regulation of *NRG1* and *TUP1* [174]. Exploiting the negative regulators as drug targets holds excellent potential for future clinical applications. There is a growing interest in applying nanoparticles on medical devices, prosthetic devices, and catheters to combat polymicrobial biofilms in clinical settings.

6. Mechanisms of Hyphal Morphogenesis

6.1. Septin Ring Formation

Although the septin subunits are static in budding yeast cells, upon hyphal induction, Cdc3, Cdc12, and Sep7 form a stable core, while the Cdc10 subunit becomes dynamic, shuttling between the septin ring and the cytoplasm [175]. Cdc3 and Cdc12 are essential, whereas Cdc10 and Cdc11 are not. However, the deletion of *CDC10* and *CDC11* leads to defects in cytokinesis. During hyphal growth, Cdc11 is phosphorylated by the cyclin-CDK (cyclin-dependent kinase) complex Ccn1-Cdc28, and another cyclin-CDK complex Hgc1-Cdc28 maintains its phosphorylated state; mutations to the phosphorylation sites in Cdc11 impair the maintenance of polarized growth [176]. Cdc11 phosphorylation by the septin ring-associated kinase Gin4 primes it for further phosphorylation by Ccn1-Cdc28 [176]. Both *cdc10Δ/Δ* and *cdc11Δ/Δ* mutants have abnormalities in septum formation during hyphal growth and form curved hyphae [177,178]. Cdc10 dynamics are dependent on Sep7 and its phosphorylation status [175]. *sep7Δ/Δ* mutants can form hyphae, but the hyphal compartments separate after cytokinesis. Ccn1-Cdc28 and Hgc1-Cdc28 phosphorylate Gin4, which in turn phosphorylates Sep7 [179,180]. Deletion of *GIN4* disrupts the formation of septin rings in germ tubes resulting in a severe cytokinesis defect; *gin4Δ/Δ* mutants form pseudohyphae constitutively and cannot form true hyphae upon serum induction [181]. Depletion of Gin4 in G1 cells blocks septin ring formation [180]. Sep7 is dephosphorylated by the protein phosphatase 2A (PP2A), mediated by the structural subunit Tpd3 and the catalytic subunit Pph21 [182]. Deletion of *PPH21* or *TPD3* or its regulators, *CDC55* or *RTS1*, leads to the hyperphosphorylation of Sep7 and the disruption of septin organization [182,183]. *cdc55Δ/Δ* mutants grow as pseudohyphae under yeast growth conditions, while *rts1Δ/Δ* mutants grow as round, enlarged multinucleated cells. Both *cdc55Δ/Δ* and *rts1Δ/Δ* mutants display hyphal defects.

The nucleus migrates out from the mother cell to the septin band within the developing hyphae, and the first nuclear division occurs in this subapical compartment [184]. One daughter nucleus migrates back to the mother cell, while the other nucleus migrates to the apical compartment. After mitosis, the protein phosphatase Cdc14, which regulates mitotic exit, localizes to the septum in yeast cells and dephosphorylates the Mob2-Cbk1 complex, allowing the transcription factor Ace2 to translocate to the nucleus and activate the transcription of genes involved in cell separation [185]. However, in hyphal cells, Cdc14 does not localize to the septum, and Mob2-Cbk1 remains at the hyphal tip [185]; thus, cytokinesis does not result in cell separation or the formation of a constriction between cells as observed in yeast or pseudohyphae, respectively. The septin ring splits into two rings with the formation of the primary septum dividing the hyphal compartments. Both rings are maintained in hyphal cells, unlike in yeast and pseudohyphal cells, where the septin rings are dissembled after cytokinesis. However, in *sep7Δ/Δ* mutants, Cdc14 can localize to the hyphal septum, activating the Ace2-dependent cell separation program, resulting in hyphal cell separation [186]. The subapical compartment of the hyphae is vacuolated and remains in the G1 phase.

The nucleosome assembly protein, Nap1, plays a role in septin ring formation and dynamics [187]. Deletion of *NAP1* results in constitutively pseudohyphal cells that can transit to true hyphae under hyphal-inducing conditions. Phosphorylation of Nap1 occurs in a cell cycle-dependent manner, which involves Gin4 and Cla4, a second septin ring-associated kinase. Phosphorylated Nap1 translocates from the cytoplasm to the emerging bud neck. In *cdc10Δ/Δ* and *cdc11Δ/Δ* mutants, Nap1 remains in the cytoplasm even though it is hyperphosphorylated. After mitosis, Nap1 is dephosphorylated in a manner that is dependent upon PP2A and Cdc14.

6.2. Polarization of the Actin Cytoskeleton

Actin cytoskeleton polarization is required for the morphogenesis of *C. albicans*, regardless of cell type. The actin cytoskeleton, made up of actin patches and cables, maintains directional growth by directing vesicular flow for tip expansion. In yeast and pseudohyphae, polarized growth is driven by the polarisome, a complex that includes the polarisome scaffold protein Spa2, the formin Bni1 that serves as the actin cable nucleator, and the formin-actin-binding protein Bud6 [188]. Actin cables, comprised of long bundles of actin filaments, converge at the apical site. During polarized growth, post-Golgi membrane-bound secretory vesicles are continuously delivered to the apical site, supplying material required to expand the plasma membrane and synthesize new cell walls. The vesicles are tethered to the actin cables by the Rab-type GTPase Sec4, activated by the GEF Sec2 [189,190], while the class V myosin, Myo2, complexed to the regulatory light chain Mlc1, provides the motive force for vesicle transport [191]. Upon arrival at the plasma membrane, the secretory vesicles dock with the exocyst before fusing with the plasma membrane. The exocyst is a complex that comprises Sec3, Sec5, Sec6, Sec8, Sec10, Sec15, Exo70, and Exo84 [192]. Sec4 mediates vesicle tethering with the exocyst through its interaction with Sec15 [189,190].

Although the polarisome and exocyst complexes also localize to the hyphal tip, polarized growth in hyphae is driven by a Spitzenkörper, a vesicle-rich structure responsible for hyphal growth directionality, which is present during all stages of the cell cycle, including septation [193]. Spa2, Bni1, and Bud6 coordinate the functions of the Spitzenkörper and the polarisome complex at the hyphal tip [191,193]. During hyphal growth, the post-Golgi secretory vesicles travel along actin cables to the Spitzenkörper, which acts as a vesicle supply center and is maintained at a fixed distance from the hyphal tip (Figure 5). The vesicle-associated proteins Sec4, Sec2, and Mlc1 are localized to the Spitzenkörper during hyphal growth [191,194]. At the Spitzenkörper, the secretory vesicles are loaded onto actin cables nucleated by the polarisome and transported to the plasma membrane, where they dock with the exocyst. Actin cables are essential in hyphal growth, as their disruption inhibits hyphal formation [193]. Loss of *BNI1* does not affect bud emergence, as germ tube formation can be initiated in *bni1Δ/Δ* mutants. However, the germ tubes are wider in diameter, and *bni1Δ/Δ* mutants cannot maintain polarized cell growth [195]. Deletion of *SPA2* leads to polarity and hyphal growth defects [196]. *spa2Δ/Δ* mutants display random budding with multiple surface protrusions. Similar to the *bni1Δ/Δ* mutants, *spa2Δ/Δ* mutants can form germ tubes. However, unlike in *bni1Δ/Δ* mutants, hyphal growth can be maintained in the *spa2Δ/Δ* mutants, albeit in the form of severely swollen and curvy hyphae. Actin depolymerizing drugs, cytochalasin A and latrunculin A, disrupt the actin cytoskeleton, thus inhibiting hyphal growth and also suppressing the expression of hyphal-specific genes [143,197,198]. Chlorpropham, a drug affecting actin microfilament organization, inhibits hyphal growth [199].

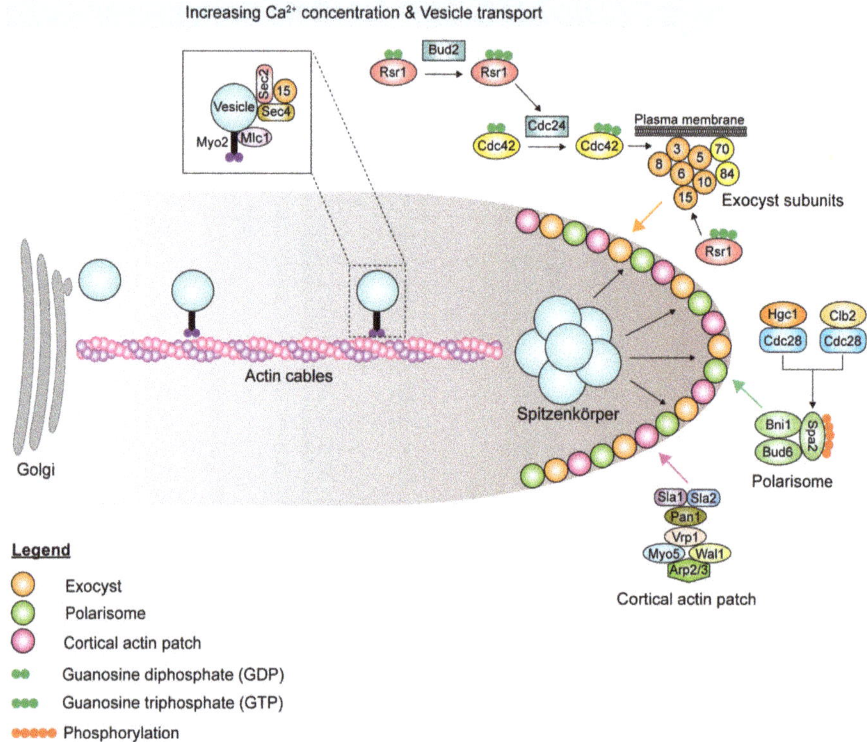

Figure 5. Schematic representation of polarized growth in *C. albicans* hyphal cells. Polarized growth is driven by Spitzenkörper, a vesicle supply center maintained at a fixed distance from the hyphal tip. Post-Golgi membrane-bound secretory vesicles are continuously delivered to the site of polarized growth. Secretory vesicles, tethered by the Rab-type GTPase Sec4 and the GEF Sec2, are transported to the hyphal tip via actin cables with the class V myosin Myo2 complexed to the regulatory light chain Mlc1, providing the motive force. The vesicles accumulate in Spitzenkörper before docking with the exocyst, which consists of Sec3, Sec5, Sec6, Sec8, Sec10, Sec15, Exo70, and Exo84 subunits, before fusing with the plasma membrane. Spa2, Bni1, and Bud6 coordinate the functions of the Spitzenkörper and the polarisome complex at the apical site of the hyphal tip. Endocytosis, endocytic recycling of polarity proteins, involves the cortical actin patches at the apical site of the hyphal tip. Actin patch organization and dynamics involve the actin cytoskeletal proteins Sla1 and Sla2, the actin skeleton-regulatory protein Pan1, and the Vpr1-Wal1-Myo5 complex, which activates the Arp2/3 complex. The landmark GTPase Rsr1, upon activation by its GEF Bud2, localizes Cdc24 to the site of tip growth, in addition to Ca²⁺ binding of the EF-hand motif in Cdc24.

The extensive exocytosis, which occurs at the apical tip and allows for rapid cell wall and membrane deposition, is counterbalanced by endocytosis. Endocytosis is essential for hyphal growth. Suppression of endocytosis suppresses hyphal elongation, and inhibition of endocytosis blocks hyphal formation, while yeast proliferation is unimpeded in both situations. Actin patches form the sites of endocytosis, which is important for maintaining polarity through the endocytic recycling of polarity proteins [200,201]. Before budding or germ tube evagination, cortical actin patches cluster at the apical site [197,202]. Actin patches are highly dynamic, with a lifetime of 5–20 s [203]. As the bud continues to enlarge in yeast cells, the cortical actin patches are redistributed isotropically throughout the bud surface [202]. However, in hyphal cells, the cortical actin patches remain clustered at the hyphal tip throughout hyphal growth [202]. Endocytosis in *C. albicans* mainly occurs via clathrin-mediated endocytosis, and various genes involved in the process have been studied. Sla1 and Sla2 are actin cytoskeletal proteins involved in actin patch organization and dynamics, as well as actin cable polarization, and necessary for normal endocytosis [204–209]. Cortical actin patches formed in *sla1Δ/Δ* mutants are depolarized and less dynamic and form short filaments [206,210]. *sla2Δ/Δ* mutants cannot undergo hyphal and pseudohyphal growth as the localization and orientation of actin patches and cables are defective [204,205]. *sla2Δ/Δ* mutants grow slower and form enlarged cells, as Swe1, the morphogenesis checkpoint kinase, delays cell cycle progression. Swe1 phosphorylates the Clb2-Cdc28 complex in response to perturbations to the actin cytoskeleton, thus delaying the normal transition from polarized growth to isotropic bud growth and delaying nuclear division. Pan1 is a clathrin-mediated endocytosis scaffold protein that is essential for endocytosis [211]. Depletion of Pan1 leads to the formation of thick and swollen cells that have abnormal filamentation. The inhibitory protein kinase Akl1 interacts with Pan1 to repress endocytosis, suppressing hyphal elongation [212]. Deletion of *AKL1* results in faster hyphal elongation rates and longer hyphae, while *AKL1* overexpression reduces hyphal elongation rates. However, overexpression of *PAN1* counteracts the effects of *AKL1* overexpression.

The myosin type I protein Myo5, the Wiskott–Aldrich Syndrome protein (WASP) homolog *WAL1*, and the WASP-interacting protein Vpr1 form a complex similar to that in *Saccharomyces cerevisiae* [213]. The Vpr1-Wal1-Myo5 complex is required for the polarized distribution of cortical actin patches. The deletion of *MYO5* leads to mislocalization of cortical actin patches, with the patches dispersed throughout the bud and the mother cell, resulting in excessive isotropic growth [214]. *myo5Δ/Δ* mutants are unable to endocytose and cannot form hyphae [214]. Deletion of *WAL1* and *VRP1* leads to defects in polarized growth [213,215]. *wal1Δ/Δ* mutants can initiate but cannot maintain hyphal growth. Instead, *wal1Δ/Δ* mutants form elongated, pseudohyphal cells under hyphae-inducing conditions. *vpr1Δ/Δ* mutants have a defect in hyphal formation that is slightly less severe than in *wal1Δ/Δ* mutants. Cortical actin patches are depolarized in both the mother cells and buds of *vpr1Δ/Δ* mutants. Myo5 and Wal1 activate the actin module Arp2/3 complex to initiate actin polymerization. Deleting *ARP2* or *ARP3* leads to an inability to form hyphae, although endocytosis is not abolished [208,209]. Deleting *RVS161* and *RVS167*, which encode Bin-Amphiphysin-Rvs (BAR) domain proteins, results in defective actin patch polarization, with the *rvs161Δ/Δ* mutants displaying a more severe defect in endocytosis and morphogenesis than the *rvs167Δ/Δ* mutants [207].

6.3. The Role of Ras- and Rho-Family GTPase

The small Ras- and Rho-family GTPases play essential roles in hyphal maintenance. The small Rho GTPase Cdc42 is the master regulator of polarized growth. Cdc42 affects hyphal growth and maintenance in at least two ways. Firstly, Cdc42 affects morphogenesis at the transcriptional level. Reduced expression levels of Cdc42 lead to decreased expression of hyphal-specific genes [216]. Secondly, decreasing cellular levels of active Cdc42 results in yeast and hyphae larger and rounder in shape, indicative of polarized growth defect [216,217]. Cdc42 cycles between GDP- and GTP-bound states. The GEF Cdc24 mediates the formation of GTP-bound Cdc42 [142,216]. [142,216]. The GAPs Rga2 and Bem3 mediate the return of Cdc42 from the GTP- to the GDP-bound form. Cdc42 and Cdc24, both required for viability, localize to the hyphal tip during hyphal growth [142,216]. Bem3 is localized to the apical zone of polarized growth, while Rga2 is localized to the septum and is phosphorylated in a hyphal-specific manner [218]. Loss of *RGA2* and *BEM3* results in the formation of a Spitzenkörper-like structure under pseudohyphal-promoting conditions, and the mutants have a morphology resembling true hyphae. Bem1 is a polarity establishment scaffolding protein that binds GTP-bound Cdc42, keeping it localized to the apical site [219].

The Ras-like GTPase Rsr1, a landmark protein that is the master regulator of the bud site selection system, regulates the amount and distribution of Cdc42 activity at the hyphal tip [220,221]. Rsr1 activity is regulated by the GEF Bud5 and the GAP Bud2. Bud5 is localized to the apical site, while Bud2 is localized to the subapical region and septin ring. *rsr1Δ/Δ* mutants have defects in polarized growth; yeast cells are larger and rounder, while the hyphae are wider than wild-type cells. Rsr1 is involved in regulating the recruitment and spatial distribution of vesicles at the hyphal tip [220,221]. Loss of Rsr1 affects the size of the fixed region to which vesicles are delivered and also affects the localization of exocyst subunits [222]. Rsr1 may play a role in limiting the competition for Cdc42 between the septum and the hyphal tip.

7. CDKs, Cyclins, and Their Roles in Hyphal Morphogenesis

Maintenance of cell signaling is important for cell cycle progression and cell growth. The cell cycle-associated cyclins and CDKs tightly regulate the small GTPases and other components of polarized growth. *C. albicans* has three G1 cyclins (Ccn1, Cln3, and Hgc1) and two B-type mitotic cyclins (Clb2, Clb4), of which only Cln3 and Clb2 are essential. The essential CDK Cdc28 serves as the master regulator that controls cell cycle progression at G1/S and G2/M phases via specific cyclin interactions that dictate the timing of the phases. Levels of the G1 and B-type mitotic cyclins oscillate during the cell cycle, and a single cyclin-Cdc28 complex can regulate multiple events within each phase of the cell cycle. Cdc28 is usually stable and present at constant levels throughout the cell cycle; however, its depletion leads to filamentous growth [223]. Ccn1 and Cln3 levels in yeast cells are high in the G1 phase, coinciding with bud emergence and apical growth, and decline in the early G2 phase. Clb2 levels peak in the early G2/M phase, while Clb4 levels reach their peak in the mid-G2/M phase [224]. Levels of both B-cyclins start to decline in the M phase and disappear during exit from mitosis [185,224].

In hyphal cells, polarized growth continues at the apical site throughout the cell cycle, indicating the decoupling of cell elongation from the cell cycle. Ccn1 and Cln3 levels are accumulated earlier and persist for a longer time during hyphal growth [224,225], extending the G1 phase in the early germ tube. Accumulation of the mitotic cyclins, Clb2 and Clb4, is delayed in hyphal cells. Although it is not required for the initiation of hyphal growth, high levels of Ccn1 are required for maintenance of hyphal growth, along with Cln3. The forkhead family transcription factor, Fkh2, usually undergoes cell cycle-dependent phosphorylation to induce the expression of genes that regulate cell cycle progression [226,227]. However, upon hyphal induction, Fkh2 is phosphorylated by Ccn1/Cln3-Cdc28 and Mob2-Cbk1 in a cell cycle-independent manner, redirecting it to enhance the expression of hyphal-specific genes such as the hyphal-specific G1 cyclin *HGC1* (Figure 6) [226,227]. *fkh2Δ/Δ* mutants grow constitutively as pseudohyphae under both yeast and hyphal-inducing conditions [226,227]. During hyphal growth, Ccn1-Cdc28 and Cln3-Cdc28 complexes phosphorylate Mob2, the activator of Cbk1, the cell wall integrity kinase, inhibiting the activation of Ace2 (Figure 6) [228]. Cln3-Cdc28 complex regulates cortical actin patches via phosphorylation of Sla1 [206].

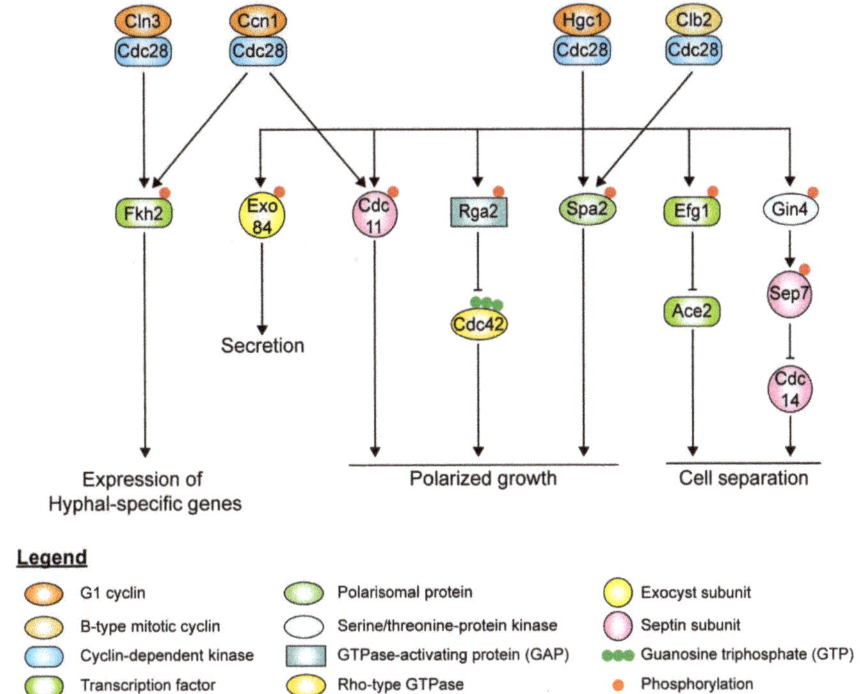

Figure 6. *C. albicans* morphogenesis is tightly regulated by the cell cycle-associated cyclins and cyclin-dependent kinase (CDK). The essential CDK Cdc28 serves as the master regulator that controls the cell cycle progression at G1/S and G2/M phases by forming CDK complexes with specific cyclins. Levels of Cdc28 are relatively stable throughout the cell cycle and deplete during hyphal growth. In contrast, levels of the G1 and B-type mitotic cyclins oscillate during the cell cycle. G1 cyclins Cln3 and Ccn1 peak in the G1 phase and decline in the early G2 phase, while B-type mitotic cyclin Clb2 peaks in the early G2/M phase and declines in the M phase. Upon hyphal induction, Fkh2 is phosphorylated by Cln3-Cdc28 and Ccn1-Cdc28 complexes in a cell cycle-dependent manner to enhance the expression of hyphal-specific genes. The Hgc1-Cdc28 complex is essential for the maintenance of hyphal growth. The exocyst subunit Exo84 is phosphorylated by the Hgc1-Cdc28 complex for the regulation of polarized secretion. Phosphorylation of the septin subunit Cdc11 (by Ccn1-Cdc28 and Hgc1-Cdc28 complexes), GAP Rga2 (by the Hgc1-Cdc28 complex), and the polarisome protein Spa2 (by Hgc1-Cdc28 and Clb2-Cdc28 complexes) promote polarized growth. Rga2 is phosphorylated and inactivated by Hgc1-Cdc28, which relives the repression of the GTPase Cdc42. Phosphorylation of the transcription factor Efg1 and protein kinase Gin4 inhibit cell separation. Phosphorylated Efg1 binds to promoters of Ace2 target genes, inhibiting their transcription. Phosphorylated Gin4 modifies the dynamics of the septin ring by subsequent phosphorylation of Sep7 and deactivating the cell separation program via inhibition of the protein phosphatase Cdc14.

The hyphal-specific G1 cyclin Hgc1 does not regulate the cell cycle but plays a critical role in hyphal morphogenesis (Figure 6) [169]. Besides suppression by Tup1 and Nrg1, the expression of *HGC1* is positively regulated by the transcription factor Ume6, which ensures that Hgc1 is expressed throughout the cell cycle as long as the inducing conditions remain [169]. Hgc1 interacts with Cdc28, forming a complex regulating by phosphorylation regulators and components of cell polarity, membrane trafficking, and cell separation, which is required to maintain hyphal growth (Figure 6). The Hgc1-Cdc28 complex phosphorylates and inactivates Rga2, sequestering it from the hyphal tip to allow Cdc42 localization at the hyphal tip to persist during polarized growth [217,229]. Hgc1-Cdc28, together with Clb2-Cdc28, phosphorylates Spa2, localizing the polarisome to the hyphal tip [230]. Hgc1-Cdc28 complex phosphorylates the exocyst subunits Exo84 and

Sec2, allowing them to be recycled at the growing hyphal tip [194,231]. The Hgc1-Cdc28 complex phosphorylates Efg1, leading to Efg1 competitively binding to promoters of Ace2 target genes, thereby repressing the expression of cell separation activators to prevent cell separation after cytokinesis [232]. The Hgc1-Cdc28 complex also plays a role in regulating the septin ring dynamics during hyphal growth via Sep7 [175].

The cyclins Pcl1 and Pcl5 and the CDK Pho85, although not essential for cell cycle progression, contribute to morphogenesis in response to environmental cues. The Pcl1-Pho85 complex is required for temperature-dependent filamentation induced by Hsp90 inhibition [233]. The transcription factor Hms1 is required for filamentation induced by high temperatures. The Pcl1-Pho85 complex phosphorylates Hms1, which then binds to hyphal-specific genes. It also regulates the degradation of the transcription factor Gcn4, which is indirectly involved in filamentous growth in response to amino acid starvation [234]. Gcn4 induces *PCL5* expression, and the Pcl5-Pho85 complex phosphorylates Gcn4, leading to its degradation [235,236].

8. Cell Cycle Perturbation Leads to Morphogenesis

Although hyphal growth is not directly controlled by the cell cycle, perturbing the cell cycle can cause significant pseudohyphal growth under non-hyphal-inducing conditions or block hyphal growth under hyphal-inducing conditions [134,135,237]. Loss of Ccn1 does not induce morphogenesis but causes a filamentation defect under serum induction [225]. Depletion of Cln3, Clb2, or Clb4 results in filamentous growth in the absence of hyphal-inducing stimuli [134,237]. Cells depleted of Cln3 undergo cell cycle arrest in the G1 phase and form filaments before the resumption of the cell cycle [135]. In the absence of mitotic cyclins, polarized growth promoted by G1 cyclins is not entirely suppressed, and filamentation occurs. Clb2-depleted cells form elongated projections during cell cycle arrest and are inviable, whereas Clb4-depleted cells grow constitutively as pseudohyphae and remain viable [224]. Depletion of Cdc28 also promotes filamentous growth [223].

Genotoxic stresses that disrupt cell cycle progression and activate DNA damage/replication checkpoints lead to filamentation [238]. Pharmacological inhibition of cell cycle progression by the DNA replication inhibitors hydroxyurea (HU) and aphidicolin (AC) or DNA damage induced by UV radiation or the alkylating agent methylmethane sulfonate (MMS) result in S phase arrest, inducing filamentous growth [133,239,240]. Checkpoint proteins play a crucial role in response to DNA replication and DNA damage stresses. The protein kinase Rad53 plays a central role in the DNA replication and DNA damage checkpoint. *rad53Δ/Δ* mutants cannot switch to filamentous growth in response to DNA replication inhibitors and DNA damage; mutations to the Rad53 FHA domains inhibit filamentation in response to DNA damage, but not cell cycle arrest [241,242]. Deletion of *RAD9*, which encodes a checkpoint protein upstream of Rad53, blocks DNA damage-induced filamentation [241]. Depleting the DNA repair protein Rad52 or deleting its gene results in the accumulation of spontaneous DNA damages that trigger the DNA damage checkpoint, resulting in filamentous growth [243]. After the stress is relieved, deactivation of the cell cycle checkpoint is necessary for the cell cycle to progress. Rad53 is dephosphorylated by the protein phosphatase 2A-like complex Pph3-Psy2. Deletion of *PPH3* that encodes the catalytic subunit, *PSY2* that encodes the regulatory subunit, or *TIP41* that encodes the regulator of the Pph3-Psy2 complex enhances MMS-induced filamentous growth and delays the filament-to-yeast transition following DNA damage stress relief [239,240]. The histone acetyltransferases Hat1 and Hat2 are required for the repair of DNA damages caused by endogenous and exogenous agents; *hat1Δ/Δ* mutants accumulate DNA damages rapidly and switch to filamentous growth [244].

Perturbations to mitosis can also lead to the switch to filamentous growth under non-hyphal-inducing conditions. The cell cycle regulatory polo-like kinase Cdc5 is required for the early stages of nuclear division and chromatin separation and mediates spindle formation during the S phase [245]. Depletion of Cdc5 leads to mitotic inhibition and blocks the cell cycle in the G2 phase, leading to hyphal-like growth; however, the cells eventually lose viability [245]. The cytoplasmic dynein, Dyn1, mediates nuclear movement during mitosis. Deletion of *DYN1* or depletion of Dyn1 results in filamentous growth, which requires the spindle position checkpoint protein Bub2 [246,247]. Pharmacological perturbation of mitosis by the microtubule inhibitor nocodazole and deletion of *MAD2* that encodes a spindle assembly checkpoint protein leads to pseudohyphal growth [248].

Degradation of cyclins and CDK inhibitors is regulated mainly by ubiquitin-proteasome-dependent proteolysis and is required for orderly cell cycle progression. Degradation of these cell cycle regulatory proteins is mediated by two multiprotein ubiquitin ligase E3 complexes, the Skp1-Cullin/Cdc53-F-box (SCF) complex and the anaphase-promoting complex/cyclosome (APC/C) complex. The multiprotein SCF complex, consisting of the linker protein Skp1, the scaffold protein Cullin/Cdc53, and a substrate recognition F-box protein, plays a central role in regulating the temporal and spatial degradation of cell cycle regulatory proteins. Depleting the essential *CDC53* or the deletion of the F-box protein genes *CDC4* and *GRR1* leads to filamentous growth. SCF^{Cdc4} is required for the degradation of the CDK inhibitor Sol1 and the transcription factors Ume6 and Gcn4 [235,236,249]. Deleting *CDC4* leads to constitutively filamentous growth with a mix of hyphal and pseudohyphal cells [250]. As Sol1 represses the Clb2-Cdc28 complex, deletion of *CDC4* stabilizes Sol1, inhibiting the switch from apical to isotropic growth, resulting in a pseudohyphal phenotype. However, deleting *SOL1* in the *cdc4*Δ/Δ mutant background still gives rise to constitutively filamentous growth [249]. SCF^{Grr1} is required for the degradation of Ccn1 and Cln3. The deletion of *GRR1* stabilizes Ccn1 and Cln3 levels, and the *grr1*Δ/Δ mutants grow constitutively as pseudohyphae under yeast conditions [251].

The APC/C complex mediates protein degradation during mitotic progression [252]. While little is known about the APC/C complex in *C. albicans*, the two co-activators, Cdc20 and Cdh1, have been identified recently. Cdc20 is essential and mediates the degradation of Clb2 and Cdc5. Depletion of Cdc20 results in the accumulation of Clb2 and Cdc5, leading to a delay in metaphase and telophase; Cdc20-depleted cells grow as long filaments over time [252]. Cdh1 likely plays a role in regulating mitotic exit by influencing Clb2 and Cdc5 degradation; deletion of *CDH1* results in a delay in Clb2 degradation and elevated levels of Cdc5 [252]. *cdh1*Δ/Δ mutants display pleiotropic phenotypes, with a mix of yeast, elongated buds, and pseudohyphae.

9. Conclusions

In summary, the extensive research findings over the years have provided us with illuminating insights into the activation and regulation of hyphal morphogenesis in *C. albicans*. The factors involved are often crucial in controlling the balance between commensalism and invasive infection by *C. albicans*. The yeast-to-hyphal transition in *C. albicans* is highly dependent on the complex interplay between internal signal transduction signaling pathways and external environmental cues that reflect the host niches. This transition is governed by a complex network of signaling pathways, including the cAMP-PKA pathway, the MAP kinase pathways, and the Cek1-mediated and PKC pathways. Activation of these pathways in response to various cues triggers the activation of specific transcription factors such as Efg1, Flo8, Ume6, Tec1, and Cph1. Crosstalk between the cAMP-PKA and MAP kinase pathways add a further layer of complexity to the existing signaling network as multiple signaling pathways can converge to the same set of transcription factors. Following hyphal initiation, subsequent hyphal development requires delicate mechanisms to maintain hyphal elongation. Polarized growth requires continuous delivery of membrane-bound secretory vesicles, along the actin cables, to the site of polarized growth. The vesicles accumulate as Spitzenkörper in the subapical region before docking with the exocyst and polarisome components. This exocytosis process is counterbalanced by the endocytosis process. Cell morphology of *C. albicans* is known to be tightly linked to cell cycle progression through cyclin-CDK complexes. One of the most important CDK complexes is the Hgc1-Cdc28 complex, which governs multiple cellular processes required for hyphal development. The Hgc1-Cdc28 complex plays important roles in polarized growth, polarized secretion, and inhibition of cell separation, which ensures the formation of long tubular cells without constriction at the septal junction. Lastly, perturbation of the cell cycle can either induce or impair the highly polarized growth in *C. albicans* under different conditions.

10. Future Directions

However, much remains to be explored and unraveled in *C. albicans* morphogenesis. While the mechanisms behind hyphal induction during cell cycle arrest have been uncovered, there are still missing gaps. Future work could uncover the genes that regulate filamentation in response to the cell cycle perturbation and elucidate how the signals are transduced to the cAMP-PKA pathway and the activated downstream transcriptional regulators. The link between nutrient and osmotic stress and filamentous growth via the PKC pathway has been uncovered, but the downstream transcriptional regulators remain elusive. Applying evolutionary tools of systems biology in combination with animal-based studies will propel discoveries in this field. The recent development of the transposon-mediated mutagenesis systems in haploid *C. albicans* strains would allow genome-wide screening for novel genes with functions that influence morphogenesis [253,254]. Future work towards identifying additional downstream transcriptional regulatory genes could open up new avenues towards antifungal therapy development.

Author Contributions: All authors contributed to the manuscript preparation. E.W.L.C. conceived the original idea and drafted the manuscript; L.M.P. assisted in the initial draft preparation; Y.W. supervised the project and made substantial scientific contributions. All authors have read and agreed to the published version of the manuscript.

Funding: This work was supported by funding from the National Research Foundation (NRF2019-NRF-ISF003-3039) and National Medical Research Council (NMRC/OFIRG/0072/2018 and NMRC/O FIRG/0055/2019) of Singapore to Y.W.

Institutional Review Board Statement: Not applicable.

Informed Consent Statement: Not applicable.

Data Availability Statement: Not applicable.

Conflicts of Interest: The authors declare no conflict of interest.

References

1. Scully, C.; El-Kabir, M.; Samaranayake, L.P. *Candida* and Oral Candidosis: A Review. *Crit. Rev. Oral Biol. Med.* **1994**, *5*, 125–157. [CrossRef] [PubMed]
2. Achkar, J.M.; Fries, B.C. *Candida* Infections of the Genitourinary Tract. *Clin. Microbiol. Rev.* **2010**, *23*, 253–273. [CrossRef] [PubMed]
3. Rosenbach, A.; Dignard, D.; Pierce, J.V.; Whiteway, M.; Kumamoto, C.A. Adaptations of *Candida albicans* for Growth in the Mammalian Intestinal Tract. *Eukaryot. Cell* **2010**, *9*, 1075–1086. [CrossRef] [PubMed]
4. Kühbacher, A.; Burger-Kentischer, A.; Rupp, S. Interaction of *Candida* Species with the Skin. *Microorganisms* **2017**, *5*, 32. [CrossRef]
5. Pappas, P.G.; Lionakis, M.S.; Arendrup, M.C.; Ostrosky-Zeichner, L.; Kullberg, B.J. Invasive candidiasis. *Nat. Rev. Dis. Primers* **2018**, *4*, 18026. [CrossRef]
6. Bongomin, F.; Gago, S.; Oladele, R.O.; Denning, D.W. Global and Multi-National Prevalence of Fungal Diseases-Estimate Precision. *J. Fungi* **2017**, *3*, 57. [CrossRef]
7. Bassetti, M.; Mikulska, M.; Viscoli, C. Bench-to-bedside review: Therapeutic management of invasive candidiasis in the intensive care unit. *Crit. Care* **2010**, *14*, 244. [CrossRef]
8. Gudlaugsson, O.; Gillespie, S.; Lee, K.; Vande Berg, J.; Hu, J.; Messer, S.; Herwaldt, L.; Pfaller, M.; Diekema, D. Attributable mortality of nosocomial candidemia, revisited. *Clin. Infect. Dis.* **2003**, *37*, 1172–1177. [CrossRef]
9. Macphail, G.L.; Taylor, G.D.; Buchanan-Chell, M.; Ross, C.; Wilson, S.; Kureishi, A. Epidemiology, treatment and outcome of candidemia: A five-year review at three Canadian hospitals. *Mycoses* **2002**, *45*, 141–145. [CrossRef]
10. Zaoutis, T.E.; Argon, J.; Chu, J.; Berlin, J.A.; Walsh, T.J.; Feudtner, C. The epidemiology and attributable outcomes of candidemia in adults and children hospitalized in the United States: A propensity analysis. *Clin. Infect. Dis. Am.* **2005**, *41*, 1232–1239. [CrossRef]
11. Morgan, J.; Meltzer, M.I.; Plikaytis, B.D.; Sofair, A.N.; Huie-White, S.; Wilcox, S.; Harrison, L.H.; Seaberg, E.C.; Hajjeh, R.A.; Teutsch, S.M. Excess mortality, hospital stay, and cost due to candidemia: A case-control study using data from population-based candidemia surveillance. *Infect. Control. Hosp. Epidemiol.* **2005**, *26*, 540–547. [CrossRef]

12. Brown, A.J.; Gow, N.A. Regulatory networks controlling *Candida albicans* morphogenesis. *Trends Microbiol.* **1999**, *7*, 333–338. [CrossRef]
13. Gow, N.A. Germ tube growth of *Candida albicans. Curr. Top. Med. Mycol.* **1997**, *8*, 43–55. [PubMed]
14. Mukaremera, L.; Lee, K.K.; Mora-Montes, H.M.; Gow, N.A.R. *Candida albicans* Yeast, Pseudohyphal, and Hyphal Morphogenesis Differentially Affects Immune Recognition. *Front. Immunol.* **2017**, *8*, 629. [CrossRef] [PubMed]
15. Nemecek, J.C.; Wüthrich, M.; Klein, B.S. Global control of dimorphism and virulence in fungi. *Science* **2006**, *312*, 583–588. [CrossRef] [PubMed]
16. Trevijano-Contador, N.; Rueda, C.; Zaragoza, O. Fungal morphogenetic changes inside the mammalian host. *Semin. Cell Dev. Biol.* **2016**, *57*, 100–109. [CrossRef] [PubMed]
17. Slutsky, B.; Buffo, J.; Soll, D.R. High-frequency switching of colony morphology in *Candida albicans. Science* **1985**, *230*, 666–669. [CrossRef]
18. Sudbery, P.; Gow, N.; Berman, J. The distinct morphogenic states of *Candida albicans. Trends Microbiol.* **2004**, *12*, 317–324. [CrossRef]
19. Braunsdorf, C.; Mailänder-Sánchez, D.; Schaller, M. Fungal sensing of host environment. *Cell Microbiol.* **2016**, *18*, 1188–1200. [CrossRef]
20. Chin, V.K.; Foong, K.J.; Maha, A.; Rusliza, B.; Norhafizah, M.; Chong, P.P. Multi-step pathogenesis and induction of local immune response by systemic *Candida albicans* infection in an intravenous challenge mouse model. *Int. J. Mol. Sci.* **2014**, *15*, 14848–14867. [CrossRef]
21. Di Carlo, P.; Di Vita, G.; Guadagnino, G.; Cocorullo, G.; D'Arpa, F.; Salamone, G.; Salvatore, B.; Gulotta, G.; Cabibi, D. Surgical pathology and the diagnosis of invasive visceral yeast infection: Two case reports and literature review. *World J. Emerg. Surg.* **2013**, *8*, 38. [CrossRef]
22. Noble, S.M.; Gianetti, B.A.; Witchley, J.N. *Candida albicans* cell-type switching and functional plasticity in the mammalian host. *Nat. Rev. Microbiol.* **2017**, *15*, 96. [CrossRef]
23. Van der Meer, J.W.M.; van de Veerdonk, F.L.; Joosten, L.A.B.; Kullberg, B.-J.; Netea, M.G. Severe *Candida* spp. infections: New insights into natural immunity. *Int. J. Antimicrob. Agents* **2010**, *36*, S58–S62. [CrossRef]
24. Erwig, L.P.; Gow, N.A.R. Interactions of fungal pathogens with phagocytes. *Nat. Rev. Microbiol.* **2016**, *14*, 163–176. [CrossRef]
25. Fradin, C.; De Groot, P.; MacCallum, D.; Schaller, M.; Klis, F.; Odds, F.C.; Hube, B. Granulocytes govern the transcriptional response, morphology and proliferation of *Candida albicans* in human blood. *Mol. Microbiol.* **2005**, *56*, 397–415. [CrossRef]
26. Lorenz, M.C.; Bender, J.A.; Fink, G.R. Transcriptional response of *Candida albicans* upon internalization by macrophages. *Eukaryot. Cell* **2004**, *3*, 1076–1087. [CrossRef]
27. Naglik, J.R.; Moyes, D.L.; Wächtler, B.; Hube, B. *Candida albicans* interactions with epithelial cells and mucosal immunity. *Microbes Infect.* **2011**, *13*, 963–976. [CrossRef]
28. Grubb, S.E.W.; Murdoch, C.; Sudbery, P.E.; Saville, S.P.; Lopez-Ribot, J.L.; Thornhill, M.H. Adhesion of *Candida albicans* to Endothelial Cells under Physiological Conditions of Flow. *Infect. Immun.* **2009**, *77*, 3872. [CrossRef]
29. Yang, W.; Yan, L.; Wu, C.; Zhao, X.; Tang, J. Fungal invasion of epithelial cells. *Microbiol. Res.* **2014**, *169*, 803–810. [CrossRef]
30. Rogiers, O.; Frising, U.C.; Kucharíková, S.; Jabra-Rizk, M.A.; van Loo, G.; Van Dijck, P.; Wullaert, A. Candidalysin Crucially Contributes to Nlrp3 Inflammasome Activation by *Candida albicans* Hyphae. *mBio* **2019**, *10*, e02221–e02318. [CrossRef]
31. Lo, H.J.; Köhler, J.R.; DiDomenico, B.; Loebenberg, D.; Cacciapuoti, A.; Fink, G.R. Nonfilamentous *C. albicans* mutants are avirulent. *Cell* **1997**, *90*, 939–949. [CrossRef]
32. Schweizer, A.; Rupp, S.; Taylor, B.N.; Röllinghoff, M.; Schröppel, K. The TEA/ATTS transcription factor CaTec1p regulates hyphal development and virulence in *Candida albicans. Mol. Microbiol.* **2000**, *38*, 435–445. [CrossRef] [PubMed]
33. Klengel, T.; Liang, W.-J.; Chaloupka, J.; Ruoff, C.; Schröppel, K.; Naglik, J.R.; Eckert, S.E.; Mogensen, E.G.; Haynes, K.; Tuite, M.F.; et al. Fungal adenylyl cyclase integrates CO2 sensing with cAMP signaling and virulence. *Curr. Biol.* **2005**, *15*, 2021–2026. [CrossRef] [PubMed]
34. Brock, M. Fungal metabolism in host niches. *Curr. Opin. Microbiol.* **2009**, *12*, 371–376. [CrossRef]
35. Xu, X.-L.; Lee, R.T.H.; Fang, H.-M.; Wang, Y.-M.; Li, R.; Zou, H.; Zhu, Y.; Wang, Y. Bacterial Peptidoglycan Triggers *Candida albicans* Hyphal Growth by Directly Activating the Adenylyl Cyclase Cyr1p. *Cell Host Microbe* **2008**, *4*, 28–39. [CrossRef]
36. Maidan, M.M.; Thevelein, J.M.; Van Dijck, P. Carbon source induced yeast-to-hypha transition in *Candida albicans* is dependent on the presence of amino acids and on the G-protein-coupled receptor Gpr1. *Biochem. Soc. Trans.* **2005**, *33*, 291–293. [CrossRef]
37. Lu, Y.; Su, C.; Wang, A.; Liu, H. Hyphal development in *Candida albicans* requires two temporally linked changes in promoter chromatin for initiation and maintenance. *PLoS Biol.* **2011**, *9*, e1001105. [CrossRef]
38. Song, W.; Wang, H.; Chen, J. *Candida albicans* Sfl2, a temperature-induced transcriptional regulator, is required for virulence in a murine gastrointestinal infection model. *FEMS Yeast Res.* **2011**, *11*, 209–222. [CrossRef]
39. Shapiro, R.S.; Uppuluri, P.; Zaas, A.K.; Collins, C.; Senn, H.; Perfect, J.R.; Heitman, J.; Cowen, L.E. Hsp90 orchestrates temperature-dependent *Candida albicans* morphogenesis via Ras1-PKA signaling. *Curr. Biol.* **2009**, *19*, 621–629. [CrossRef]
40. Taschdjian, C.L.; Burchall, J.J.; Kozinn, P.J. Rapid identification of *Candida albicans* by filamentation on serum and serum substitutes. *AMA J. Dis. Child.* **1960**, *99*, 212–215. [CrossRef]
41. Tan, C.T.; Xu, X.; Qiao, Y.; Wang, Y. A peptidoglycan storm caused by beta-lactam antibiotic's action on host microbiota drives *Candida albicans* infection. *Nat. Commun.* **2021**, *12*, 2560. [CrossRef]

42. Hall, R.A.; De Sordi, L.; Maccallum, D.M.; Topal, H.; Eaton, R.; Bloor, J.W.; Robinson, G.K.; Levin, L.R.; Buck, J.; Wang, Y.; et al. CO2 acts as a signalling molecule in populations of the fungal pathogen *Candida albicans*. *PLoS Pathog.* **2010**, *6*, e1001193. [CrossRef]

43. Miwa, T.; Takagi, Y.; Shinozaki, M.; Yun, C.W.; Schell, W.A.; Perfect, J.R.; Kumagai, H.; Tamaki, H. Gpr1, a putative G-protein-coupled receptor, regulates morphogenesis and hypha formation in the pathogenic fungus *Candida albicans*. *Eukaryot. Cell* **2004**, *3*, 919–931. [CrossRef]

44. D'Souza, C.A.; Heitman, J. Conserved cAMP signaling cascades regulate fungal development and virulence. *FEMS Microbiol. Rev.* **2001**, *25*, 349–364. [CrossRef]

45. Maidan, M.M.; De Rop, L.; Serneels, J.; Exler, S.; Rupp, S.; Tournu, H.; Thevelein, J.M.; Van Dijck, P. The G protein-coupled receptor Gpr1 and the Galpha protein Gpa2 act through the cAMP-protein kinase A pathway to induce morphogenesis in *Candida albicans*. *Mol. Biol. Cell* **2005**, *16*, 1971–1986. [CrossRef]

46. Zou, H.; Fang, H.M.; Zhu, Y.; Wang, Y. *Candida albicans* Cyr1, Cap1 and G-actin form a sensor/effector apparatus for activating cAMP synthesis in hyphal growth. *Mol. Microbiol.* **2010**, *75*, 579–591. [CrossRef]

47. Lopes, J.P.; Stylianou, M.; Backman, E.; Holmberg, S.; Jass, J.; Claesson, R.; Urban, C.F. Evasion of Immune Surveillance in Low Oxygen Environments Enhances *Candida albicans* Virulence. *mBio* **2018**, *9*, e02120–e02218. [CrossRef]

48. Desai, P.R.; van Wijlick, L.; Kurtz, D.; Juchimiuk, M.; Ernst, J.F. Hypoxia and Temperature Regulated Morphogenesis in *Candida albicans*. *PLoS Genet.* **2015**, *11*, e1005447. [CrossRef]

49. Giusani, A.D.; Vinces, M.; Kumamoto, C.A. Invasive filamentous growth of *Candida albicans* is promoted by Czf1p-dependent relief of Efg1p-mediated repression. *Genetics* **2002**, *160*, 1749–1753. [CrossRef]

50. Mulhern, S.M.; Logue, M.E.; Butler, G. *Candida albicans* transcription factor Ace2 regulates metabolism and is required for filamentation in hypoxic conditions. *Eukaryot. Cell* **2006**, *5*, 2001–2013. [CrossRef]

51. Saputo, S.; Kumar, A.; Krysan, D.J. Efg1 directly regulates ACE2 expression to mediate cross talk between the cAMP/PKA and RAM pathways during *Candida albicans* morphogenesis. *Eukaryot. Cell* **2014**, *13*, 1169–1180. [CrossRef] [PubMed]

52. Mayer, F.L.; Wilson, D.; Hube, B. *Candida albicans* pathogenicity mechanisms. *Virulence* **2013**, *4*, 119–128. [CrossRef] [PubMed]

53. Davis, D.; Edwards, J.E., Jr.; Mitchell, A.P.; Ibrahim, A.S. *Candida albicans* RIM101 pH response pathway is required for host-pathogen interactions. *Infect. Immun.* **2000**, *68*, 5953–5959. [CrossRef] [PubMed]

54. Davis, D.; Wilson, R.B.; Mitchell, A.P. RIM101-dependent and-independent pathways govern pH responses in *Candida albicans*. *Mol. Cell. Biol.* **2000**, *20*, 971–978. [CrossRef]

55. El Barkani, A.; Kurzai, O.; Fonzi, W.A.; Ramon, A.; Porta, A.; Frosch, M.; Mühlschlegel, F.A. Dominant Active Alleles of RIM101 (PRR2) Bypass the pH Restriction on Filamentation of *Candida albicans*. *Mol. Cell. Biol.* **2000**, *20*, 4635–4647. [CrossRef]

56. Davis, D.A. How human pathogenic fungi sense and adapt to pH: The link to virulence. *Curr. Opin. Microbiol.* **2009**, *12*, 365–370. [CrossRef]

57. Vylkova, S.; Carman, A.J.; Danhof, H.A.; Collette, J.R.; Zhou, H.; Lorenz, M.C. The fungal pathogen *Candida albicans* autoinduces hyphal morphogenesis by raising extracellular pH. *mBio* **2011**, *2*, e00055–e00111. [CrossRef]

58. Vesely, E.M.; Williams, R.B.; Konopka, J.B.; Lorenz, M.C. N-Acetylglucosamine Metabolism Promotes Survival of *Candida albicans* in the Phagosome. *mSphere* **2017**, *2*. [CrossRef]

59. Pande, K.; Chen, C.; Noble, S.M. Passage through the mammalian gut triggers a phenotypic switch that promotes *Candida albicans* commensalism. *Nat. Genet.* **2013**, *45*, 1088–1091. [CrossRef]

60. Naseem, S.; Konopka, J.B. N-acetylglucosamine Regulates Virulence Properties in Microbial Pathogens. *PLoS Pathog.* **2015**, *11*, e1004947. [CrossRef]

61. Sudbery, P.E. Growth of *Candida albicans* hyphae. *Nat. Rev. Microbiol.* **2011**, *9*, 737–748. [CrossRef]

62. Alvarez, F.J.; Konopka, J.B. Identification of an N-Acetylglucosamine Transporter That Mediates Hyphal Induction in *Candida albicans*. *Mol. Biol. Cell* **2007**, *18*, 965–975. [CrossRef]

63. Naseem, S.; Gunasekera, A.; Araya, E.; Konopka, J.B. N-acetylglucosamine (GlcNAc) induction of hyphal morphogenesis and transcriptional responses in *Candida albicans* are not dependent on its metabolism. *J. Biol. Chem.* **2011**, *286*, 28671–28680. [CrossRef]

64. Naseem, S.; Min, K.; Spitzer, D.; Gardin, J.; Konopka, J.B. Regulation of Hyphal Growth and N-Acetylglucosamine Catabolism by Two Transcription Factors in *Candida albicans*. *Genetics* **2017**, *206*, 299–314. [CrossRef]

65. Su, C.; Lu, Y.; Liu, H. N-acetylglucosamine sensing by a GCN5-related N-acetyltransferase induces transcription via chromatin histone acetylation in fungi. *Nat. Commun.* **2016**, *7*, 12916. [CrossRef]

66. Min, K.; Biermann, A.; Hogan, D.A.; Konopka, J.B. Genetic Analysis of NDT80 Family Transcription Factors in *Candida albicans* Using New CRISPR-Cas9 Approaches. *mSphere* **2018**, *3*, e00545–e00618. [CrossRef]

67. Rocha, C.R.; Schröppel, K.; Harcus, D.; Marcil, A.; Dignard, D.; Taylor, B.N.; Thomas, D.Y.; Whiteway, M.; Leberer, E. Signaling through adenylyl cyclase is essential for hyphal growth and virulence in the pathogenic fungus *Candida albicans*. *Mol. Biol. Cell* **2001**, *12*, 3631–3643. [CrossRef]

68. Parrino, S.M.; Si, H.; Naseem, S.; Groudan, K.; Gardin, J.; Konopka, J.B. cAMP-independent signal pathways stimulate hyphal morphogenesis in *Candida albicans*. *Mol. Microbiol.* **2017**, *103*, 764–779. [CrossRef]

69. Min, K.; Naseem, S.; Konopka, J.B. *N*-Acetylglucosamine Regulates Morphogenesis and Virulence Pathways in Fungi. *J. Fungi* **2020**, *6*, 8. [CrossRef]

70. Dabrowa, N.; Taxer, S.S.; Howard, D.H. Germination of *Candida albicans* induced by proline. *Infect. Immun.* **1976**, *13*, 830–835. [CrossRef]

71. Land, G.A.; McDonald, W.C.; Stjernholm, R.L.; Friedman, T.L. Factors affecting filamentation in *Candida albicans*: Relationship of the uptake and distribution of proline to morphogenesis. *Infect. Immun.* **1975**, *11*, 1014–1023. [CrossRef]

72. Ghosh, S.; Navarathna, D.H.; Roberts, D.D.; Cooper, J.T.; Atkin, A.L.; Petro, T.M.; Nickerson, K.W. Arginine-induced germ tube formation in *Candida albicans* is essential for escape from murine macrophage line RAW 264.7. *Infect. Immun.* **2009**, *77*, 1596–1605. [CrossRef]

73. Miramon, P.; Lorenz, M.C. The SPS amino acid sensor mediates nutrient acquisition and immune evasion in *Candida albicans*. *Cell Microbiol.* **2016**, *18*, 1611–1624. [CrossRef]

74. Martinez, P.; Ljungdahl, P.O. Divergence of Stp1 and Stp2 transcription factors in *Candida albicans* places virulence factors required for proper nutrient acquisition under amino acid control. *Mol. Cell. Biol.* **2005**, *25*, 9435–9446. [CrossRef]

75. Miramon, P.; Pountain, A.W.; van Hoof, A.; Lorenz, M.C. The Paralogous Transcription Factors Stp1 and Stp2 of *Candida albicans* Have Distinct Functions in Nutrient Acquisition and Host Interaction. *Infect. Immun.* **2020**, *88*. [CrossRef]

76. Brega, E.; Zufferey, R.; Mamoun, C.B. *Candida albicans* Csy1p is a nutrient sensor important for activation of amino acid uptake and hyphal morphogenesis. *Eukaryot. Cell* **2004**, *3*, 135–143. [CrossRef]

77. Abdel-Sater, F.; Jean, C.; Merhi, A.; Vissers, S.; Andre, B. Amino acid signaling in yeast: Activation of Ssy5 protease is associated with its phosphorylation-induced ubiquitylation. *J. Biol. Chem.* **2011**, *286*, 12006–12015. [CrossRef]

78. Martinez, P.; Ljungdahl, P.O. An ER packaging chaperone determines the amino acid uptake capacity and virulence of *Candida albicans*. *Mol. Microbiol.* **2004**, *51*, 371–384. [CrossRef] [PubMed]

79. Silao, F.G.S.; Ward, M.; Ryman, K.; Wallström, A.; Brindefalk, B.; Udekwu, K.; Ljungdahl, P.O. Mitochondrial proline catabolism activates Ras1/cAMP/PKA-induced filamentation in *Candida albicans*. *PLoS Genet.* **2019**, *15*, e1007976. [CrossRef] [PubMed]

80. Grahl, N.; Demers, E.G.; Lindsay, A.K.; Harty, C.E.; Willger, S.D.; Piispanen, A.E.; Hogan, D.A. Mitochondrial Activity and Cyr1 Are Key Regulators of Ras1 Activation of *C. albicans* Virulence Pathways. *PLoS Pathog.* **2015**, *11*, e1005133. [CrossRef] [PubMed]

81. Davis-Hanna, A.; Piispanen, A.E.; Stateva, L.I.; Hogan, D.A. Farnesol and dodecanol effects on the *Candida albicans* Ras1-cAMP signalling pathway and the regulation of morphogenesis. *Mol. Microbiol.* **2008**, *67*, 47–62. [CrossRef]

82. Hall, R.A.; Turner, K.J.; Chaloupka, J.; Cottier, F.; De Sordi, L.; Sanglard, D.; Levin, L.R.; Buck, J.; Mühlschlegel, F.A. The quorum-sensing molecules farnesol/homoserine lactone and dodecanol operate via distinct modes of action in *Candida albicans*. *Eukaryot. Cell* **2011**, *10*, 1034–1042. [CrossRef]

83. Chauhan, N.M.; Raut, J.S.; Karuppayil, S.M. A morphogenetic regulatory role for ethyl alcohol in *Candida albicans*. *Mycoses* **2011**, *54*, e697–e703. [CrossRef]

84. Martins, M.; Henriques, M.; Azeredo, J.; Rocha, S.M.; Coimbra, M.A.; Oliveira, R. Morphogenesis control in *Candida albicans* and *Candida dubliniensis* through signaling molecules produced by planktonic and biofilm cells. *Eukaryot. Cell* **2007**, *6*, 2429–2436. [CrossRef]

85. Lindsay, A.K.; Morales, D.K.; Liu, Z.; Grahl, N.; Zhang, A.; Willger, S.D.; Myers, L.C.; Hogan, D.A. Analysis of *Candida albicans* mutants defective in the Cdk8 module of mediator reveal links between metabolism and biofilm formation. *PLoS Genet.* **2014**, *10*, e1004567. [CrossRef]

86. Alem, M.A.; Oteef, M.D.; Flowers, T.H.; Douglas, L.J. Production of tyrosol by *Candida albicans* biofilms and its role in quorum sensing and biofilm development. *Eukaryot. Cell* **2006**, *5*, 1770–1779. [CrossRef]

87. Chen, H.; Fujita, M.; Feng, Q.; Clardy, J.; Fink, G.R. Tyrosol is a quorum-sensing molecule in *Candida albicans*. *Proc. Natl. Acad. Sci. USA* **2004**, *101*, 5048–5052. [CrossRef]

88. Ghosh, S.; Kebaara, B.W.; Atkin, A.L.; Nickerson, K.W. Regulation of aromatic alcohol production in *Candida albicans*. *Appl. Environ. Microbiol.* **2008**, *74*, 7211–7218. [CrossRef]

89. Chauhan, N.M.; Mohan Karuppayil, S. Dual identities for various alcohols in two different yeasts. *Mycology* **2020**, *12*, 25–38. [CrossRef]

90. Kebaara, B.W.; Langford, M.L.; Navarathna, D.H.M.L.P.; Dumitru, R.; Nickerson, K.W.; Atkin, A.L. *Candida albicans* Tup1 Is Involved in Farnesol-Mediated Inhibition of Filamentous-Growth Induction. *Eukaryot. Cell* **2008**, *7*, 980–987. [CrossRef]

91. Lindsay, A.K.; Hogan, D.A. *Candida albicans*: Molecular interactions with *Pseudomonas aeruginosa* and *Staphylococcus aureus*. *Fungal Biol. Rev.* **2014**, *28*, 85–96. [CrossRef]

92. Hogan, D.A.; Vik, Å.; Kolter, R. A *Pseudomonas aeruginosa* quorum-sensing molecule influences *Candida albicans* morphology. *Mol. Microbiol.* **2004**, *54*, 1212–1223. [CrossRef]

93. Haiko, J.; Saeedi, B.; Bagger, G.; Karpati, F.; Özenci, V. Coexistence of *Candida* species and bacteria in patients with cystic fibrosis. *Eur. J. Clin. Microbiol. Infect. Dis.* **2019**, *38*, 1071–1077. [CrossRef]

94. Morales, D.K.; Grahl, N.; Okegbe, C.; Dietrich, L.E.P.; Jacobs, N.J.; Hogan, D.A. Control of *Candida albicans* Metabolism and Biofilm Formation by *Pseudomonas aeruginosa* Phenazines. *mBio* **2013**, *4*, e00526–e00612. [CrossRef] [PubMed]

95. Ryan, R.P.; An, S.-Q.; Allan, J.H.; McCarthy, Y.; Dow, J.M. The DSF Family of Cell-Cell Signals: An Expanding Class of Bacterial Virulence Regulators. *PLoS Pathog.* **2015**, *11*, e1004986. [CrossRef] [PubMed]

96. de Rossi, B.P.; García, C.; Alcaraz, E.; Franco, M. Stenotrophomonas maltophilia interferes via the DSF-mediated quorum sensing system with *Candida albicans* filamentation and its planktonic and biofilm modes of growth. *Rev. Argent. Microbiol.* **2014**, *46*, 288–297. [CrossRef]

97. Yang, D.; Hu, Y.; Yin, Z.; Gao, Q.; Zhang, Y.; Chan, F.Y.; Zeng, G.; Weng, L.; Wang, L.; Wang, Y. *Candida albicans* Ubiquitin and Heat Shock Factor-Type Transcriptional Factors Are Involved in 2-Dodecenoic Acid-Mediated Inhibition of Hyphal Growth. *Microorganisms* **2020**, *8*, 75. [CrossRef] [PubMed]

98. Nett, J.E.; Marchillo, K.; Spiegel, C.A.; Andes, D.R. Development and validation of an in vivo *Candida albicans* biofilm denture model. *Infect. Immun.* **2010**, *78*, 3650–3659. [CrossRef]

99. Balasubramanian, V.; Vashisht, D.; Cletus, J.; Sakthivel, N. Plant β-1,3-glucanases: Their biological functions and transgenic expression against phytopathogenic fungi. *Biotechnol. Lett.* **2012**, *34*, 1983–1990. [CrossRef]

100. Wu, Q.; Dou, X.; Wang, Q.; Guan, Z.; Cai, Y.; Liao, X. Isolation of β-1,3-Glucanase-Producing Microorganisms from *Poria cocos* Cultivation Soil via Molecular Biology. *Molecules* **2018**, *23*, 1555. [CrossRef]

101. Adams, D.J. Fungal cell wall chitinases and glucanases. *Microbiology* **2004**, *150*, 2029–2035. [CrossRef]

102. Tsai, P.W.; Yang, C.Y.; Chang, H.T.; Lan, C.Y. Characterizing the role of cell-wall beta-1,3-exoglucanase Xog1p in *Candida albicans* adhesion by the human antimicrobial peptide LL-37. *PLoS ONE* **2011**, *6*, e21394. [CrossRef]

103. Xu, H.; Nobile, C.J.; Dongari-Bagtzoglou, A. Glucanase Induces Filamentation of the Fungal Pathogen *Candida albicans*. *PLoS ONE* **2013**, *8*, e63736. [CrossRef]

104. Calera, J.A.; Zhao, X.J.; Calderone, R. Defective hyphal development and avirulence caused by a deletion of the SSK1 response regulator gene in *Candida albicans*. *Infect. Immun.* **2000**, *68*, 518–525. [CrossRef]

105. Lee, K.L.; Buckley, H.R.; Campbell, C.C. An amino acid liquid synthetic medium for the development of mycelial and yeast forms of *Candida albicans*. *Sabouraudia* **1975**, *13*, 148–153. [CrossRef]

106. Liu, H.; Kohler, J.; Fink, G. Suppression of hyphal formation in *Candida albicans* by mutation of a STE12 homolog. *Science* **1994**, *266*, 1723–1726. [CrossRef]

107. Biswas, K.; Morschhäuser, J. The Mep2p ammonium permease controls nitrogen starvation-induced filamentous growth in *Candida albicans*. *Mol. Microbiol.* **2005**, *56*, 649–669. [CrossRef]

108. Dabas, N.; Schneider, S.; Morschhäuser, J. Mutational Analysis of the *Candida albicans* Ammonium Permease Mep2p Reveals Residues Required for Ammonium Transport and Signaling. *Eukaryot. Cell* **2009**, *8*, 147–160. [CrossRef]

109. Schrevens, S.; Van Zeebroeck, G.; Riedelberger, M.; Tournu, H.; Kuchler, K.; Van Dijck, P. Methionine is required for cAMP-PKA-mediated morphogenesis and virulence of *Candida albicans*. *Mol. Microbiol.* **2018**, *108*, 258–275. [CrossRef]

110. Cao, C.; Wu, M.; Bing, J.; Tao, L.; Ding, X.; Liu, X.; Huang, G. Global regulatory roles of the cAMP/PKA pathway revealed by phenotypic, transcriptomic and phosphoproteomic analyses in a null mutant of the PKA catalytic subunit in *Candida albicans*. *Mol. Microbiol.* **2017**, *105*, 46–64. [CrossRef]

111. Hogan, D.A.; Sundstrom, P. The Ras/cAMP/PKA signaling pathway and virulence in *Candida albicans*. *Future Microbiol.* **2009**, *4*, 1263–1270. [CrossRef]

112. Bockmühl, D.P.; Ernst, J.F. A potential phosphorylation site for an A-type kinase in the Efg1 regulator protein contributes to hyphal morphogenesis of *Candida albicans*. *Genetics* **2001**, *157*, 1523–1530. [CrossRef]

113. Cao, F.; Lane, S.; Raniga, P.P.; Lu, Y.; Zhou, Z.; Ramon, K.; Chen, J.; Liu, H. The Flo8 transcription factor is essential for hyphal development and virulence in *Candida albicans*. *Mol. Biol. Cell* **2006**, *17*, 295–307. [CrossRef]

114. Fang, H.M.; Wang, Y. RA domain-mediated interaction of Cdc35 with Ras1 is essential for increasing cellular cAMP level for *Candida albicans* hyphal development. *Mol. Microbiol.* **2006**, *61*, 484–496. [CrossRef]

115. Taylor, S.S.; Buechler, J.A.; Yonemoto, W. cAMP-dependent protein kinase: Framework for a diverse family of regulatory enzymes. *Annu. Rev. Biochem.* **1990**, *59*, 971–1005. [CrossRef]

116. Bockmühl, D.P.; Krishnamurthy, S.; Gerads, M.; Sonneborn, A.; Ernst, J.F. Distinct and redundant roles of the two protein kinase A isoforms Tpk1p and Tpk2p in morphogenesis and growth of *Candida albicans*. *Mol. Microbiol.* **2001**, *42*, 1243–1257. [CrossRef]

117. Cloutier, M.; Castilla, R.; Bolduc, N.; Zelada, A.; Martineau, P.; Bouillon, M.; Magee, B.B.; Passeron, S.; Giasson, L.; Cantore, M.L. The two isoforms of the cAMP-dependent protein kinase catalytic subunit are involved in the control of dimorphism in the human fungal pathogen *Candida albicans*. *Fungal Genet. Biol.* **2003**, *38*, 133–141. [CrossRef]

118. Hoyer, L.L.; Cieslinski, L.B.; McLaughlin, M.M.; Torphy, T.J.; Shatzman, A.R.; Livi, G.P. A *Candida albicans* cyclic nucleotide phosphodiesterase: Cloning and expression in *Saccharomyces cerevisiae* and biochemical characterization of the recombinant enzyme. *Microbiology* **1994**, *1994*, 1533–1542. [CrossRef]

119. Jung, W.H.; Stateva, L.I. The cAMP phosphodiesterase encoded by CaPDE2 is required for hyphal development in *Candida albicans*. *Microbiology* **2003**, *149*, 2961–2976. [CrossRef]

120. Bahn, Y.S.; Staab, J.; Sundstrom, P. Increased high-affinity phosphodiesterase PDE2 gene expression in germ tubes counteracts CAP1-dependent synthesis of cyclic AMP, limits hypha production and promotes virulence of *Candida albicans*. *Mol. Microbiol.* **2003**, *50*, 391–409. [CrossRef] [PubMed]

121. Wilson, D.; Tutulan-Cunita, A.; Jung, W.; Hauser, N.C.; Hernandez, R.; Williamson, T.; Piekarska, K.; Rupp, S.; Young, T.; Stateva, L. Deletion of the high-affinity cAMP phosphodiesterase encoded by PDE2 affects stress responses and virulence in *Candida albicans*. *Mol. Microbiol.* **2007**, *65*, 841–856. [CrossRef] [PubMed]

122. Wilson, D.; Fiori, A.; Brucker, K.D.; Dijck, P.V.; Stateva, L. *Candida albicans* Pde1p and Gpa2p comprise a regulatory module mediating agonist-induced cAMP signalling and environmental adaptation. *Fungal Genet. Biol.* **2010**, *47*, 742–752. [CrossRef] [PubMed]

123. Harcus, D.; Nantel, A.; Marcil, A.; Rigby, T.; Whiteway, M. Transcription profiling of cyclic AMP signaling in *Candida albicans*. *Mol. Biol. Cell* **2004**, *15*, 4490–4499. [CrossRef]
124. Wang, Y. Fungal Adenylyl Cyclase Acts As a Signal Sensor and Integrator and Plays a Central Role in Interaction with Bacteria. *PLoS Pathog.* **2013**, *9*, e1003612. [CrossRef]
125. Huang, G.; Huang, Q.; Wei, Y.; Wang, Y.; Du, H. Multiple roles and diverse regulation of the Ras/cAMP/protein kinase A pathway in *Candida albicans*. *Mol. Microbiol.* **2019**, *111*, 6–16. [CrossRef]
126. Boguski, M.S.; McCormick, F. Proteins regulating Ras and its relatives. *Nature* **1993**, *366*, 643–654. [CrossRef]
127. Bahn, Y.S.; Sundstrom, P. CAP1, an adenylate cyclase-associated protein gene, regulates bud-hypha transitions, filamentous growth, and cyclic AMP levels and is required for virulence of *Candida albicans*. *J. Bacteriol.* **2001**, *183*, 3211–3223. [CrossRef]
128. Feng, Q.; Summers, E.; Guo, B.; Fink, G. Ras signaling is required for serum-induced hyphal differentiation in *Candida albicans*. *J. Bacteriol.* **1999**, *181*, 6339–6346. [CrossRef]
129. Leberer, E.; Harcus, D.; Dignard, D.; Johnson, L.; Ushinsky, S.; Thomas, D.Y.; Schröppel, K. Ras links cellular morphogenesis to virulence by regulation of the MAP kinase and cAMP signalling pathways in the pathogenic fungus *Candida albicans*. *Mol. Microbiol.* **2001**, *42*, 673–687. [CrossRef]
130. Brown, A.J.; Barelle, C.J.; Budge, S.; Duncan, J.; Harris, S.; Lee, P.R.; Leng, P.; Macaskill, S.; Abdul Murad, A.M.; Ramsdale, M.; et al. Gene regulation during morphogenesis in *Candida albicans*. *Contrib. Microbiol.* **2000**, *5*, 112–125. [CrossRef]
131. Leach, M.D.; Farrer, R.A.; Tan, K.; Miao, Z.; Walker, L.A.; Cuomo, C.A.; Wheeler, R.T.; Brown, A.J.; Wong, K.H.; Cowen, L.E. Hsf1 and Hsp90 orchestrate temperature-dependent global transcriptional remodelling and chromatin architecture in *Candida albicans*. *Nat. Commun.* **2016**, *7*, 11704. [CrossRef] [PubMed]
132. Shapiro, R.S.; Zaas, A.K.; Betancourt-Quiroz, M.; Perfect, J.R.; Cowen, L.E. The Hsp90 co-chaperone Sgt1 governs *Candida albicans* morphogenesis and drug resistance. *PLoS ONE* **2012**, *7*, e44734. [CrossRef] [PubMed]
133. Bachewich, C.; Nantel, A.; Whiteway, M. Cell cycle arrest during S or M phase generates polarized growth via distinct signals in *Candida albicans*. *Mol. Microbiol.* **2005**, *57*, 942–959. [CrossRef] [PubMed]
134. Bachewich, C.; Whiteway, M. Cyclin Cln3p links G1 progression to hyphal and pseudohyphal development in *Candida albicans*. *Eukaryot. Cell* **2005**, *4*, 95–102. [CrossRef]
135. Chen, C.; Zeng, G.; Wang, Y. G1 and S phase arrest in *Candida albicans* induces filamentous growth via distinct mechanisms. *Mol. Microbiol.* **2018**, *110*, 191–203. [CrossRef]
136. Chen, R.E.; Thorner, J. Function and regulation in MAPK signaling pathways: Lessons learned from the yeast *Saccharomyces cerevisiae*. *Biochim. Biophys. Acta* **2007**, *1773*, 1311–1340. [CrossRef]
137. Monge, R.A.; Román, E.; Nombela, C.; Pla, J. The MAP kinase signal transduction network in *Candida albicans*. *Microbiology* **2006**, *152*, 905–912. [CrossRef]
138. Román, E.; Arana, D.M.; Nombela, C.; Alonso-Monge, R.; Pla, J. MAP kinase pathways as regulators of fungal virulence. *Trends Microbiol.* **2007**, *15*, 181–190. [CrossRef]
139. Eisman, B.; Alonso-Monge, R.; Román, E.; Arana, D.; Nombela, C.; Pla, J. The Cek1 and Hog1 Mitogen-Activated Protein Kinases Play Complementary Roles in Cell Wall Biogenesis and Chlamydospore Formation in the Fungal Pathogen *Candida albicans*. *Eukaryot. Cell* **2006**, *5*, 347–358. [CrossRef]
140. Galán-Díez, M.; Arana, D.M.; Serrano-Gómez, D.; Kremer, L.; Casasnovas, J.M.; Ortega, M.; Cuesta-Domínguez, A.; Corbí, A.L.; Pla, J.; Fernández-Ruiz, E. *Candida albicans* beta-glucan exposure is controlled by the fungal CEK1-mediated mitogen-activated protein kinase pathway that modulates immune responses triggered through dectin-1. *Infect. Immun.* **2010**, *78*, 1426–1436. [CrossRef]
141. Köhler, J.R.; Fink, G.R. *Candida albicans* strains heterozygous and homozygous for mutations in mitogen-activated protein kinase signaling components have defects in hyphal development. *Proc. Natl. Acad. Sci. USA* **1996**, *93*, 13223–13228. [CrossRef]
142. Bassilana, M.; Blyth, J.; Arkowitz, R.A. Cdc24, the GDP-GTP exchange factor for Cdc42, is required for invasive hyphal growth of *Candida albicans*. *Eukaryot. Cell* **2003**, *2*, 9–18. [CrossRef]
143. Hazan, I.; Liu, H. Hyphal tip-associated localization of Cdc42 is F-actin dependent in *Candida albicans*. *Eukaryot. Cell* **2002**, *1*, 856–864. [CrossRef]
144. Chen, H.; Zhou, X.; Ren, B.; Cheng, L. The regulation of hyphae growth in *Candida albicans*. *Virulence* **2020**, *11*, 337–348. [CrossRef]
145. Huang, G. Regulation of phenotypic transitions in the fungal pathogen *Candida albicans*. *Virulence* **2012**, *3*, 251–261. [CrossRef]
146. Herrero de Dios, C.; Román, E.; Diez, C.; Alonso-Monge, R.; Pla, J. The transmembrane protein Opy2 mediates activation of the Cek1 MAP kinase in *Candida albicans*. *Fungal Genet. Biol.* **2013**, *50*, 21–32. [CrossRef]
147. Román, E.; Cottier, F.; Ernst, J.F.; Pla, J. Msb2 signaling mucin controls activation of Cek1 mitogen-activated protein kinase in *Candida albicans*. *Eukaryot. Cell* **2009**, *8*, 1235–1249. [CrossRef]
148. Hope, H.; Schmauch, C.; Arkowitz, R.A.; Bassilana, M. The *Candida albicans* ELMO homologue functions together with Rac1 and Dck1, upstream of the MAP Kinase Cek1, in invasive filamentous growth. *Mol. Microbiol.* **2010**, *76*, 1572–1590. [CrossRef]
149. Hope, H.; Bogliolo, S.; Arkowitz, R.A.; Bassilana, M. Activation of Rac1 by the Guanine Nucleotide Exchange Factor Dck1 Is Required for Invasive Filamentous Growth in the Pathogen *Candida albicans*. *Mol. Biol. Cell* **2008**, *19*, 3638–3651. [CrossRef]
150. Bassilana, M.; Arkowitz, R.A. Rac1 and Cdc42 Have Different Roles in *Candida albicans* Development. *Eukaryot. Cell* **2006**, *5*, 321–329. [CrossRef]

151. Riggle, P.J.; Andrutis, K.A.; Chen, X.; Tzipori, S.R.; Kumamoto, C.A. Invasive Lesions Containing Filamentous Forms Produced by a *Candida albicans* Mutant That Is Defective in Filamentous Growth in Culture. *Infect. Immun.* **1999**, *67*, 3649–3652. [CrossRef] [PubMed]

152. Levin, D.E. Regulation of cell wall biogenesis in *Saccharomyces cerevisiae*: The cell wall integrity signaling pathway. *Genetics* **2011**, *189*, 1145–1175. [CrossRef] [PubMed]

153. Perez, P.; Rincón, S.A. Rho GTPases: Regulation of cell polarity and growth in yeasts. *Biochem. J.* **2010**, *426*, 243–253. [CrossRef] [PubMed]

154. Xie, J.L.; Grahl, N.; Sless, T.; Leach, M.D.; Kim, S.H.; Hogan, D.A.; Robbins, N.; Cowen, L.E. Signaling through Lrg1, Rho1 and Pkc1 Governs *Candida albicans* Morphogenesis in Response to Diverse Cues. *PLoS Genet.* **2016**, *12*, e1006405. [CrossRef]

155. Biswas, S.; Van Dijck, P.; Datta, A. Environmental Sensing and Signal Transduction Pathways Regulating Morphopathogenic Determinants of *Candida albicans*. *Microbiol. Mol. Biol. Rev.* **2007**, *71*, 348–376. [CrossRef]

156. Braun, B.R.; Johnson, A.D. Control of Filament Formation in *Candida albicans* by the Transcriptional Repressor TUP1. *Science* **1997**, *277*, 105–109. [CrossRef]

157. Murad, A.M.; Leng, P.; Straffon, M.; Wishart, J.; Macaskill, S.; MacCallum, D.; Schnell, N.; Talibi, D.; Marechal, D.; Tekaia, F.; et al. NRG1 represses yeast-hypha morphogenesis and hypha-specific gene expression in *Candida albicans*. *EMBO J.* **2001**, *20*, 4742–4752. [CrossRef]

158. Kadosh, D.; Johnson, A.D. Rfg1, a protein related to the *Saccharomyces cerevisiae* hypoxic regulator Rox1, controls filamentous growth and virulence in *Candida albicans*. *Mol. Cell. Biol.* **2001**, *21*, 2496–2505. [CrossRef]

159. Braun, B.R.; Johnson, A.D. TUP1, CPH1 and EFG1 make independent contributions to filamentation in *Candida albicans*. *Genetics* **2000**, *155*, 57–67. [CrossRef]

160. Kadosh, D.; Johnson, A.D. Induction of the *Candida albicans* filamentous growth program by relief of transcriptional repression: A genome-wide analysis. *Mol. Biol. Cell* **2005**, *16*, 2903–2912. [CrossRef]

161. Banerjee, M.; Thompson, D.S.; Lazzell, A.; Carlisle, P.L.; Pierce, C.; Monteagudo, C.; López-Ribot, J.L.; Kadosh, D. UME6, a novel filament-specific regulator of *Candida albicans* hyphal extension and virulence. *Mol. Biol. Cell* **2008**, *19*, 1354–1365. [CrossRef]

162. Braun, B.R.; Kadosh, D.; Johnson, A.D. NRG1, a repressor of filamentous growth in *C. albicans*, is down-regulated during filament induction. *EMBO J.* **2001**, *20*, 4753–4761. [CrossRef]

163. Jensen, E.C.; Hornby, J.M.; Pagliaccetti, N.E.; Wolter, C.M.; Nickerson, K.W.; Atkin, A.L. Farnesol restores wild-type colony morphology to 96% of *Candida albicans* colony morphology variants recovered following treatment with mutagens. *Genome* **2006**, *49*, 346–353. [CrossRef]

164. Mosel, D.D.; Dumitru, R.; Hornby, J.M.; Atkin, A.L.; Nickerson, K.W. Farnesol concentrations required to block germ tube formation in *Candida albicans* in the presence and absence of serum. *Appl. Environ. Microbiol.* **2005**, *71*, 4938–4940. [CrossRef]

165. Ramage, G.; Saville, S.P.; Wickes, B.L.; López-Ribot, J.L. Inhibition of *Candida albicans* biofilm formation by farnesol, a quorum-sensing molecule. *Appl. Environ. Microbiol.* **2002**, *68*, 5459–5463. [CrossRef]

166. Lu, Y.; Su, C.; Unoje, O.; Liu, H. Quorum sensing controls hyphal initiation in *Candida albicans* through Ubr1-mediated protein degradation. *Proc. Natl. Acad. Sci. USA* **2014**, *111*, 1975–1980. [CrossRef]

167. Martin, R.; Moran, G.P.; Jacobsen, I.D.; Heyken, A.; Domey, J.; Sullivan, D.J.; Kurzai, O.; Hube, B. The *Candida albicans*-Specific Gene EED1 Encodes a Key Regulator of Hyphal Extension. *PLoS ONE* **2011**, *6*, e18394. [CrossRef]

168. Mendelsohn, S.; Pinsky, M.; Weissman, Z.; Kornitzer, D. Regulation of the *Candida albicans* Hypha-Inducing Transcription Factor Ume6 by the CDK1 Cyclins Cln3 and Hgc1. *mSphere* **2017**, *2*, e00248–e00316. [CrossRef]

169. Zheng, X.; Wang, Y.; Wang, Y. Hgc1, a novel hypha-specific G1 cyclin-related protein regulates *Candida albicans* hyphal morphogenesis. *EMBO J.* **2004**, *23*, 1845–1856. [CrossRef]

170. Carlisle, P.L.; Kadosh, D. *Candida albicans* Ume6, a filament-specific transcriptional regulator, directs hyphal growth via a pathway involving Hgc1 cyclin-related protein. *Eukaryot. Cell* **2010**, *9*, 1320–1328. [CrossRef]

171. Lu, Y.; Su, C.; Solis, N.V.; Filler, S.G.; Liu, H. Synergistic Regulation of Hyphal Elongation by Hypoxia, CO2, and Nutrient Conditions Controls the Virulence of *Candida albicans*. *Cell Host Microbe* **2013**, *14*, 499–509. [CrossRef] [PubMed]

172. Lu, Y.; Su, C.; Ray, S.; Yuan, Y.; Liu, H. CO2 Signaling through the Ptc2-Ssn3 Axis Governs Sustained Hyphal Development of *Candida albicans* by Reducing Ume6 Phosphorylation and Degradation. *mBio* **2019**, *10*. [CrossRef] [PubMed]

173. Hamdy, R.; Soliman, S.S.M.; Alsaadi, A.I.; Fayed, B.; Hamoda, A.M.; Elseginy, S.A.; Husseiny, M.I.; Ibrahim, A.S. Design and synthesis of new drugs inhibitors of *Candida albicans* hyphae and biofilm formation by upregulating the expression of *TUP1* transcription repressor gene. *Eur. J. Pharm. Sci.* **2020**, *148*, 105327. [CrossRef] [PubMed]

174. Padmavathi, A.R.; Das, A.; Priya, A.; Sushmitha, T.J.; Pandian, S.K.; Toleti, S.R. Impediment to growth and yeast-to-hyphae transition in *Candida albicans* by copper oxide nanoparticles. *Biofouling* **2020**, *36*, 56–72. [CrossRef]

175. González-Novo, A.; Correa-Bordes, J.; Labrador, L.; Sánchez, M.; Vázquez de Aldana, C.R.; Jiménez, J. Sep7 is essential to modify septin ring dynamics and inhibit cell separation during *Candida albicans* hyphal growth. *Mol. Biol. Cell* **2008**, *19*, 1509–1518. [CrossRef]

176. Sinha, I.; Wang, Y.M.; Philp, R.; Li, C.R.; Yap, W.H.; Wang, Y. Cyclin-dependent kinases control septin phosphorylation in *Candida albicans* hyphal development. *Dev. Cell* **2007**, *13*, 421–432. [CrossRef]

177. Sudbery, P.E. The germ tubes of *Candida albicans* hyphae and pseudohyphae show different patterns of septin ring localization. *Mol. Microbiol.* **2001**, *41*, 19–31. [CrossRef]

178. Warenda, A.J.; Konopka, J.B. Septin function in *Candida albicans* morphogenesis. *Mol. Biol. Cell* **2002**, *13*, 2732–2746. [CrossRef]
179. Au Yong, J.Y.; Wang, Y.M.; Wang, Y. The Nim1 kinase Gin4 has distinct domains crucial for septin assembly, phospholipid binding and mitotic exit. *J. Cell Sci.* **2016**, *129*, 2744–2756. [CrossRef]
180. Li, C.R.; Au Yong, J.Y.; Wang, Y.M.; Wang, Y. CDK regulates septin organization through cell-cycle-dependent phosphorylation of the Nim1-related kinase Gin4. *J. Cell Sci.* **2012**, *125*, 2533–2543. [CrossRef]
181. Wightman, R.; Bates, S.; Amornrrattanapan, P.; Sudbery, P. In *Candida albicans*, the Nim1 kinases Gin4 and Hsl1 negatively regulate pseudohypha formation and Gin4 also controls septin organization. *J Cell Biol.* **2004**, *164*, 581–591. [CrossRef]
182. Liu, Q.; Han, Q.; Wang, N.; Yao, G.; Zeng, G.; Wang, Y.; Huang, Z.; Sang, J.; Wang, Y. Tpd3-Pph21 phosphatase plays a direct role in Sep7 dephosphorylation in *Candida albicans*. *Mol. Microbiol.* **2016**, *101*, 109–121. [CrossRef]
183. Han, Q.; Pan, C.; Wang, Y.; Wang, N.; Wang, Y.; Sang, J. The PP2A regulatory subunits, Cdc55 and Rts1, play distinct roles in *Candida albicans'* growth, morphogenesis, and virulence. *Fungal Genet. Biol.* **2019**, *131*, 103240. [CrossRef]
184. Finley, K.R.; Berman, J. Microtubules in *Candida albicans* hyphae drive nuclear dynamics and connect cell cycle progression to morphogenesis. *Eukaryot. Cell* **2005**, *4*, 1697–1711. [CrossRef]
185. Clemente-Blanco, A.; Gonzalez-Novo, A.; Machin, F.; Caballero-Lima, D.; Aragon, L.; Sanchez, M.; de Aldana, C.R.; Jimenez, J.; Correa-Bordes, J. The Cdc14p phosphatase affects late cell-cycle events and morphogenesis in *Candida albicans*. *J. Cell Sci.* **2006**, *119*, 1130–1143. [CrossRef]
186. González-Novo, A.; Vázquez de Aldana, C.R.; Jiménez, J. Fungal septins: One ring to rule it all? *Cent. Eur. J. Biol.* **2009**, *4*, 274–289. [CrossRef]
187. Huang, Z.X.; Zhao, P.; Zeng, G.S.; Wang, Y.M.; Sudbery, I.; Wang, Y. Phosphoregulation of Nap1 plays a role in septin ring dynamics and morphogenesis in *Candida albicans*. *mBio* **2014**, *5*, e00915-13. [CrossRef]
188. Xie, Y.; Loh, Z.Y.; Xue, J.; Zhou, F.; Sun, J.; Qiao, Z.; Jin, S.; Deng, Y.; Li, H.; Wang, Y.; et al. Orchestrated actin nucleation by the *Candida albicans* polarisome complex enables filamentous growth. *J. Biol. Chem.* **2020**, *295*, 14840–14854. [CrossRef]
189. Caballero-Lima, D.; Hautbergue, G.M.; Wilson, S.A.; Sudbery, P.E. In *Candida albicans* hyphae, Sec2p is physically associated with SEC2 mRNA on secretory vesicles. *Mol. Microbiol.* **2014**, *94*, 828–842. [CrossRef]
190. Caballero-Lima, D.; Kaneva, I.N.; Watton, S.P.; Sudbery, P.E.; Craven, C.J. The spatial distribution of the exocyst and actin cortical patches is sufficient to organize hyphal tip growth. *Eukaryot. Cell* **2013**, *12*, 998–1008. [CrossRef]
191. Jones, L.A.; Sudbery, P.E. Spitzenkorper, exocyst, and polarisome components in *Candida albicans* hyphae show different patterns of localization and have distinct dynamic properties. *Eukaryot. Cell* **2010**, *9*, 1455–1465. [CrossRef]
192. TerBush, D.R.; Maurice, T.; Roth, D.; Novick, P. The Exocyst is a multiprotein complex required for exocytosis in *Saccharomyces cerevisiae*. *EMBO J.* **1996**, *15*, 6483–6494. [CrossRef]
193. Crampin, H.; Finley, K.; Gerami-Nejad, M.; Court, H.; Gale, C.; Berman, J.; Sudbery, P. *Candida albicans* hyphae have a Spitzenkorper that is distinct from the polarisome found in yeast and pseudohyphae. *J. Cell Sci.* **2005**, *118*, 2935–2947. [CrossRef]
194. Bishop, A.; Lane, R.; Beniston, R.; Chapa-y-Lazo, B.; Smythe, C.; Sudbery, P. Hyphal growth in *Candida albicans* requires the phosphorylation of Sec2 by the Cdc28-Ccn1/Hgc1 kinase. *EMBO J.* **2010**, *29*, 2930–2942. [CrossRef] [PubMed]
195. Martin, R.; Walther, A.; Wendland, J. Ras1-induced hyphal development in *Candida albicans* requires the formin Bni1. *Eukaryot. Cell* **2005**, *4*, 1712–1724. [CrossRef]
196. Zheng, X.D.; Wang, Y.M.; Wang, Y. CaSPA2 is important for polarity establishment and maintenance in *Candida albicans*. *Mol. Microbiol.* **2003**, *49*, 1391–1405. [CrossRef]
197. Akashi, T.; Kanbe, T.; Tanaka, K. The role of the cytoskeleton in the polarized growth of the germ tube in *Candida albicans*. *Microbiology* **1994**, *140*, 271–280. [CrossRef]
198. Wolyniak, M.J.; Sundstrom, P. Role of actin cytoskeletal dynamics in activation of the cyclic AMP pathway and HWP1 gene expression in *Candida albicans*. *Eukaryot. Cell* **2007**, *6*, 1824–1840. [CrossRef]
199. Yokoyama, K.; Kaji, H.; Nishimura, K.; Miyaji, M. The role of microfilaments and microtubules in apical growth and dimorphism of *Candida albicans*. *J. Gen. Microbiol.* **1990**, *136*, 1067–1075. [CrossRef]
200. Pruyne, D.; Bretscher, A. Polarization of cell growth in yeast. *J. Cell Sci.* **2000**, *113*, 571–585. [CrossRef]
201. Kaksonen, M.; Sun, Y.; Drubin, D.G. A pathway for association of receptors, adaptors, and actin during endocytic internalization. *Cell* **2003**, *115*, 475–487. [CrossRef]
202. Anderson, J.M.; Soll, D.R. Differences in actin localization during bud and hypha formation in the yeast *Candida albicans*. *J. Gen. Microbiol.* **1986**, *132*, 2035–2047. [CrossRef] [PubMed]
203. Smith, M.G.; Swamy, S.R.; Pon, L.A. The life cycle of actin patches in mating yeast. *J. Cell Sci.* **2001**, *114*, 1505–1513. [CrossRef]
204. Asleson, C.M.; Bensen, E.S.; Gale, C.A.; Melms, A.S.; Kurischko, C.; Berman, J. *Candida albicans INT1*-induced filamentation in Saccharomyces cerevisiae depends on Sla2p. *Mol. Cell. Biol.* **2001**, *21*, 1272–1284. [CrossRef]
205. Gale, C.A.; Leonard, M.D.; Finley, K.R.; Christensen, L.; McClellan, M.; Abbey, D.; Kurischko, C.; Bensen, E.; Tzafrir, I.; Kauffman, S.; et al. *SLA2* mutations cause *SWE1*-mediated cell cycle phenotypes in *Candida albicans* and *Saccharomyces cerevisiae*. *Microbiology* **2009**, *155*, 3847–3859. [CrossRef]
206. Zeng, G.; Wang, Y.M.; Wang, Y. Cdc28-Cln3 phosphorylation of Sla1 regulates actin patch dynamics in different modes of fungal growth. *Mol. Biol. Cell* **2012**, *23*, 3485–3497. [CrossRef]
207. Douglas, L.M.; Martin, S.W.; Konopka, J.B. BAR domain proteins Rvs161 and Rvs167 contribute to *Candida albicans* endocytosis, morphogenesis, and virulence. *Infect. Immun.* **2009**, *77*, 4150–4160. [CrossRef]

208. Epp, E.; Nazarova, E.; Regan, H.; Douglas, L.M.; Konopka, J.B.; Vogel, J.; Whiteway, M. Clathrin- and Arp2/3-independent endocytosis in the fungal pathogen *Candida albicans*. *mBio* **2013**, *4*, e00476. [CrossRef]

209. Epp, E.; Walther, A.; Lépine, G.; Leon, Z.; Mullick, A.; Raymond, M.; Wendland, J.; Whiteway, M. Forward genetics in *Candida albicans* that reveals the Arp2/3 complex is required for hyphal formation, but not endocytosis. *Mol. Microbiol.* **2010**, *75*, 1182–1198. [CrossRef]

210. Reijnst, P.; Jorde, S.; Wendland, J. *Candida albicans* SH3-domain proteins involved in hyphal growth, cytokinesis, and vacuolar morphology. *Curr. Genet.* **2010**, *56*, 309–319. [CrossRef]

211. Martin, R.; Hellwig, D.; Schaub, Y.; Bauer, J.; Walther, A.; Wendland, J. Functional analysis of *Candida albicans* genes whose *Saccharomyces cerevisiae* homologues are involved in endocytosis. *Yeast* **2007**, *24*, 511–522. [CrossRef]

212. Bar-Yosef, H.; Gildor, T.; Ramirez-Zavala, B.; Schmauch, C.; Weissman, Z.; Pinsky, M.; Naddaf, R.; Morschhauser, J.; Arkowitz, R.A.; Kornitzer, D. A Global Analysis of Kinase Function in *Candida albicans* Hyphal Morphogenesis Reveals a Role for the Endocytosis Regulator Akl1. *Front. Cell Infect. Microbiol.* **2018**, *8*, 17. [CrossRef]

213. Borth, N.; Walther, A.; Reijnst, P.; Jorde, S.; Schaub, Y.; Wendland, J. *Candida albicans* Vrp1 is required for polarized morphogenesis and interacts with Wal1 and Myo5. *Microbiology* **2010**, *156*, 2962–2969. [CrossRef]

214. Oberholzer, U.; Marcil, A.; Leberer, E.; Thomas, D.Y.; Whiteway, M. Myosin I is required for hypha formation in *Candida albicans*. *Eukaryot. Cell* **2002**, *1*, 213–228. [CrossRef]

215. Walther, A.; Wendland, J. Polarized hyphal growth in *Candida albicans* requires the Wiskott-Aldrich Syndrome protein homolog Wal1p. *Eukaryot Cell* **2004**, *3*, 471–482. [CrossRef]

216. Bassilana, M.; Hopkins, J.; Arkowitz, R.A. Regulation of the Cdc42/Cdc24 GTPase module during *Candida albicans* hyphal growth. *Eukaryot. Cell* **2005**, *4*, 588–603. [CrossRef]

217. Ushinsky, S.C.; Harcus, D.; Ash, J.; Dignard, D.; Marcil, A.; Morchhauser, J.; Thomas, D.Y.; Whiteway, M.; Leberer, E. CDC42 is required for polarized growth in human pathogen *Candida albicans*. *Eukaryot. Cell* **2002**, *1*, 95–104. [CrossRef]

218. Court, H.; Sudbery, P. Regulation of Cdc42 GTPase activity in the formation of hyphae in *Candida albicans*. *Mol. Biol. Cell* **2007**, *18*, 265–281. [CrossRef]

219. Kozubowski, L.; Saito, K.; Johnson, J.M.; Howell, A.S.; Zyla, T.R.; Lew, D.J. Symmetry-Breaking Polarization Driven by a Cdc42p GEF-PAK Complex. *Curr. Biol.* **2008**, *18*, 1719–1726. [CrossRef]

220. Hausauer, D.L.; Gerami-Nejad, M.; Kistler-Anderson, C.; Gale, C.A. Hyphal guidance and invasive growth in *Candida albicans* require the Ras-like GTPase Rsr1p and its GTPase-activating protein Bud2p. *Eukaryot. Cell* **2005**, *4*, 1273–1286. [CrossRef]

221. Pulver, R.; Heisel, T.; Gonia, S.; Robins, R.; Norton, J.; Haynes, P.; Gale, C.A. Rsr1 focuses Cdc42 activity at hyphal tips and promotes maintenance of hyphal development in *Candida albicans*. *Eukaryot. Cell* **2013**, *12*, 482–495. [CrossRef] [PubMed]

222. Guo, P.P.; Yong, J.Y.A.; Wang, Y.M.; Li, C.R. Sec15 links bud site selection to polarised cell growth and exocytosis in *Candida albicans*. *Sci. Rep.* **2016**, *6*, 26464. [CrossRef] [PubMed]

223. Umeyama, T.; Kaneko, A.; Niimi, M.; Uehara, Y. Repression of *CDC28* reduces the expression of the morphology-related transcription factors, Efg1p, Nrg1p, Rbf1p, Rim101p, Fkh2p and Tec1p and induces cell elongation in *Candida albicans*. *Yeast* **2006**, *23*, 537–552. [CrossRef] [PubMed]

224. Bensen, E.S.; Clemente-Blanco, A.; Finley, K.R.; Correa-Bordes, J.; Berman, J. The mitotic cyclins Clb2p and Clb4p affect morphogenesis in *Candida albicans*. *Mol. Biol. Cell* **2005**, *16*, 3387–3400. [CrossRef]

225. Loeb, J.D.; Sepulveda-Becerra, M.; Hazan, I.; Liu, H. A G1 cyclin is necessary for maintenance of filamentous growth in *Candida albicans*. *Mol. Biol. Cell* **1999**, *19*, 4019–4027. [CrossRef]

226. Bensen, E.S.; Filler, S.G.; Berman, J. A forkhead transcription factor is important for true hyphal as well as yeast morphogenesis in *Candida albicans*. *Eukaryot. Cell* **2002**, *1*, 787–798. [CrossRef]

227. Greig, J.A.; Sudbery, I.M.; Richardson, J.P.; Naglik, J.R.; Wang, Y.; Sudbery, P.E. Cell Cycle-Independent Phospho-Regulation of Fkh2 during Hyphal Growth Regulates *Candida albicans* Pathogenesis. *PLoS Pathog.* **2015**, *11*, e1004630. [CrossRef]

228. Gutiérrez-Escribano, P.; González-Novo, A.; Suárez, M.B.; Li, C.R.; Wang, Y.; de Aldana, C.R.; Correa-Bordes, J. CDK-dependent phosphorylation of Mob2 is essential for hyphal development in *Candida albicans*. *Mol. Biol. Cell* **2011**, *22*, 2458–2469. [CrossRef]

229. Zheng, X.-D.; Lee, R.T.H.; Wang, Y.-M.; Lin, Q.-S.; Wang, Y. Phosphorylation of Rga2, a Cdc42 GAP, by CDK/Hgc1 is crucial for *Candida albicans* hyphal growth. *EMBO J.* **2007**, *26*, 3760–3769. [CrossRef]

230. Wang, H.; Huang, Z.X.; Au Yong, J.Y.; Zou, H.; Zeng, G.; Gao, J.; Wang, Y.; Wong, A.H.; Wang, Y. CDK phosphorylates the polarisome scaffold Spa2 to maintain its localization at the site of cell growth. *Mol. Microbiol.* **2016**, *101*, 250–264. [CrossRef]

231. Caballero-Lima, D.; Sudbery, P.E. In *Candida albicans*, phosphorylation of Exo84 by Cdk1-Hgc1 is necessary for efficient hyphal extension. *Mol. Biol. Cell* **2014**, *25*, 1097–1110. [CrossRef]

232. Wang, A.; Raniga, P.P.; Lane, S.; Lu, Y.; Liu, H. Hyphal chain formation in *Candida albicans*: Cdc28-Hgc1 phosphorylation of Efg1 represses cell separation genes. *Mol. Biol. Cell* **2009**, *29*, 4406–4416. [CrossRef]

233. Shapiro, R.S.; Sellam, A.; Tebbji, F.; Whiteway, M.; Nantel, A.; Cowen, L.E. Pho85, Pcl1, and Hms1 signaling governs *Candida albicans* morphogenesis induced by high temperature or Hsp90 compromise. *Curr. Biol.* **2012**, *22*, 461–470. [CrossRef]

234. Tripathi, G.; Wiltshire, C.; Macaskill, S.; Tournu, H.; Budge, S.; Brown, A.J. Gcn4 co-ordinates morphogenetic and metabolic responses to amino acid starvation in *Candida albicans*. *EMBO J.* **2002**, *21*, 5448–5456. [CrossRef]

235. Gildor, T.; Shemer, R.; Atir-Lande, A.; Kornitzer, D. Coevolution of cyclin Pcl5 and its substrate Gcn4. *Eukaryot. Cell* **2005**, *4*, 310–318. [CrossRef]

236. Shemer, R.; Meimoun, A.; Holtzman, T.; Kornitzer, D. Regulation of the transcription factor Gcn4 by Pho85 cyclin *PCL5*. *Mol. Biol. Cell* **2002**, *22*, 5395–5404. [CrossRef]

237. Chapa y Lazo, B.; Bates, S.; Sudbery, P. The G1 cyclin Cln3 regulates morphogenesis in *Candida albicans*. *Eukaryot. Cell* **2005**, *4*, 90–94. [CrossRef]

238. Berman, J. Morphogenesis and cell cycle progression in *Candida albicans*. *Curr. Opin. Microbiol.* **2006**, *9*, 595–601. [CrossRef]

239. Wang, H.; Gao, J.; Li, W.; Wong, A.H.; Hu, K.; Chen, K.; Wang, Y.; Sang, J. Pph3 dephosphorylation of Rad53 is required for cell recovery from MMS-induced DNA damage in *Candida albicans*. *PLoS ONE* **2012**, *7*, e37246. [CrossRef]

240. Feng, J.; Duan, Y.; Sun, W.; Qin, Y.; Zhuang, Z.; Zhu, D.; Sun, X.; Jiang, L. CaTip41 regulates protein phosphatase 2A activity, CaRad53 deactivation and the recovery of DNA damage-induced filamentation to yeast form in *Candida albicans*. *FEMS Yeast Res.* **2016**, *16*, fow009. [CrossRef]

241. Shi, Q.M.; Wang, Y.M.; Zheng, X.D.; Lee, R.T.; Wang, Y. Critical role of DNA checkpoints in mediating genotoxic-stress-induced filamentous growth in *Candida albicans*. *Mol. Biol. Cell* **2007**, *18*, 815–826. [CrossRef]

242. Loll-Krippleber, R.; d'Enfert, C.; Feri, A.; Diogo, D.; Perin, A.; Marcet-Houben, M.; Bougnoux, M.E.; Legrand, M. A study of the DNA damage checkpoint in *Candida albicans*: Uncoupling of the functions of Rad53 in DNA repair, cell cycle regulation and genotoxic stress-induced polarized growth. *Mol. Microbiol.* **2014**, *91*, 452–471. [CrossRef]

243. Andaluz, E.; Ciudad, T.; Gomez-Raja, J.; Calderone, R.; Larriba, G. Rad52 depletion in *Candida albicans* triggers both the DNA-damage checkpoint and filamentation accompanied by but independent of expression of hypha-specific genes. *Mol. Microbiol.* **2006**, *59*, 1452–1472. [CrossRef] [PubMed]

244. Tscherner, M.; Stappler, E.; Hnisz, D.; Kuchler, K. The histone acetyltransferase Hat1 facilitates DNA damage repair and morphogenesis in *Candida albicans*. *Mol. Microbiol.* **2012**, *86*, 1197–1214. [CrossRef] [PubMed]

245. Bachewich, C.; Thomas, D.Y.; Whiteway, M. Depletion of a polo-like kinase in *Candida albicans* activates cyclase-dependent hyphal-like growth. *Mol. Biol. Cell* **2003**, *14*, 2163–2180. [CrossRef] [PubMed]

246. Finley, K.R.; Bouchonville, K.J.; Quick, A.; Berman, J. Dynein-dependent nuclear dynamics affect morphogenesis in *Candida albicans* by means of the Bub2p spindle checkpoint. *J. Cell Sci.* **2008**, *121*, 466–476. [CrossRef]

247. Martin, R.; Walther, A.; Wendland, J. Deletion of the dynein heavy-chain gene *DYN1* leads to aberrant nuclear positioning and defective hyphal development in *Candida albicans*. *Eukaryot. Cell* **2004**, *3*, 1574–1588. [CrossRef] [PubMed]

248. Bai, C.; Ramanan, N.; Wang, Y.M.; Wang, Y. Spindle assembly checkpoint component CaMad2p is indispensable for *Candida albicans* survival and virulence in mice. *Mol. Microbiol.* **2002**, *45*, 31–44. [CrossRef]

249. Atir-Lande, A.; Gildor, T.; Kornitzer, D. Role for the SCFCDC4 ubiquitin ligase in *Candida albicans* morphogenesis. *Mol. Microbiol. Cell* **2005**, *16*, 2772–2785. [CrossRef]

250. Shieh, J.C.; White, A.; Cheng, Y.C.; Rosamond, J. Identification and functional characterization of *Candida albicans CDC4*. *J. Biomed. Sci.* **2005**, *12*, 913–924. [CrossRef]

251. Li, W.J.; Wang, Y.M.; Zheng, X.D.; Shi, Q.M.; Zhang, T.T.; Bai, C.; Li, D.; Sang, J.L.; Wang, Y. The F-box protein Grr1 regulates the stability of Ccn1, Cln3 and Hof1 and cell morphogenesis in *Candida albicans*. *Mol. Microbiol.* **2006**, *62*, 212–226. [CrossRef]

252. Chou, H.; Glory, A.; Bachewich, C. Orthologues of the anaphase-promoting complex/cyclosome coactivators Cdc20p and Cdh1p are important for mitotic progression and morphogenesis in *Candida albicans*. *Eukaryot. Cell* **2011**, *10*, 696–709. [CrossRef]

253. Gao, J.; Wang, H.; Li, Z.; Wong, A.H.; Wang, Y.Z.; Guo, Y.; Lin, X.; Zeng, G.; Liu, H.; Wang, Y.; et al. *Candida albicans* gains azole resistance by altering sphingolipid composition. *Nat. Commun.* **2018**, *9*, 4495. [CrossRef]

254. Segal, E.S.; Gritsenko, V.; Levitan, A.; Yadav, B.; Dror, N.; Steenwyk, J.L.; Silberberg, Y.; Mielich, K.; Rokas, A.; Gow, N.A.R.; et al. Gene Essentiality Analyzed by In Vivo Transposon Mutagenesis and Machine Learning in a Stable Haploid Isolate of *Candida albicans*. *mBio* **2018**, *9*. [CrossRef]

 *pathogens*

Review

On and Off: Epigenetic Regulation of *C. albicans* Morphological Switches

Elise Iracane †, **Samuel Vega-Estévez** † and **Alessia Buscaino** *

Kent Fungal Group, School of Biosciences, University of Kent, Canterbury, CT2 7NJ, UK;
E.Iracane@kent.ac.uk (E.I.); S.Vega-Estevez@kent.ac.uk (S.V.-E.)
* Correspondence: A.Buscaino@kent.ac.uk
† These authors contributed equally to this work.

Abstract: The human fungal pathogen *Candida albicans* is a dimorphic opportunistic pathogen that colonises most of the human population without creating any harm. However, this fungus can also cause life-threatening infections in immunocompromised individuals. The ability to successfully colonise different host niches is critical for establishing infections and pathogenesis. *C. albicans* can live and divide in various morphological forms critical for its survival in the host. Indeed, *C. albicans* can grow as both yeast and hyphae and can form biofilms containing hyphae. The transcriptional regulatory network governing the switching between these different forms is complex but well understood. In contrast, non-DNA based epigenetic modulation is emerging as a crucial but still poorly studied regulatory mechanism of morphological transition. This review explores our current understanding of chromatin-mediated epigenetic regulation of the yeast to hyphae switch and biofilm formation. We highlight how modification of chromatin structure and non-coding RNAs contribute to these morphological transitions.

Keywords: *Candida albicans*; epigenetic; yeast; chromatin; biofilm; hyphae

Citation: Iracane, E.; Vega-Estévez, S.; Buscaino, A. On and Off: Epigenetic Regulation of *C. albicans* Morphological Switches. *Pathogens* **2021**, *10*, 1463. https://doi.org/10.3390/pathogens10111463

Academic Editors: Jonathan Richardson and Lawrence S. Young

Received: 4 August 2021
Accepted: 5 November 2021
Published: 11 November 2021

Publisher's Note: MDPI stays neutral with regard to jurisdictional claims in published maps and institutional affiliations.

1. Introduction

Epigenetics is a popular term first defined by Conrad Waddington in the early 1940s as "the process by which the genotype brings the phenotype into being" [1]. Since then, the meaning of epigenetics has significantly changed. Arthur Riggs defined epigenetics as the study of mitotically and/or meiotically heritable changes in the gene function that are not explained by changes in the DNA sequence [2]. Riggs' definition focuses on heritability: the ability of an epigenetic mark to be passed to subsequent generations of cells and/or organisms. There is no doubt that heritable epigenetics is an important regulatory mechanism. However, this definition excludes many important, not heritable mechanisms often labelled as "epigenetic". For example, post-translation modifications of histone proteins and their effect on gene expression are often described as an epigenetic regulatory mechanism. However, chromatin marks are, in the majority of the cases, transient and not heritable. Likewise, Riggs' definition excludes the role of non-coding RNAs (ncRNAs) in transcription and other DNA-based organisms. To overcome this conundrum, Adrian Bird redefined epigenetics as "the structural adaptation of chromosomal regions to register, signal or perpetuate altered activity states" [3]. This definition focuses on changes in gene function that are independent of changes in the underlying DNA sequence. Importantly, these changes can be heritable or not. Epigenetic regulatory mechanisms include changes in gene expression and chromosome function triggered by chromatin modification, chromatin remodelling and ncRNAs activity [2–8]. In this review, we will adopt Adrian Bird's definition.

Human fungal pathogens are microbial organisms that kill more than 1.5 million people annually and reduce the quality of life of >1 billion people [9]. Additionally, the recent staggering escalation in the number of invasive fungal infections and the emergence of antifungal drug resistance poses an ever-increasing threat to human health. Fungal pathogens grow in association with their host, and establishing how these organisms adapt to hostile host environments is key to understanding how they cause life-threatening infections and develop resistance to antifungal drugs.

Chief among human fungal pathogens is *Candida albicans,* a CTG(Ser1)-clade organism in which the CTG codon is translated as serine rather than leucine [10,11]. *C. albicans* colonises almost every organ in the human body, and therefore, it is exposed to rapid environmental changes [12]. Indeed, *C. albicans* is a harmless commensal yeast found in the skin, gut, oral cavity and mucosa [13]. However, this fungal pathogen can become virulent, establishing an extensive range of mucosal and systemic infections. For example, *C. albicans* can cause vulvovaginal candidiasis (VVC), an infection estimated to afflict 75% of all women at least once in their lifetime [14] or candidiasis, systemic infections that can be life-threatening in immunocompromised individuals and are associated with high mortality rates (up to 50%) [9]. Phenotypic plasticity is a critical regulatory mechanism that drives rapid adaptation to hostile host environments. Indeed, environmental changes can induce dramatic morphological changes, and phenotypic switches are critical host adaptation and virulence drivers. For example, *C. albicans* can grow as a single rounded yeast cell or as multicellular hyphae. Yeast cells are critical for host colonisation, early infection and dissemination, while hyphae facilitate tissue invasion and damage [15,16]. Filamentous cells are also crucial for biofilm formation, a highly organised structure that confers resistance to antimicrobial therapies and the host immune response [17]. *C. albicans* cells can also switch between a white and opaque state. White and opaque cells have different appearances, gene expression profiles and mating behaviours [18].

Epigenetic regulatory mechanisms are emerging as essential modulators of *C. albicans'* phenotypic plasticity. Indeed, epigenetic regulation can sense environmental changes leading to the rapid and reversible modulation of gene expression and adaptation to hostile environments. Recently, Qasim et al. [19] reviewed the role of epigenetics in the white–opaque switch extensively. This review will discuss the contribution of epigenetics to *C. albicans* phenotypic plasticity by focusing on the gene-regulation changes in the yeast -hyphae switch and biofilm formation.

2. *C. albicans'* Chromatin Structure: The Basics

In eukaryotes, DNA is packed around specific histone proteins within the nucleus to form a compact structure called chromatin. The basic unit of chromatin is the nucleosome, formed by 147 base pairs (bp) of DNA wrapped around an octamer of histones. This octamer is composed of two dimers of the histones H2A–H2B and the histone tetramer $(H3)_2(H4)_2$. Nucleosomes are organised into arrays that are further packaged by histone H1, promoting chromatin folding into compact fibres [20] (Figure 1). The diploid *C. albicans* genome contains two homologous pairs of divergently transcribed histones H2A (*HTA1* (orf19.6924)) and H2B (*HTB1* (orf19.6925)), as well as histone H3 (*HHT2* (orf19.1853) and *HHT21* (orf19.1061)) and H4 (*HHF1* (orf19.1059) and *HHF2* (orf19.1854)) genes. A putative histone H1 (*HHO1* (orf19.5137.1)) can also be identified [21]. Although histones are slow-evolving proteins, variability in histone proteins has been documented in most eukaryotes and histone variants play critical biological roles [22–24]. For example, the histone H3 variant, CENP-A^{Cse4}, epigenetically defines centromeres in each chromosome. *C. albicans* CENP-A^{Cse4} marks regional centromeres associated with its eight chromosomes [25].

Chromatin allows the packaging of DNA into a compact structure that can fit inside the nucleus while permitting efficient accessibility to DNA-binding proteins. However, chromatin is also an obstacle to all DNA-templated biological processes, including transcription, replication, recombination and repair [26]. Consequently, changes in chromatin structure can have a profound impact on nuclear processes, and chromatin is a crucial regulator of DNA-based activities. For example, chromatin can be assembled into two functionally and structurally different chromatin structures. Gene-rich regions and non-repetitive DNA are assembled into euchromatin, an open chromatin state that is permissive to transcription. In contrast, heterochromatin is a transcriptionally silent chromatin state that is associated with gene-poor and repetitive regions of the genome [27]. Chromatin structure can be modulated by three distinct mechanisms: (i) post-translation modification of histone proteins, (ii) chromatin remodelling and (iii) ncRNAs (Figure 2).

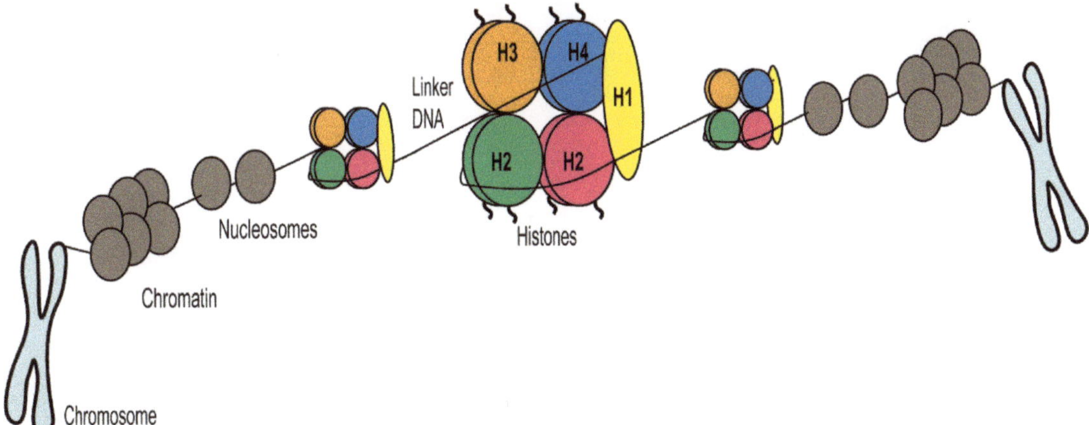

Figure 1. Chromatin organisation and compaction in eukaryotes. Chromatin, a DNA–protein complex, forms chromosomes within the nucleus of eukaryotic cells. The central unit of chromatin is the nucleosome composed of a histone octamer and DNA. Histone H1 promotes further chromatin folding in some eukaryotes.

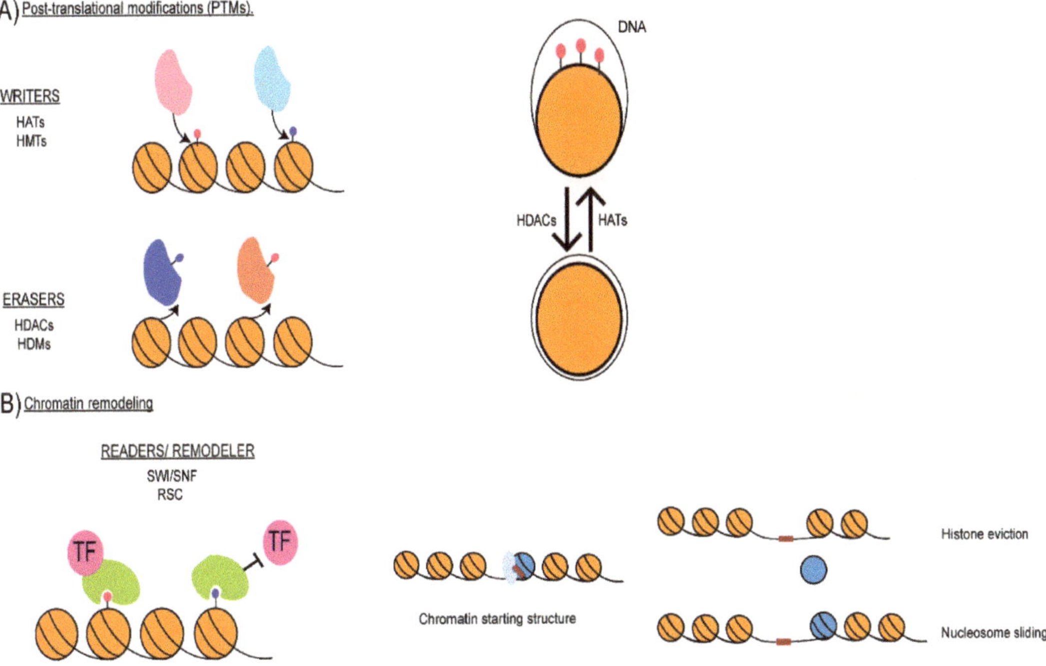

Figure 2. *Cont.*

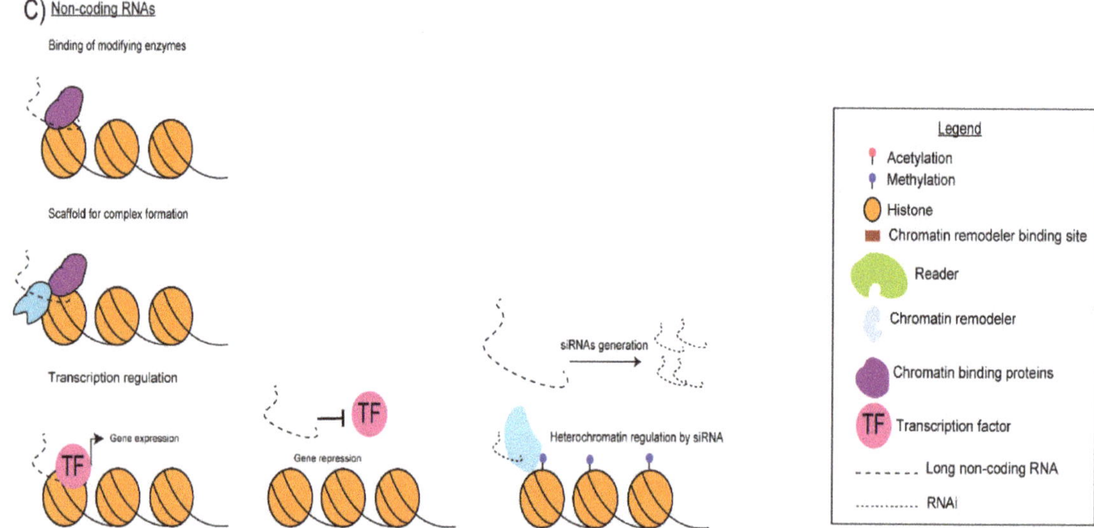

Figure 2. Mechanisms of modulation in chromatin structures. (**A**) Left: schematics of chromatin writers and erasers: writers, such as HATs and HMTs, add epigenetic marks to histone proteins, while erasers, such as HDACs and HDMs, remove epigenetic marks from histone proteins. Right: schematic of how post-translation modifications, such as histone acetylation, can affect DNA-histone interactions. (**B**) Modes of action of readers and chromatin remodelers. Left: reader proteins, components of chromatin remodelling complex, bind modified histone tails recruiting or blocking transcription factors. Right: schematics of chromatin remodellers' activity, including histone eviction and nucleosome sliding. (**C**) Mode of action of non-coding RNAs (ncRNAs). Top: ncRNAs can recruit modifiers to chromatin or act as a scaffold promoting the formation of protein complexes. Middle: ncRNAs can activate or repress transcription. ncRNAs can also be processed into siRNAs by the RNAi machinery. siRNAs can seed heterochromatin formation. HAT: histone acetyltransferase; HMT: histone methyltransferase; HDAC: histone deacetylase; HDM: histone demethylase; RSC: remodels the structure of chromatin.

3. Histone Post-Translational Modifications

Histone proteins are formed by a globular core and unstructured basic amino-terminal tails that can be post-translationally modified. The most common post-translational modifications (PTMs), also known as histone marks, include methylation, acetylation, ubiquitination, ADP-ribosylation and the sumoylation of lysine (K) residues; the methylation of arginine (R) residues and the phosphorylation of serine (S) and threonine (T) residues. In addition, the same amino acid can be affected by multiple modifications (i.e., mono, di- or tri-methylated) [28].

Histone marks are differentially associated with euchromatin and heterochromatin regions. At euchromatic transcriptionally active regions, genes promoters are assembled into a chromatin state containing acetylated histones that are tri-methylated on H3K4 (H3K4me^3), while histone H3 methylated on K36 (H3K36me) is found at gene bodies [26]. Likewise, enhancers and super-enhancers are marked by the mono-methylation of histone H3 on K4 (H3K4me^1) and the acetylation of histone H3 on K27 (H3K27Ac) [29,30]. Genome-wide chromatin profiling demonstrates that the *C. albicans* transcriptionally active genome is packaged into canonical euchromatin, where gene promoters of active genes are enriched in H3K4me^3 while gene bodies are marked byacetylated histone H3 (H3K9Ac) and H4 (H4K16Ac) (Figure 3) [31]. In many eukaryotic organisms heterochromatic regions are enriched in repressive histone marks, such as the methylation of K9 on histone H3 (H3K9me) or the methylation of K27 on histone H3 (H3K27me). Furthermore, high levels of DNA methylation on position five of cytosines (5mC) are associated with heterochromatin [26,32]. Similarly to *S. cerevisiae*, *C. albicans* is devoid of H3K9me and H3K27me [33]. Although 5mC mark has been detected in *C. albicans*, it is unclear whether DNA methylation is associated with heterochromatic regions in this organism [34]. Instead, chromatin profiling studies have demonstrated that *C. albicans* heterochromatic regions are characterised by low levels of both histone acetylation and methylation [31] (Figure 3).

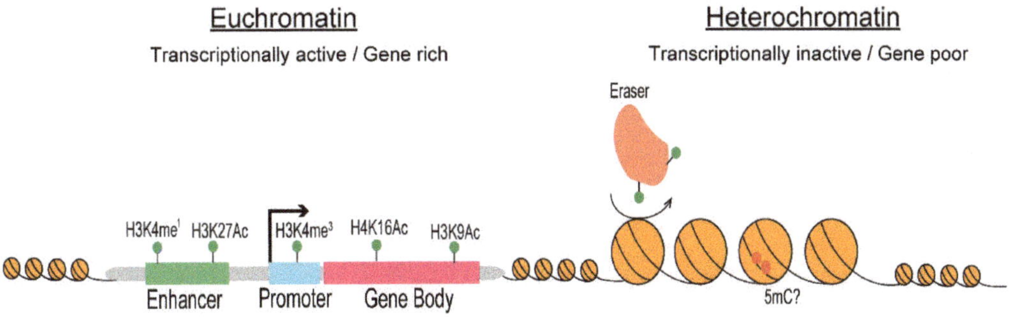

Figure 3. Histone modifications and transcriptional activity in *C, albicans*. (**Left**): schematic of histone modifications associated with active genes that are assembled into euchromatin. Coding regions (gene body) and regulatory regions (enhancer and promoter) are shown. (**Right**): schematic of chromatin marks associated with gene-poor heterochromatic regions.

The post-translation modification of histone proteins is a dynamic and reversible process catalysed by "writer" and "eraser" enzymes that add and remove epigenetic marks (Figure 2). For example, the additions of acetyl groups to histone tails are carried out by histone acetyltransferases (HATs), whereas their removal is conducted by histone deacetylases (HDACs); methylation marks are added by histone methyltransferases (HMTs) and removed by histone demethylases (HDMs). Additionally, HAT and HDAC can also catalyse the addition and removal of acyl groups different from those in acetylation, such as crotonyl, succinyl, β-hydroxybutyryl, and propionyl [35]. The primary histone modifiers found in *C. albicans* and their orthologs found in *S. cerevisiae*, *S. pombe* and humans are listed in Table 1.

Table 1. *Candida albicans* histone modifiers and RNA interference actors and their orthologs in *S. cerevisiae*, *S. pombe* and humans.

	C. albicans Gene and Known Function Candida Genome [21]	*S. cerevisiae* Ortholog SGD [36]	*S. pombe* Ortholog Pombase [37]	Human Ortholog Alliance of Genome Resources [38]
Histone Acetyltransferase	*ESA1* (orf19.5416) NuA4 HAT complex acts on H4K5, H4K12	*ESA1 (YOR244W) (alias: TAS1, KAT5)*	*MST1 (SPAC637.12c)*	*TIP60 (alias: KAT5)*
	SAS2 (orf19.2087) SAS HAT complex acts on H4K16	*SAS2 (YMR127C) (alias: KAT8)*	*MST2 (SPAC17G8.13c)*	
	GCN5 (orf19.705) SAGA/ADA complex	*GCN5 (YGR252W) (alias: ADA4, SW19, AAS104, KAT2)*	*GCN5 (SPAC1952.05)*	*KAT2B (alias: CAF)* *KAT2A (alias: GCN5)*
	ADA2 (orf19.2331) SAGA/ADA complex acts on H3K9	*ADA2 (YDR448W) (alias: SWI8)*	*ADA2 (SPCC24B10.08c)*	*TADA2B (alias: ADA2B)* *TADA2A (alias: ADA2A)*
	RTT109 (orf19.7491) acts on H3K56	*RTT109 (YLL002W) (alias: KIM2, REM50, KAT11)*	rtt109 (SPBC342.06c)	No ortholog
	YNG2 (orf19.878) *(alias: NBN1)* NuA4 HAT complex acts on nucleosomal H4	*YNG2 (YHR090C) (alias: EAF4, NBN1)*	*png1 (SPAC3G9.08)*	*ING3 (alias: Eaf4, ING2)* *ING2 ± (alias: ING1L)* *ING4 ±* *ING5 ±*

Table 1. *Cont.*

	C. albicans Gene and Known Function Candida Genome [21]	*S. cerevisiae* Ortholog SGD [36]	*S. pombe* Ortholog Pombase [37]	Human Ortholog Alliance of Genome Resources [38]
	NAT4 (orf19.4664)	NAT4 (YMR069W)	naa40 (SPCC825.04c)	NAA40 (alias: NAT11, PATT1)
	HAT1 (orf19.779)	HAT1 (YPL001W) (alias: KAT1)	hat1 (SPAC139.06)	HAT1 (alias: KAT1)
	SAS3 (orf19.2540)	SAS3 (YBL052C) (alias: KAT6)	MST2 (SPAC17G8.13c)	KAT7 (alias: HBO1, MYST-2)
				KAT6A ± (alias: MYST-3)
				KAT6B ± (alias: MYST-4)
				KAT8 ± (alias: MYST-1)
Histone Deacetylase	HDA1 (orf19.2606) in a complex with Hda2 and Hda3	HDA1 (YNL021W)	clr3 (SPBC800.03)	HDAC10 (alias: HD10)
				HDAC6 (alias: HD6)
	SET3 (orf19.7221) SET3 HDAC complex with Hos2, Snt1 and Sif2	SET3 (YKR029C) / SET4 (YJL105W) (SET3 paralog)	set3 (SPAC22E12.11c)	KMT2E *
				SETD5 *
	RPD3 (orf19.2834)	RPD3 (YNL330C)	clr6 (SPBC36.05c)	HDAC1 (alias: KDAC1, RPD3)
	RPD31 (orf19.6801)			HDAC2 (alias: KDAC2, RPD3)
	SIR2 (orf19.1992) (alias: SIR21)	HST1 (YOL068C) (SIR2 paralog) / SIR2 (YDL042C)	sir2 (SPBC16D10.07c)	SIRT1 (alias: SIR2)
	HST1 (orf19.4761) (alias: SIR22)	HST1 (YOL068C) SIR2 (YDL042C)		
	HST2 (orf19.2580)	HST2 (YPL015C)	hst2 (SPCC132.02)	SIRT3
				SIRT2
	HST3 (orf19.1934) Acts on H3K56	HST3 (YOR025W)	hst4 (SPAC1783.04c)	No ortholog
Histone Methyltransferase	SET1 (orf19.6009) Acts on H3K4	SET1 (YHR119W) (alias: KMT2)	set1 (SPCC306.04c)	SETD1B (alias: KMT2G)
				SETD1A (alias: KMT2F)
Histone Demethylase	RPH1 (orf19.2743)	RPH1 (YER169W)	jmj3 (SPBC83.07)	KDM4A
				KDM4B
				KDM4C
				KDM4D
				KDM4E
Chromatin Remodeler	SWR1 (orf19.1871) SWR1 complex	SWR1 (YDR334W)	swr1 (SPAC11E3.01c)	SRCAP (alias: SWR1)
	SWI1 (orf19.5657) SWI/SNF complex	SWI1 (YPL016W) (alias: ADR6)	sol1 (SPBC30B4.04c)	ARID5A (alias: MRF1)
				ARID5B (alias: MRF2)
	SNF2 (orf19.1526) SWI/SNF complex	SNF2 (YOR290C)	snf21 (SPAC1250.01)	SMARCA2
				SMARCA4
	STH1 (orf19.239) RSC complex	STH1 (YIL126W)		SMARCA2
				SMARCA4
RNA interference	DCR1 (orf19.3796)	No ortholog	dcr1 (SPCC188.13c)	DICER1
	AGO1 (orf19.2903) RISC complex	No ortholog	ago1 (SPCC736.11)	PIWIL1
				PIWIL2
				PIWIL3
				PIWIL4

(*) human orthologs are histone methyltransferase; (±) *S. pombe* ortholog only.

Histone marks alter chromatin architecture and its function via two main distinct mechanisms. Firstly, PMTs can alter histone-DNA interaction modulating higher-order chromatin structure and affecting gene expression and regulation [39]. For example, histone acetylation reduces the net positive charge of histone tails, and therefore, will weaken histone–DNA interaction, resulting in an open chromatin conformation that is permissive to transcription (Figure 2A) [39]. Histone crotonylation activates transcription more potently than histone acetylation [40]. This is because the crotonyl group is more hydrophobic and rigid than acetyl groups, disrupting histone-DNA interactions [40]. Histone marks can also be recognised by "reader" proteins, which can influence chromatin dynamics and function via promoting or blocking the recruitment of transcription factors and/or other chromatin-modifying factors (Figure 2B) [41]. For example, bromodomain-containing proteins specifically bind acetylated histones, chromodomain containing proteins recognise specific methylation marks and the YEATS domain recognises the crotonyl marks [42].

4. Chromatin Remodelling Regulates Gene Expression and Chromatin Structure

Nucleosomes deposited on DNA can be a physical barrier, reducing chromatin accessibility and gene expression. Chromatin remodelling is the regulatory process that changes the interactions between DNA and histone proteins leading to complete or partial disassembly of the nucleosomes (histone eviction) or nucleosome reposition (nucleosome sliding) [43] (Figure 2C). Chromatin remodelling is catalysed by ATP-dependent multi-subunit protein complexes known as chromatin remodelers [43]. ATP-dependent chromatin remodelers belong to four subfamilies: switch/sucrose non-fermentable (SWI/SNF), imitation switch (ISWI), chromodomain helicase DNA-binding (CHD/NuRD/Mi-2) and inositol-requiring 80 (INO80) [44]. Among those, the SWI/SNF subfamily is the primary remodeler catalysing nucleosome sliding and eviction. Initially identified in budding yeast, SWI/SNF complexes are highly conserved across eukaryotes [45]. The SWI/SNF complex can be targeted to acetylated transcriptionally active chromatin, as it can bind acetylated histones (and non-histone proteins) through a bromodomain subunit [44]. Therefore, SWI/SNF activity generally correlates with transcriptional activation even if the complex has also been linked to transcriptional repression [46–50].

Different yeast species contain a second remodelling complex similar to SWI/SNF, the RSC (remodels the structure of chromatin) complex [44]. This complex is essential for survival in *S. cerevisiae*, although it is not required for growth in *S. pombe* [51]. The RSC complex binds promoters and intergenic regions and is specifically recruited to RNA polymerase II to tune gene transcription [52]. Four chromatin remodeler catalytic subunits have been described in *C. albicans*: *STH1*, *SNF2*, *SWR1* and *SWI1*. Sth1 is the catalytic subunit of the RSC complex, which in *C. albicans* is composed of a total of 13 subunits, including two CTG (Ser1)-clade-specific (Nri1 and Nri2) [53]. Snf2 and Swi1 are catalytic subunits of the SWI/SNF complex [54], and Swr1 is the major subunit of the SWR1 complex [55].

5. Non-Coding Transcription and Non-Coding RNAs

Large fractions of eukaryotic genomes are extensively transcribed but not translated into functional proteins. The act of non-coding transcription and its associated histone modifications and changes in nucleosome density can interfere with the activity of nearby genes [56]. However, ncRNAs can also regulate gene expression by interacting with DNA, RNA and proteins and modulating chromatin structure [57]. Finally, ncRNAs can be processed into small silencing RNAs by RNA interference (RNAi) machinery. The RNAse III-like enzyme dicer (Dcr) and the PIWI domain-containing protein Argonaute (Ago) are at the core of the RNAi machinery and responsible for the generation of the three major branches of small ncRNAs—short interference RNAs (siRNA), micro RNAs (miRNAs) and PIWI-interacting RNAs (piRNAs)—that differ in their biogenesis and mechanisms of action (Figure 4).

In the siRNA pathway, RNAi is triggered by a dsRNA precursor that can arise endogenously by transcription of repetitive DNA and by convergent transcription. This precursor dsRNA is processed into a 20–24-nucleotide (nt) siRNA duplex by Dcr [58]. One strand of the duplex is loaded into Ago, an effector complex. Ago uses base-pairing interaction to target cognate RNAs for inactivation. siRNA-mediated silencing can be co-transcriptional by seeding the assembly of repressive heterochromatin or post-transcriptional RNA cleavage using Ago-slicer endonuclease activity. siRNA-directed heterochromatin assembly has been best described in the fission yeast *S. pombe*, wherein RNAi machinery triggers the formation of transcriptionally silenced hypoacetylated chromatin, methylated on lysine 9 of histone H3 (H3K9) [59]. In the miRNA pathway, short stem-loop dsRNA precursors are pre-processed by a nuclear RNAse III complex (Drosha–Pasha) before final processing into miRNAs in the cytoplasm by Dcr. miRNAs are loaded into an Ago-containing protein complex and targeted to the 3′ untranslated region (3′-UTR) of the target mRNA blocking its translation [60]. The piRNA pathway silences transposable elements in the germline of many animal species [61]. In contrast to the other pathways, piRNAs are not generated by dsRNAs precursors and their biogenesis is independent of Dcr [62]. A single transcript is generated from a piRNA cluster and processed into piRNAs by PIWI-domain containing proteins. The piRNA pathway controls transposons through several distinct but interlinked mechanisms. Whereas cytoplasmic PIWI proteins silence their targets post-transcriptionally through piRNA direct cleavage, nuclear Piwi–piRNA complexes function at the transcriptional level via heterochromatin assembly [63]. Although we still know very little about the nature and the putative function of *C. albicans* non-coding RNAs, RNA profiling analyses have identified many non-coding transcripts whose expression differ under distinct growth conditions [64].

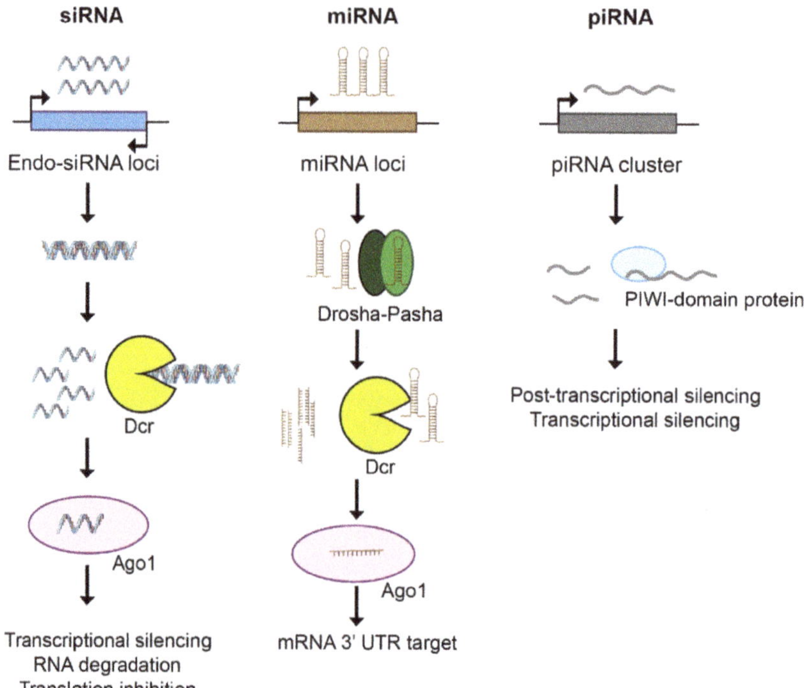

Figure 4. RNA interference pathways. In the siRNA pathway, a precursor dsRNA is processed into a siRNA duplex by dicer (Dcr), and one strand of the duplex is loaded into argonaute (Ago1), which targets it against complementary RNAs for inactivation. In the miRNA pathway, loop dsRNA precursors are processed, first, in the nucleus by drosha–pasha and later in the cytoplasm by dicer to produce miRNAs. These are loaded into an Ago and targeted to the 3′-UTR region of the target mRNA to block translation. In the piRNA pathway, a single transcript is generated from a piRNA cluster and processed into piRNAs by PIWI-domain-containing proteins that silence their targets post-transcriptionally in the cytoplasm through piRNA direct cleavage or, transcriptionally, in the nucleus via heterochromatin assembly.

Furthermore, it has been shown that *C. albicans* contains active RNAi machinery in vitro and in a heterologous yeast system. CaDcr1 is a non-canonical enzyme that can generate small RNAs and catalyse the 35S ribosomal RNA [65]. Future studies will establish the impact of non-coding RNAs in *C. albicans* biology.

6. Chromatin-Mediated Regulation of the Yeast to Hypha Morphological Switch

Modulation of the yeast to hypha morphological transition relies on a complex interplay of a transcriptional regulator and chromatin modifiers [66,67]. Hyphal growth can be divided into two different stages: initiation and maintenance. In yeast cells, the transcriptional repressors Nrg1 and Tup1 inhibit hyphal morphogenesis by blocking the expression of a subset of filament-specific genes [68,69]. During the initiation stage, Nrg1 protein levels decrease sharply, and the Nrg1-mediated repression is cleared. After that, during the maintenance phase, Nrg1 protein levels recover rapidly, but Nrg1 binding to promoters of hypha-specific genes is inhibited [70,71].

Hyphal growth is induced by a broad range of environmental and host factors, including serum, nutrient starvation, hypoxia and high CO_2 concentration [72,73]. These different host signals are integrated by redundant sensing pathways that modulate the activity of transcriptional regulators (such as Chp1, Egf1 and Flo8), resulting in the transcriptional upregulation of hundreds of genes such as genes encoding for cell wall proteins, adhesins and secreted aspartyl proteinases (SAP) [74]. Chromatin modifiers are emerging as important regulators of the yeast-to-hypha transcriptional programme [75–77] (Figure 5A). For example, the concerted and opposite activities of the NuA4 HAT complex and the Hda1 HDAC are necessary for the initiation–maintenance transition and for activating the hyphal-specific transcriptional programme. Upon hyphae induction, the NuA4 complex is recruited to the promoters of hyphae-specific genes. Dynamic acetylation of histone H4 and the NuA4 components Yng2 have been proposed to be necessary for NuA4-dependent hypha induction [75,78]. Consequently, deletion of the *ESA1* gene, encoding for the catalytic subunit of the NuA4 complex, hinders filamentous growth [75]. The HDAC Hda1 promotes hypha maintenance by deacetylating Yng2, and this modification is critical to sustaining hyphal maintenance blocking Nrg1 binding to hyphae-specific promoters in response to serum or nutrient limitation [79]. Importantly, HDA1 is not required for hyphae maintenance or elongation in hypoxia or the presence of elevated CO_2, demonstrating the complexity of the hyphae regulatory programme [80,81].

Several other histone modifiers are important for the yeast–hyphae switch. For example, the HAT Gcn5 is a positive regulator of hyphal growth, while the HATs Sas2 and Hat1 are negative regulators of hyphae formation [75,82,83]. The Set3/Hos2 histone deacetylase complex negatively regulates the yeast-to-hyphae switch by modulating the kinetics of the filamentous transcriptional programme [76,84]. Furthermore, the catalytic activity of the HDAC Sir2 modulates hyphae formation, as the number of hyphae is reduced in a *C. albicans* strain expressing catalytic inactive Sir2 [77].

It is largely unknown how chromatin modifiers modulate the yeast-to-hyphae switch, as the critical substrates necessary for this morphological transition have not been identified yet. Identifying these substrates will be essential to unveil the role of protein post-translation modifications in filamentous growth as chromatin modifiers modify histones and non-histones proteins [85,86]. Furthermore, histone crotonylation is emerging as a crucial post-translation modification regulating *C. albicans* filamentous growth [87]. As HATs can catalyse both acetylation and crotonylation, it will be essential to dissect which modifications are the key regulators of filamentous growth.

Several studies demonstrate that chromatin remodelling controls the yeast-to-hyphae transition. Indeed, the *C. albicans* SWI/SNF and RSC chromatin-remodelling complexes are required for filamentation growth [54,88,89]. However, the molecular mechanism(s) of the SWI/SNF-mediated regulation of hypha formation is still unclear. Indeed, it has been shown that, upon hyphal induction, the SWI/SNF catalytic subunit Snf2 binds the promoters of the hyphae-specific genes *HWP1*, *ALS3* and *ECE1* [78]. This observation suggests that SWI/SNF directly controls the filamentous transcriptional programme by chromatin remodelling of hyphae-specific genes. However, genome-wide chromatin profiling of a different SWI/SNF component, Snf6, did not detect any specific interaction with hyphae-specific genes (including *HWP1*, *ALS3* and *ECE1*). Furthermore, RNA sequencing analyses of wild type (WT) and *SNF6* deletion strains suggest that SWI/SNF is a general transcriptional regulator in both yeast and hyphal cells and that SWI/SNF controls filamentation indirectly [88]. Future studies will determine whether Snf2 plays a role in hyphae formation independently of other SWI/SNF components.

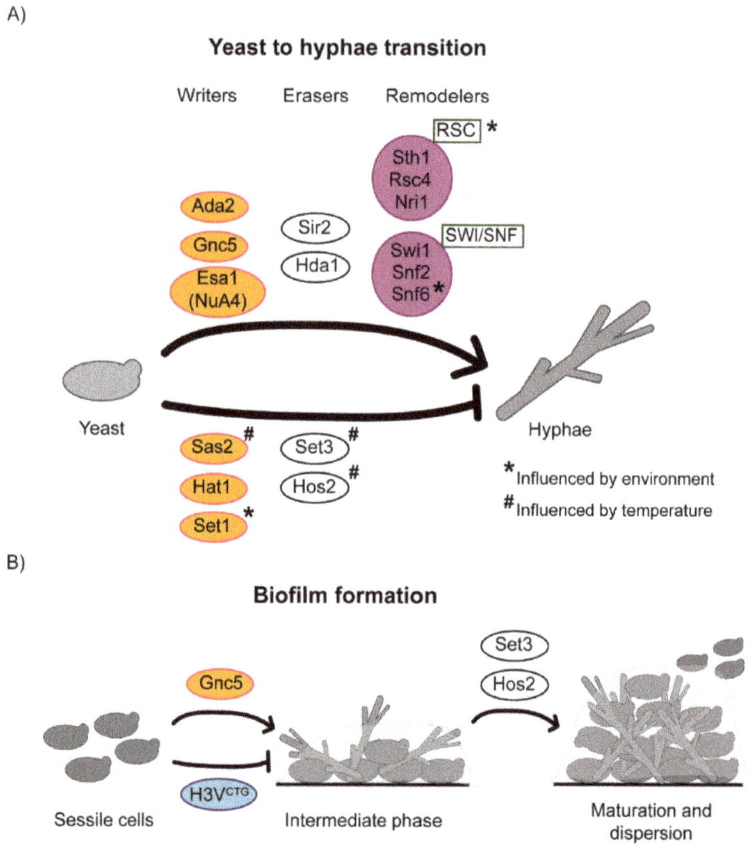

Figure 5. Influence of chromatin modifiers and remodelers on (**A**) the yeast–hyphae transition and (**B**) biofilm formation in *C. albicans*. Proteins promoting the yeast-to-hyphae transition or biofilm formation are represented on top. Proteins repressing the yeast-to-hyphae transition or biofilm formation are represented below. Proteins influenced by the environmental conditions or temperature are represented with (*) or (#) respectively. Writers, erasers, remodelers and histone variants are represented in orange, white, magenta and blue bubbles, respectively.

Although the role of non-coding RNAs in the modulation of hyphal transition is largely unexplored, genome-wide gene-expression profiling studies have identified several novel ncRNAs specifically expressed in hyphae-inducing growth conditions [64,90]. It is still unknown whether these ncRNAs have a function, but it is interesting to note that some of these non-coding RNAs have expression profiles similar to the expression profile of hyphae-specific genes. In the future, it will be essential to determine the function of these ncRNAs.

7. Chromatin-Mediated Regulation of the Planktonic-Biofilm Transition

C. albicans biofilm consists of a layer of yeast cells overlaid by filamentous hyphal and pseudo-hyphal cells surrounded by an extracellular matrix formed by polysaccharides and proteins. The formation of biofilms is a multi-step process consisting of four stages: (1) the adherence and colonisation of yeast cells to the surface, (2) yeast cell proliferation forming the basal layer, (3) the growth of hyphae and pseudo-hyphae with the formation of the extracellular matrix and complex three-dimensional architecture, (4) the dissemination of progeny biofilm cells to seed new sites [91]. Seven master regulators (Bcr1, Brg1, Efg1, Flo8, Ndt80, Tec1 and Rob1) are critical for normal biofilm formation in vivo and in vitro [92,93]. Of these seven regulators, Bcr1, Efg1 and Ndt80 are important modulators of biofilm formation in non-albicans *Candida* species that are evolutionarily distant from *C. albicans* [94]. The biofilm master regulators are transcriptional regulators controlling the expression of thousands of genes expressed differentially between yeast and biofilm cells [95].

An increasing body of evidence suggests that chromatin modifiers and chromatin remodelling regulate different stages of biofilm formation (Figure 5B). Firstly, a specific chromatin state, marked by the histone H3 variant H3V^{CTG} (*ORF19.6791*), acts as a negative regulator of biofilm formation in planktonic cells [96]. H3V^{CTG} contains three variant amino acids (Ser31, Thr32 and Thr80) replaced by Val31, Ser32 and Ser80. Val31 and Ser32 are essential for the variant function. H3V^{CTG} binds promoters of biofilm-related genes in planktonic cells, but it does not mark these gene promoters in biofilm-inducing growth conditions. Additionally, H3V^{CTG} mutant strains produce more robust biofilms than WT cells in vivo and in vitro, suggesting that H3V^{CTG} represses biofilm formation [96]. The role of H3V^{CTG} in other CTG-clade yeast species is unknown. H3V^{CTG} likely regulates biological processes distinct from biofilm formation as this histone variant is expressed in CTG-clade organisms such as *Scheffersomyces stipitis* and *Debaryomyces hansenii* that do not form biofilm under several biofilm-inducing conditions [94,96].

Hyphae formation is important in the biofilm process. Therefore, it is likely that chromatin modifiers regulating filamentous growth are also required for biofilm maturation. Accordingly, deletion of the HAT *GCN5* leads to a strong decrease of adhesion and a dysregulation of Als1-mediated adhesion, which hints at the role of Gcn5 in biofilm establishment [84]. Additionally, it has been shown that chromatin-mediated transcriptional regulation is important for regulating biofilm dispersal. Indeed, the HDAC Set3/Hos2 is a positive regulator of biofilm dispersal. *C. albicans* strain deleted for the *SET3* gene are hyper filamentous and have a reduced number of yeast cells leading to a reduced biofilm dispersal [84].

It is still unknown whether non-coding RNAs regulate biofilm formation. However, ncRNAs might play an crucial regulatory role in biofilm formation as specific ncRNAs are differentially expressed in biofilm cells compared to planktonic cells [64].

8. Conclusions

An increasing body of evidence demonstrates that chromatin modifiers and chromatin remodelers modulate the gene expression programmes associated with the yeast to hyphae switch and with biofilm formation, two interconnected processes playing important roles for host adaption and pathogenesis, as well as the white-opaque switch (Figure 6). Despite the emerging central role of chromatin-mediated regulation in controlling *C. albicans* biology, our understanding of these regulatory processes is still in its infancy. This is because we lack the fundamental knowledge and understanding of how chromatin structures change in different host hostile environments and whether chromatin modulation differs among *C. albicans* clinical isolates. To start filling this gap in knowledge, chromatin profiling of different *C. albicans* morphological forms should be performed. Similarly, histone and non-histone substrates of chromatin modifiers should be identified using biochemical approaches. In addition, it will be exciting to dissect the role of non-coding RNAs and the RNAi machinery to *C. albicans* morphological switches.

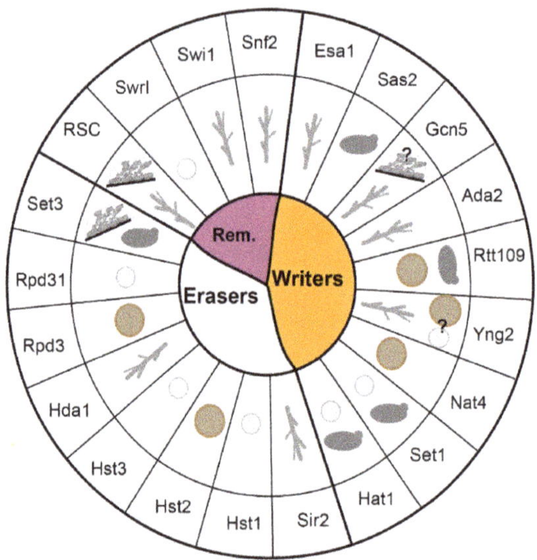

Required for the establishment of:

Figure 6. Summary of the involvement of the main erasers, writers and remodelers (Rem.) in the yeast–hyphae transition, the white–opaque switch and biofilm formation. The question mark indicates a possible involvement not confirmed yet.

Author Contributions: Conceptualisation: E.I., S.V.-E. and A.B. writing—original draft preparation: E.I., S.V.-E.; writing—review and editing: A.B.; visualisation: E.I., S.V.-E.; supervision: A.B.; project administration: A.B.; funding acquisition: A.B. All authors have read and agreed to the published version of the manuscript.

Funding: This research was funded by a Leverhulme Trust Project Grant (RPG-2020-186) to E.I. and A.B and by a BBSRC Responsive Mode Grant (BB/T006315/1) to S.V.E. and A.B.

Institutional Review Board Statement: Not applicable.

Informed Consent Statement: Not applicable.

Data Availability Statement: No new data were created or analyzed in this study. Data sharing is not applicable to this article.

Acknowledgments: We would like to thank members of the Buscaino Laboratory and the Kent Fungal Group for support, discussion and feedback.

Conflicts of Interest: The authors declare no conflict of interest. The funders had no role in the design of the study; in the collection, analyses, or interpretation of data; in the writing of the manuscript, or in the decision to publish the results.

References

1. Waddington, C.H. The Epigenotype. *Int. J. Epidemiol.* **2012**, *41*, 10–13. [CrossRef] [PubMed]
2. Russo, V.E.; Martienssen, R.A.; Riggs, A.D. *Epigenetic Mechanisms of Gene Regulation*; Cold Spring Harbor Laboratory Press: Woodbury, NY, USA, 1996.
3. Bird, A. Perceptions of epigenetics. *Nature* **2007**, *447*, 396–398. [CrossRef] [PubMed]
4. Nanney, D.L. Epigenetic control systems. *Proc. Natl. Acad. Sci. USA* **1958**, *44*, 712–717. [CrossRef] [PubMed]
5. Holliday, R. Epigenetics: An overview. *Dev. Genet.* **1994**, *15*, 453–457. [CrossRef] [PubMed]

6.   Aghcheh, R.K.; Kubicek, C.P. Epigenetics as an emerging tool for improvement of fungal strains used in biotechnology. *Appl. Microbiol. Biotechnol.* **2015**, *99*, 6167–6181. [CrossRef] [PubMed]

7.   Lappalainen, T.; Greally, J.M. Associating cellular epigenetic models with human phenotypes. *Nat. Rev. Genet.* **2017**, *18*, 441–451. [CrossRef]

8.   Cavalli, G.; Heard, E. Advances in epigenetics link genetics to the environment and disease. *Nature* **2019**, *571*, 489–499. [CrossRef] [PubMed]

9.   Brown, G.D.; Denning, D.W.; Gow, N.A.R.; Levitz, S.M.; Netea, M.G.; White, T.C. Hidden Killers: Human Fungal Infections. *Sci. Transl. Med.* **2012**, *4*, 165rv13. [CrossRef] [PubMed]

10. Ohama, T.; Suzuki, T.; Mori, M.; Osawa, S.; Ueda, T.; Watanabe, K.; Nakase, T. Non-universal decoding of the leucine codon CUG in several *Candida* species. *Nucleic Acids Res.* **1993**, *21*, 4039–4045. [CrossRef]

11. Krassowski, T.; Coughlan, A.Y.; Shen, X.-X.; Zhou, X.; Kominek, J.; Opulente, D.A.; Riley, R.; Grigoriev, I.V.; Maheshwari, N.; Shields, D.C.; et al. Evolutionary instability of CUG-Leu in the genetic code of budding yeasts. *Nat. Commun.* **2018**, *9*, 1887. [CrossRef] [PubMed]

12. Lionakis, M.S.; Lim, J.K.; Lee, C.-C.R.; Murphy, P.M. Organ-Specific Innate Immune Responses in a Mouse Model of Invasive Candidiasis. *J. Innate Immun.* **2011**, *3*, 180–199. [CrossRef] [PubMed]

13. da Silva Dantas, A.; Lee, K.K.; Raziunaite, I.; Schaefer, K.; Wagener, J.; Yadav, B.; Gow, N.A. Cell biology of Candida albicans–host interactions. *Curr. Opin. Microbiol.* **2016**, *34*, 111–118. [CrossRef] [PubMed]

14. Gonçalves, B.; Ferreira, C.; Alves, C.T.; Henriques, M.; Azeredo, J.; Silva, S. Vulvovaginal candidiasis: Epidemiology, microbiology and risk factors. *Crit. Rev. Microbiol.* **2016**, *42*, 905–927. [CrossRef] [PubMed]

15. Mayer, F.L.; Wilson, D.; Hube, B. Candida albicans pathogenicity mechanisms. *Virulence* **2013**, *4*, 119–128. [CrossRef] [PubMed]

16. Desai, J. Candida albicans Hyphae: From Growth Initiation to Invasion. *J. Fungi* **2018**, *4*, 10. [CrossRef] [PubMed]

17. Taff, H.T.; Mitchell, K.F.; Edward, J.A.; Andes, D.R. Mechanisms of Candida biofilm drug resistance. *Future Microbiol.* **2013**, *8*, 1325–1337. [CrossRef]

18. Beekman, C.N.; Cuomo, C.A.; Bennett, R.J.; Ene, I.V. Comparative genomics of white and opaque cell states supports an epigenetic mechanism of phenotypic switching in Candida albicans. *G3 GenesGenomesGenetics* **2021**, *11*, jkab001. [CrossRef] [PubMed]

19. Qasim, M.N.; Valle Arevalo, A.; Nobile, C.J.; Hernday, A.D. The Roles of Chromatin Accessibility in Regulating the Candida albicans White-Opaque Phenotypic Switch. *J. Fungi* **2021**, *7*, 37. [CrossRef]

20. Luger, K.; Dechassa, M.L.; Tremethick, D.J. New insights into nucleosome and chromatin structure: An ordered state or a disordered affair? *Nat. Rev. Mol. Cell Biol.* **2012**, *13*, 436–447. [CrossRef] [PubMed]

21. Skrzypek, M.S.; Binkley, J.; Binkley, G.; Miyasato, S.R.; Simison, M.; Sherlock, G. The *Candida* Genome Database (CGD): Incorporation of Assembly 22, systematic identifiers and visualization of high throughput sequencing data. *Nucleic Acids Res.* **2017**, *45*, D592–D596. [CrossRef]

22. Gu, M.; Naiyachit, Y.; Wood, T.J.; Millar, C.B. H2A.Z marks antisense promoters and has positive effects on antisense transcript levels in budding yeast. *BMC Genom.* **2015**, *16*, 99. [CrossRef] [PubMed]

23. Martire, S.; Banaszynski, L.A. The roles of histone variants in fine-tuning chromatin organization and function. *Nat. Rev. Mol. Cell Biol.* **2020**, *21*, 522–541. [CrossRef] [PubMed]

24. Zofall, M.; Fischer, T.; Zhang, K.; Zhou, M.; Cui, B.; Veenstra, T.D.; Grewal, S.I.S. Histone H2A.Z cooperates with RNAi and heterochromatin factors to suppress antisense RNAs. *Nature* **2009**, *461*, 419–422. [CrossRef] [PubMed]

25. Freire-Benéitez, V.; Price, R.J.; Buscaino, A. The Chromatin of Candida albicans Pericentromeres Bears Features of Both Euchromatin and Heterochromatin. *Front. Microbiol.* **2016**, *7*, 759. [CrossRef] [PubMed]

26. Li, B.; Carey, M.; Workman, J.L. The Role of Chromatin during Transcription. *Cell* **2007**, *128*, 707–719. [CrossRef] [PubMed]

27. Strålfors, A.; Ekwall, K. Heterochromatin and Euchromatin—Organization, Boundaries, and Gene Regulation. In *Reviews in Cell Biology and Molecular Medicine*; American Cancer Society: Atlanta, GA, USA, 2011; ISBN 978-3-527-60090-8.

28. Bannister, A.J.; Kouzarides, T. Regulation of chromatin by histone modifications. *Cell Res.* **2011**, *21*, 381–395. [CrossRef] [PubMed]

29. Calo, E.; Wysocka, J. Modification of enhancer chromatin: What, how and why? *Mol. Cell* **2013**, *49*, 825–837. [CrossRef]

30. Wang, X.; Cairns, M.J.; Yan, J. Super-enhancers in transcriptional regulation and genome organization. *Nucleic Acids Res.* **2019**, *47*, 11481–11496. [CrossRef]

31. Price, R.J.; Weindling, E.; Berman, J.; Buscaino, A. Chromatin Profiling of the Repetitive and Non-repetitive Genomes of the Human Fungal Pathogen Candida albicans. *mBio* **2019**, *10*, e01376-19. [CrossRef] [PubMed]

32. Saksouk, N.; Simboeck, E.; Déjardin, J. Constitutive heterochromatin formation and transcription in mammals. *Epigenetics Chromatin* **2015**, *8*, 3. [CrossRef] [PubMed]

33. Erlendson, A.A.; Friedman, S.; Freitag, M. A Matter of Scale and Dimensions: Chromatin of Chromosome Landmarks in the Fungi. *Microbiol. Spectr.* **2017**, *5*, 571–597. [CrossRef]

34. Mishra, P.K.; Baum, M.; Carbon, J. DNA methylation regulates phenotype-dependent transcriptional activity in Candida albicans. *Proc. Natl. Acad. Sci. USA* **2011**, *108*, 11965–11970. [CrossRef] [PubMed]

35. Li, X.; Egervari, G.; Wang, Y.; Berger, S.L.; Lu, Z. Regulation of chromatin and gene expression by metabolic enzymes and metabolites. *Nat. Rev. Mol. Cell Biol.* **2018**, *19*, 563–578. [CrossRef] [PubMed]

36. Engel, S.R.; Dietrich, F.S.; Fisk, D.G.; Binkley, G.; Balakrishnan, R.; Costanzo, M.C.; Dwight, S.S.; Hitz, B.C.; Karra, K.; Nash, R.S.; et al. The Reference Genome Sequence of *Saccharomyces cerevisiae*: Then and Now. *G3 GenesGenomesGenetics* **2014**, *4*, 389–398. [CrossRef] [PubMed]

37. Wood, V.; Harris, M.A.; McDowall, M.D.; Rutherford, K.; Vaughan, B.W.; Staines, D.M.; Aslett, M.; Lock, A.; Bahler, J.; Kersey, P.J.; et al. PomBase: A comprehensive online resource for fission yeast. *Nucleic Acids Res.* **2012**, *40*, D695–D699. [CrossRef]

38. The Alliance of Genome Resources Consortium; Agapite, J.; Albou, L.-P.; Aleksander, S.; Argasinska, J.; Arnaboldi, V.; Attrill, H.; Bello, S.M.; Blake, J.A.; Blodgett, O.; et al. Alliance of Genome Resources Portal: Unified model organism research platform. *Nucleic Acids Res.* **2020**, *48*, D650–D658. [CrossRef]

39. Tolsma, T.O.; Hansen, J.C. Post-translational modifications and chromatin dynamics. *Essays Biochem.* **2019**, *63*, 89–96. [CrossRef] [PubMed]

40. Sabari, B.R.; Tang, Z.; Huang, H.; Yong-Gonzalez, V.; Molina, H.; Kong, H.E.; Dai, L.; Shimada, M.; Cross, J.R.; Zhao, Y.; et al. Intracellular Crotonyl-CoA Stimulates Transcription through p300-Catalyzed Histone Crotonylation. *Mol. Cell* **2015**, *58*, 203–215. [CrossRef]

41. Musselman, C.A.; Lalonde, M.-E.; Côté, J.; Kutateladze, T.G. Perceiving the epigenetic landscape through histone readers. *Nat. Struct. Mol. Biol.* **2012**, *19*, 1218–1227. [CrossRef] [PubMed]

42. Andrews, F.H.; Shinsky, S.A.; Shanle, E.K.; Bridgers, J.B.; Gest, A.; Tsun, I.K.; Krajewski, K.; Shi, X.; Strahl, B.D.; Kutateladze, T.G. The Taf14 YEATS domain is a reader of histone crotonylation. *Nat. Chem. Biol.* **2016**, *12*, 396–398. [CrossRef]

43. Becker, P.B.; Hörz, W. ATP-Dependent Nucleosome Remodeling. *Annu. Rev. Biochem.* **2002**, *71*, 247–273. [CrossRef] [PubMed]

44. Tyagi, M.; Imam, N.; Verma, K.; Patel, A.K. Chromatin remodelers: We are the drivers!! *Nucleus* **2016**, *7*, 388–404. [CrossRef] [PubMed]

45. Tang, L.; Nogales, E.; Ciferri, C. Structure and Function of SWI/SNF Chromatin Remodeling Complexes and Mechanistic Implications for Transcription. *Prog. Biophys. Mol. Biol.* **2010**, *102*, 122–128. [CrossRef] [PubMed]

46. Lin, A.; Du, Y.; Xiao, W. Yeast chromatin remodeling complexes and their roles in transcription. *Curr. Genet.* **2020**, *66*, 657–670. [CrossRef] [PubMed]

47. Biggar, S.R.; Crabtree, G.R. Continuous and widespread roles for the Swi-Snf complex in transcription. *EMBO J.* **1999**, *18*, 2254–2264. [CrossRef] [PubMed]

48. Peterson, C.L.; Herskowitz, I. Characterization of the yeast SWI1, SWI2, and SWI3 genes, which encode a global activator of transcription. *Cell* **1992**, *68*, 573–583. [CrossRef]

49. Sudarsanam, P.; Cao, Y.; Wu, L.; Laurent, B.C.; Winston, F. The nucleosome remodeling complex, Snf/Swi, is required for the maintenance of transcription in vivo and is partially redundant with the histone acetyltransferase, Gcn5. *EMBO J.* **1999**, *18*, 3101–3106. [CrossRef] [PubMed]

50. Martens, J.A.; Winston, F. Evidence that Swi/Snf directly represses transcription in S. cerevisiae. *Genes Dev.* **2002**, *16*, 2231–2236. [CrossRef]

51. Monahan, B.J.; Villén, J.; Marguerat, S.; Bähler, J.; Gygi, S.P.; Winston, F. Fission yeast SWI/SNF and RSC complexes show compositional and functional differences from budding yeast. *Nat. Struct. Mol. Biol.* **2008**, *15*, 873–880. [CrossRef] [PubMed]

52. Lorch, Y.; Kornberg, R.D. Chromatin-remodeling and the initiation of transcription. *Q. Rev. Biophys.* **2015**, *48*, 465–470. [CrossRef]

53. Balachandra, V.K.; Verma, J.; Shankar, M.; Tucey, T.M.; Traven, A.; Schittenhelm, R.B.; Ghosh, S.K. The RSC (Remodels the Structure of Chromatin) complex of Candida albicans shows compositional divergence with distinct roles in regulating pathogenic traits. *PLoS Genet.* **2020**, *16*, e1009071. [CrossRef] [PubMed]

54. Mao, X.; Cao, F.; Nie, X.; Liu, H.; Chen, J. The Swi/Snf chromatin remodeling complex is essential for hyphal development in Candida albicans. *FEBS Lett.* **2006**, *580*, 2615–2622. [CrossRef]

55. Guan, Z.; Liu, H. Overlapping Functions between SWR1 Deletion and H3K56 Acetylation in Candida albicans. *Eukaryot. Cell* **2015**, *14*, 578–587. [CrossRef] [PubMed]

56. Kaikkonen, M.U.; Lam, M.T.Y.; Glass, C.K. Non-coding RNAs as regulators of gene expression and epigenetics. *Cardiovasc. Res.* **2011**, *90*, 430–440. [CrossRef] [PubMed]

57. Till, P.; Mach, R.L.; Mach-Aigner, A.R. A current view on long non-coding RNAs in yeast and filamentous fungi. *Appl. Microbiol. Biotechnol.* **2018**, *102*, 7319–7331. [CrossRef] [PubMed]

58. Paturi, S.; Deshmukh, M.V. A Glimpse of "Dicer Biology" Through the Structural and Functional Perspective. *Front. Mol. Biosci.* **2021**, *8*, 643657. [CrossRef] [PubMed]

59. Smialowska, A.; Djupedal, I.; Wang, J.; Kylsten, P.; Swoboda, P.; Ekwall, K. RNAi mediates post-transcriptional repression of gene expression in fission yeast Schizosaccharomyces pombe. *Biochem. Biophys. Res. Commun.* **2014**, *444*, 254–259. [CrossRef] [PubMed]

60. O'Brien, J.; Hayder, H.; Zayed, Y.; Peng, C. Overview of MicroRNA Biogenesis, Mechanisms of Actions, and Circulation. *Front. Endocrinol.* **2018**, *9*, 402. [CrossRef] [PubMed]

61. Izumi, N.; Shoji, K.; Suzuki, Y.; Katsuma, S.; Tomari, Y. Zucchini consensus motifs determine the mechanism of pre-piRNA production. *Nature* **2020**, *578*, 311–316. [CrossRef] [PubMed]

62. Huang, X.; Fejes Tóth, K.; Aravin, A.A. piRNA Biogenesis in Drosophila melanogaster. *Trends Genet.* **2017**, *33*, 882–894. [CrossRef] [PubMed]

63. Batki, J.; Schnabl, J.; Wang, J.; Handler, D.; Andreev, V.I.; Stieger, C.E.; Novatchkova, M.; Lampersberger, L.; Kauneckaite, K.; Xie, W.; et al. The nascent RNA binding complex SFiNX licenses piRNA-guided heterochromatin formation. *Nat. Struct. Mol. Biol.* **2019**, *26*, 720–731. [CrossRef]

64. Sellam, A.; Hogues, H.; Askew, C.; Tebbji, F.; van Het Hoog, M.; Lavoie, H.; Kumamoto, C.A.; Whiteway, M.; Nantel, A. Experimental annotation of the human pathogen Candida albicans coding and non-coding transcribed regions using high-resolution tiling arrays. *Genome Biol.* **2010**, *11*, R71. [CrossRef] [PubMed]

65. Bernstein, D.A.; Vyas, V.K.; Weinberg, D.E.; Drinnenberg, I.A.; Bartel, D.P.; Fink, G.R. Candida albicans Dicer (CaDcr1) is required for efficient ribosomal and spliceosomal RNA maturation. *Proc. Natl. Acad. Sci. USA* **2012**, *109*, 523–528. [CrossRef]

66. Fuchs, B.B.; Eby, J.; Nobile, C.J.; El Khoury, J.B.; Mitchell, A.P.; Mylonakis, E. Role of filamentation in Galleria mellonella killing by Candida albicans. *Microbes Infect.* **2010**, *12*, 488–496. [CrossRef]

67. Noble, S.M.; Gianetti, B.A.; Witchley, J.N. Candida albicans cell-type switching and functional plasticity in the mammalian host. *Nat. Rev. Microbiol.* **2017**, *15*, 96–108. [CrossRef]

68. Murad, A.M.A. NRG1 represses yeast-hypha morphogenesis and hypha-specific gene expression in Candida albicans. *EMBO J.* **2001**, *20*, 4742–4752. [CrossRef] [PubMed]

69. Braun, B.R.; Johnson, A.D. TUP1, CPH1 and EFG1 make independent contributions to filamentation in candida albicans. *Genetics* **2000**, *155*, 57–67. [CrossRef] [PubMed]

70. Banerjee, M.; Thompson, D.S.; Lazzell, A.; Carlisle, P.L.; Pierce, C.; Monteagudo, C.; López-Ribot, J.L.; Kadosh, D. *UME6*, a Novel Filament-specific Regulator of *Candida albicans* Hyphal Extension and Virulence. *Mol. Biol. Cell* **2008**, *19*, 1354–1365. [CrossRef]

71. Shapiro, R.S.; Robbins, N.; Cowen, L.E. Regulatory Circuitry Governing Fungal Development, Drug Resistance, and Disease. *Microbiol. Mol. Biol. Rev.* **2011**, *75*, 213–267. [CrossRef] [PubMed]

72. Ernst, J.F. Transcription factors in Candida albicans—Environmental control of morphogenesis. *Microbiology* **2000**, *146*, 1763–1774. [CrossRef] [PubMed]

73. Dunker, C.; Polke, M.; Schulze-Richter, B.; Schubert, K.; Rudolphi, S.; Gressler, A.E.; Pawlik, T.; Prada Salcedo, J.P.; Niemiec, M.J.; Slesiona-Künzel, S.; et al. Rapid proliferation due to better metabolic adaptation results in full virulence of a filament-deficient Candida albicans strain. *Nat. Commun.* **2021**, *12*, 3899. [CrossRef] [PubMed]

74. Nadeem, S.G.; Shafiq, A.; Hakim, S.T.; Anjum, Y.; Kazm, S.U. Effect of Growth Media, pH and Temperature on Yeast to Hyphal Transition in Candida albicans. *Open J. Med. Microbiol.* **2013**, *3*, 185–192. [CrossRef]

75. Wang, X.; Chang, P.; Ding, J.; Chen, J. Distinct and Redundant Roles of the Two MYST Histone Acetyltransferases Esa1 and Sas2 in Cell Growth and Morphogenesis of Candida albicans. *Eukaryot. Cell* **2013**, *12*, 438–449. [CrossRef] [PubMed]

76. Hnisz, D.; Majer, O.; Frohner, I.E.; Komnenovic, V.; Kuchler, K. The Set3/Hos2 Histone Deacetylase Complex Attenuates cAMP/PKA Signaling to Regulate Morphogenesis and Virulence of Candida albicans. *PLOS Pathog.* **2010**, *6*, e1000889. [CrossRef] [PubMed]

77. Zhao, G.; Rusche, L.N. Genetic Analysis of Sirtuin Deacetylases in Hyphal Growth of *Candida albicans. mSphere* **2021**, *6*, e00053-2. [CrossRef] [PubMed]

78. Lu, Y.; Su, C.; Mao, X.; Raniga, P.P.; Liu, H.; Chen, J. Efg1-mediated Recruitment of NuA4 to Promoters Is Required for Hypha-specific Swi/Snf Binding and Activation in Candida albicans. *Mol. Biol. Cell* **2008**, *19*, 4260–4272. [CrossRef] [PubMed]

79. Lu, Y.; Su, C.; Wang, A.; Liu, H. Hyphal Development in Candida albicans Requires Two Temporally Linked Changes in Promoter Chromatin for Initiation and Maintenance. *PLoS Biol.* **2011**, *9*, e1001105. [CrossRef]

80. Lu, Y.; Su, C.; Solis, N.V.; Filler, S.G.; Liu, H. Synergistic Regulation of Hyphal Elongation by Hypoxia, CO2, and Nutrient Conditions Controls the Virulence of Candida albicans. *Cell Host Microbe* **2013**, *14*, 499–509. [CrossRef] [PubMed]

81. Kadosh, D.; Lopez-Ribot, J.L. Candida albicans: Adapting to Succeed. *Cell Host Microbe* **2013**, *14*, 483–485. [CrossRef] [PubMed]

82. Shivarathri, R.; Tscherner, M.; Zwolanek, F.; Singh, N.K.; Chauhan, N.; Kuchler, K. The Fungal Histone Acetyl Transferase Gcn5 Controls Virulence of the Human Pathogen Candida albicans through Multiple Pathways. *Sci. Rep.* **2019**, *9*, 9445. [CrossRef]

83. Chang, P.; Fan, X.; Chen, J. Function and subcellular localization of Gcn5, a histone acetyltransferase in Candida albicans. *Fungal Genet. Biol.* **2015**, *81*, 132–141. [CrossRef] [PubMed]

84. Nobile, C.J.; Fox, E.P.; Hartooni, N.; Mitchell, K.F.; Hnisz, D.; Andes, D.R.; Kuchler, K.; Johnson, A.D. A Histone Deacetylase Complex Mediates Biofilm Dispersal and Drug Resistance in Candida albicans. *mBio* **2014**, *5*, e01201-14. [CrossRef]

85. Chen, J.; Liu, Q.; Zeng, L.; Huang, X. Protein Acetylation/Deacetylation: A Potential Strategy for Fungal Infection Control. *Front. Microbiol.* **2020**, *11*, 574736. [CrossRef]

86. Li, Y.; Li, H.; Sui, M.; Li, M.; Wang, J.; Meng, Y.; Sun, T.; Liang, Q.; Suo, C.; Gao, X.; et al. Fungal acetylome comparative analysis identifies an essential role of acetylation in human fungal pathogen virulence. *Commun. Biol.* **2019**, *2*, 154. [CrossRef] [PubMed]

87. Wang, Q.; Verma, J.; Vidan, N.; Wang, Y.; Tucey, T.M.; Lo, T.L.; Harrison, P.F.; See, M.; Swaminathan, A.; Kuchler, K.; et al. The YEATS Domain Histone Crotonylation Readers Control Virulence-Related Biology of a Major Human Pathogen. *Cell Rep.* **2020**, *31*, 107528. [CrossRef]

88. Tebbji, F.; Chen, Y.; Sellam, A.; Whiteway, M. The Genomic Landscape of the Fungus-Specific SWI/SNF Complex Subunit, Snf6, in Candida albicans. *mSphere* **2017**, *2*, e00497-17. [CrossRef]

89. Pukkila-Worley, R.; Peleg, A.Y.; Tampakakis, E.; Mylonakis, E. Candida albicans Hyphal Formation and Virulence Assessed Using a Caenorhabditis elegans Infection Model. *Eukaryot. Cell* **2009**, *8*, 1750–1758. [CrossRef] [PubMed]

90. Bruno, V.M.; Wang, Z.; Marjani, S.L.; Euskirchen, G.M.; Martin, J.; Sherlock, G.; Snyder, M. Comprehensive annotation of the transcriptome of the human fungal pathogen Candida albicans using RNA-seq. *Genome Res.* **2010**, *20*, 1451–1458. [CrossRef] [PubMed]
91. Rai, L.S.; Singha, R.; Brahma, P.; Sanyal, K. Epigenetic determinants of phenotypic plasticity in Candida albicans. *Fungal Biol. Rev.* **2018**, *32*, 10–19. [CrossRef]
92. Nobile, C.J.; Fox, E.P.; Nett, J.E.; Sorrells, T.R.; Mitrovich, Q.M.; Hernday, A.D.; Tuch, B.B.; Andes, D.R.; Johnson, A.D. A Recently Evolved Transcriptional Network Controls Biofilm Development in Candida albicans. *Cell* **2012**, *148*, 126–138. [CrossRef]
93. Fox, E.P.; Bui, C.K.; Nett, J.E.; Hartooni, N.; Mui, M.C.; Andes, D.R.; Nobile, C.J.; Johnson, A.D. An expanded regulatory network temporally controls Candida albicans biofilm formation. *Mol. Microbiol.* **2015**, *96*, 1226–1239. [CrossRef] [PubMed]
94. Mancera, E.; Nocedal, I.; Hammel, S.; Gulati, M.; Mitchell, K.F.; Andes, D.R.; Nobile, C.J.; Butler, G.; Johnson, A.D. Evolution of the complex transcription network controlling biofilm formation in Candida species. *eLife* **2021**, *10*, e64682. [CrossRef] [PubMed]
95. Gulati, M.; Nobile, C.J. Candida albicans biofilms: Development, regulation, and molecular mechanisms. *Microbes Infect.* **2016**, *18*, 310–321. [CrossRef] [PubMed]
96. Rai, L.S.; Singha, R.; Sanchez, H.; Chakraborty, T.; Chand, B.; Bachellier-Bassi, S.; Chowdhury, S.; d'Enfert, C.; Andes, D.R.; Sanyal, K. The Candida albicans biofilm gene circuit modulated at the chromatin level by a recent molecular histone innovation. *PLoS Biol.* **2019**, *17*, e3000422. [CrossRef] [PubMed]

Review

Amino Acid Sensing and Assimilation by the Fungal Pathogen *Candida albicans* in the Human Host

Fitz Gerald S. Silao and Per O. Ljungdahl *

Department of Molecular Biosciences, Wenner-Gren Institute, SciLifeLab, Stockholm University, 114 19 Stockholm, Sweden; fitzgerald.silao@scilifelab.se
* Correspondence: per.ljungdahl@scilifelab.se

Abstract: Nutrient uptake is essential for cellular life and the capacity to perceive extracellular nutrients is critical for coordinating their uptake and metabolism. Commensal fungal pathogens, e.g., *Candida albicans*, have evolved in close association with human hosts and are well-adapted to using diverse nutrients found in discrete host niches. Human cells that cannot synthesize all amino acids require the uptake of the "essential amino acids" to remain viable. Consistently, high levels of amino acids circulate in the blood. Host proteins are rich sources of amino acids but their use depends on proteases to cleave them into smaller peptides and free amino acids. *C. albicans* responds to extracellular amino acids by pleiotropically enhancing their uptake and derive energy from their catabolism to power opportunistic virulent growth. Studies using *Saccharomyces cerevisiae* have established paradigms to understand metabolic processes in *C. albicans*; however, fundamental differences exist. The advent of CRISPR/Cas9-based methods facilitate genetic analysis in *C. albicans*, and state-of-the-art molecular biological techniques are being applied to directly examine growth requirements in vivo and in situ in infected hosts. The combination of divergent approaches can illuminate the biological roles of individual cellular components. Here we discuss recent findings regarding nutrient sensing with a focus on amino acid uptake and metabolism, processes that underlie the virulence of *C. albicans*.

Keywords: *Candida albicans*; human fungal pathogen; nutrient sensing; amino acid metabolism; proline catabolism; mitochondria; SPS-sensor; nitrogen catabolite repression; glucose repression

Citation: Silao, F.G.S.; Ljungdahl, P.O. Amino Acid Sensing and Assimilation by the Fungal Pathogen *Candida albicans* in the Human Host. *Pathogens* **2022**, *11*, 5. https://doi.org/10.3390/pathogens11010005

Academic Editor: Jonathan Richardson

Received: 17 November 2021
Accepted: 19 December 2021
Published: 22 December 2021

Publisher's Note: MDPI stays neutral with regard to jurisdictional claims in published maps and institutional affiliations.

1. Introduction

All organisms require nutrients to live, grow and successfully reproduce. The ability of an organism to assimilate nutrients in a given ecological niche is dependent on its ability to sense and respond to the availability of nutrients and on intrinsic cellular properties. Defining the key signaling events activated by nutrient sensing systems and the metabolic capacities of an organism provides a compelling description that reflects the organism's role in the niche. For opportunistic human pathogens, acquiring nutrients to support commensal or pathogenic growth is not a trivial task as the availability of key nutrients is dependent on several extrinsic host-specific factors. Such factors include host defense activities that often are linked to substantial and rapid changes in biophysical parameters, e.g., extracellular pH and the generation of reactive oxygen species (ROS), different nutrient and metabolic activities of host tissues at sites of fungal cell colonization, and the presence of competing microorganisms.

Of the fungal pathogens capable of infecting humans, *Candida albicans* is considered to be the most important, and arguably the most successful. *C. albicans* is a natural commensal of humans, capable of colonizing virtually all anatomical sites (Figure 1). This fungus can switch from harmless commensal to pathogenic growth and thereby cause a spectrum of pathologies, ranging from mucosal to life-threatening systemic infections, collectively termed candidiasis. It is imperative to distinguish the difference between commensal and invasive virulent growth; invasion is distinct from superficial colonization as the former is accompanied by inflammatory signals, resulting from the activated immune response. Clinical cases presenting *C. albicans* infections of the urogenital tract [1–8], kidney [1,7,9–13], liver [14,15], lungs [16–18], spleen [19,20] and even the heart [21–26] have been reported. In rare cases, mostly in neonates, *C. albicans* can traverse the blood–brain barrier, resulting in infections of the brain [27–30]. These observations indicate that *C. albicans* can successfully establish and grow in different host niches. Consequently, *C. albicans* must possess the means to successfully adapt in order to obtain and use a wide range of host-derived nutrients. Given their opportunistic character, the question remains open as to how *C. albicans* cells fine-tune their nutrient acquisition machineries to support commensal and pathogenic growth under apparently disparate environmental conditions.

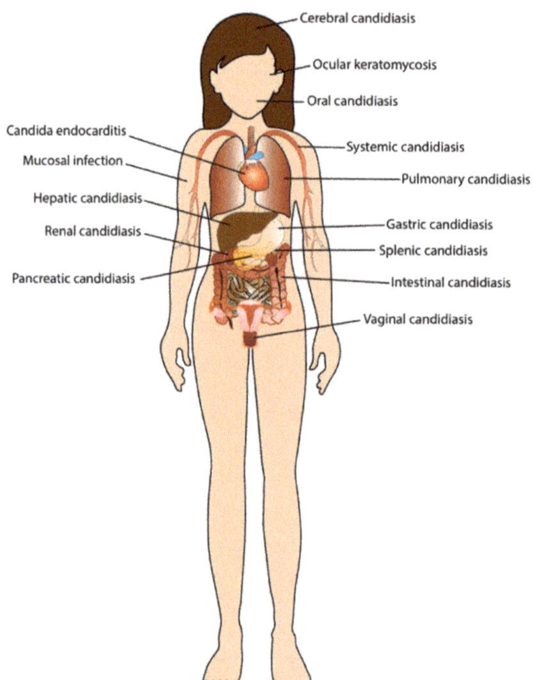

Figure 1. *C. albicans* **can infect virtually all anatomical sites in the body.** Infections can either be superficial, mainly affecting the skin or mucous membrane, or invasive, involving fungal entry into the blood (candidemia) and dissemination to other internal organs. Disseminated growth happens when *C. albicans*, colonizing an anatomical site (usually the gut), catheters or other medical implants, enter the blood and then disseminate to other organs such as the lungs (pulmonary), liver (hepatic), spleen (splenic), pancreas (pancreatic) and kidney (renal). These infections can be localized or can re-enter the bloodstream again, allowing them to reach additional anatomical sites, and in rare cases the brain (cerebral). *C. albicans* in the blood can enter the urine via the kidney, resulting in candiduria (yeast in the urine). Other complications of *Candida* infections include the appearance of fungus balls in certain sites, resulting in obstruction in these areas and the formation of abscesses. In women, vaginal candidiasis is an important concern since the vagina serves as a main reservoir for *C. albicans*. In addition, infection of the genitourinary tract is more often diagnosed in women than in men due to anatomical structural differences such as a shorter urethra and the close proximity of the vagina and anus.

C. albicans requires a source of nitrogen to synthesize proteins needed to carry out necessary cellular functions and to generate nucleotides for DNA and RNA synthesis. There is a plethora of nitrogen sources that *C. albicans* can theoretically utilize in the host, for example, amino acids, urea, peptides, proteins and N-acetyl glucosamine (GlcNAc) and even ammonia. Of the nitrogen sources available in the host, amino acids are preferred as they can be easily assimilated and used as both nitrogen and carbon sources [31,32]. Most of the current understanding of nutrient assimilation in *C. albicans* is inferred from extensive studies in the non-pathogenic yeast *Saccharomyces cerevisiae*. However, although many of the regulatory mechanisms operating between the two species are conserved, it is becoming clear that substantial differences exist, likely reflecting the different environments in which these fungi evolved. Although *S. cerevisiae* is readily found freely in nature, *C. albicans* has evolved in close association with mammalian hosts as a commensal. Furthermore, although *S. cerevisiae* prioritizes the ability to ferment sugar even in the presence of oxygen, *C. albicans* relies on mitochondrial oxidative phosphorylation to generate the energy necessary to survive in hosts. Aside from being able to thrive better at higher temperatures, i.e., 37 °C, a striking and important difference between *C. albicans* and *S. cerevisiae* is that *C. albicans* possesses mitochondria with all four proton-pumping and energy-conserving complexes of the respiratory chain, including NADH dehydrogenase (Complex I). In *S. cerevisiae* mitochondria, NADH is oxidized by matrix NADH dehydrogenases that are not coupled to energy-conserving proton-pumping mechanisms; hence, the oxidation of NADH in yeast does not directly contribute to ATP production [33].

The clear differences in metabolism between the established yeast model and the fungal pathogen *C. albicans* need to be considered when analyzing its virulence properties. In this review we focus on amino acid sensing and metabolism with an emphasis on proline catabolism. We begin by introducing some basic concepts regarding nitrogen source utilization and assimilation and then present a more focused discussion regarding amino acid metabolism, the generation of ammonia and associated consequences, and the central role of mitochondria in the production of energy for virulent growth.

2. Amino Acids as Nitrogen Sources in Host Environments

As an opportunistic pathogen, *C. albicans* can sense a multitude of environmental signals, including changes in the availability of diverse nitrogen sources, including amino acids. The signaling pathways that are induced regulate the activity of downstream effector transcription factors that engage programs of gene expression, some that are required for virulent growth (reviewed in [34,35]). A limited number of amino acid sensors have been characterized in *C. albicans* (Figure 2), the best understood being the plasma membrane-localized SPS sensor of extracellular amino acids [36–38]. Strains lacking a functional SPS sensor have a diminished capacity to take up amino acids, do not filament in the presence of serum and are less virulent during systemic infection in mice. These results provide a clear example of how an ordinary basic physiological process, such as nitrogen (amino acid) acquisition, can become an "accidental" but important virulence trait of an opportunistic human pathogen. Additional sensors of external amino acids in *C. albicans* have been reported, such as Gpr1, a G-protein-coupled receptor proposed to sense methionine [39], and Gap2, a general amino acid transporter that is the functional ortholog of Gap1 in *S. cerevisiae* (ScGap1) that could function as a transceptor [40]. Proteins that sense the availability of other well-characterized nitrogen sources, e.g., Mep2 that responds to ammonium [41], can also provide important regulatory signals governing amino acid use. Although *C. albicans* cells have been shown to respond to the presence of GlcNAc [42,43] and urea [44], active sensing mechanisms for these nitrogen sources have not been described. Nitrogen-containing compounds (amino acids) are taken up from the extracellular environment through a number of distinct transporters localized at the plasma membrane (Figure 2). Activation of the SPS sensor enhances the capacity of cells to take up and assimilate diverse nitrogen substances. This is accomplished as the signals derived from the activated SPS sensor induce the expression of several genes encoding amino acid permeases, secreted aspartyl protease *SAP2*, peptide and oligopeptide transport proteins [36,45,46].

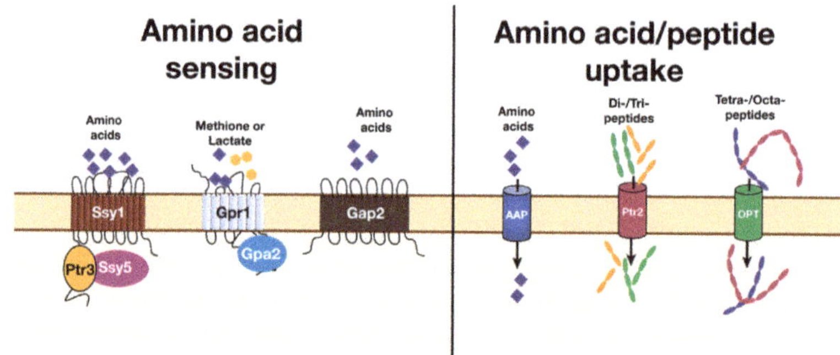

Figure 2. Amino acid sensors and transporters localized at the plasma membrane of *C. albicans*. (**Left**) the SPS sensor responds to the presence of extracellular amino acids. The main SPS sensor is composed of Ssy1, a plasma membrane-bound receptor homologous to amino acid permeases but without a capacity to transport amino acids; Ptr3, a scaffold protein that mediates intracomplex protein–protein interactions; and Ssy5, an endoprotease that proteolytically activates downstream transcription factors. The G-protein-coupled receptor 1 (Gpr1), together with intracellular cognate protein Gpa2, has been implicated in both amino acid (methionine) and lactate sensing. Gap2 is an ortholog of *S. cerevisiae* Gap1, which is thought to function as a transceptor, i.e., a functional transporter capable of initiating downstream signaling events independently of transport. (**Right**) uptake of extracellular amino acids is facilitated by a number of genetically distinct amino acid permeases (AAP) that have either broad or narrow substrate specificities. Amino acids can also enter the cell as oligopeptides taken up by Ptr2 for di-/tri-peptides, or a family of oligopeptide transporters (OPT) for oligopeptides comprising between 4 and 8 residues.

3. Nitrogen Catabolite Repression (NCR)

In addition to sensing the availability of extracellular sources of nitrogen, yeast cells can gauge the quality of internalized sources of nitrogen and respond appropriately to adjust metabolism. Nitrogen catabolite repression (NCR) is a supra-pathway that controls nitrogen source utilization through the repression of genes required for the utilization of secondary nitrogen sources when preferred ones are available. Most of the assumptions with respect to NCR in *C. albicans* are derived from extensive studies examining NCR in *S. cerevisiae*. Understanding the differences between these organisms is essential to accurately describing the hierarchy of nitrogen source assimilation and use by *C. albicans* during pathogenic growth.

In *S. cerevisiae*, NCR is controlled by four GATA transcription factors: Gln3, Gat1, Dal80 and Gzf3, all of which possess zinc-finger DNA-binding motifs that recognize a conserved GATAAG consensus sequence present in the promoters of target genes (reviewed in [47]) (Figure 3). Gln3 and Gat1 act as positive regulators of gene expression, whereas Dal80 and Gzf3 act in a negative manner to repress target gene expression. The ability of the GATA factors to compete for binding GATAAG sequences is influenced by nitrogen source availability and is even modulated by events within the nucleus [48,49]. In the presence of preferred nitrogen sources, such as ammonium and certain amino acids, Gln3 and Gat1 are tethered in the cytosol, restricting their translocation into the nucleus. For Gln3, nuclear exclusion is maintained by binding to the phosphorylated species of its interacting partner, Ure2. Gln3 and likely Gat1 can be phosphorylated, but the phosphorylation status of Gln3 does not affect its capacity to bind Ure2. Gln3 targets to and is retained in the nucleus in cells carrying deletion or mutationally inactivated alleles of Ure2, resulting in the constitutive expression of NCR-sensitive genes. Unlike Gln3, Gat1 is not entirely dependent on Ure2 for retention in the cytosol, and therefore other regulatory components apparently contribute to controlling Gat1 movement and inducing activity. Contrary to Gln3 and Gat1, Dal80 and Gzf3 are constitutively localized in the nucleus. Furthermore, in contrast to *GLN3*, *GAT1*, *GZF3* and *DAL80* are expressed under the control of promoters containing multiple

GATAAG sequences, placing their expression under NCR [50–53]. Consequently, these factors participate in regulating each other's expression (cross-regulation) and in certain instances exhibit partial autogenous regulation [48,52–54] (Figure 3).

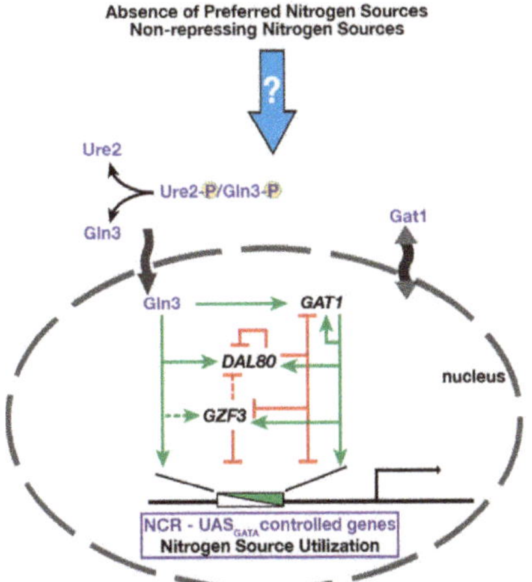

Figure 3. Schematic diagram of nitrogen catabolite repression (NCR) in yeast (adapted from Ljungdahl and Daignan-Fornier, 2012 [47]). Gln3 and Gat1 are two positive GATA effectors of NCR that are normally excluded from the nucleus under preferred nitrogen replete conditions. Nuclear exclusion is thought to occur via the interaction of Gln3 with the phosphorylated version of the Ure2 protein.

NCR has been described in *C. albicans*; however, the information available is limited to studies using strains lacking Gln3 and/or Gat1 [55–57]. Strains lacking either or both of these GATA transcription factors are unable to efficiently utilize a number of alternative nitrogen sources. This has been shown to be linked to the lack of derepression of genes necessary for their catabolism. Results from the Fonzi laboratory have shown that Gln3 and Gat1 appear to exert both specific and overlapping functions, depending on the available nitrogen sources [56]. Certain amino acids traditionally classified as poor, such as proline in *S. cerevisiae* [47,58], are readily utilized by *C. albicans* lacking Gln3 and Gat1, clearly indicating that proline utilization is not subject to tight NCR control [56]. Consistently, recent work in our group and others have shown that enzymes of the proline catabolic pathway (e.g., Put1 and Put2) can be induced in the presence of preferred nitrogen sources (e.g., ammonium or amino acids) [59–61] and even in a strain lacking Gln3 and Gat1 [59]. In addition, the gene encoding glutamate dehydrogenase (*GDH2*), a key player in central nitrogen metabolism, is robustly expressed when there is an abundance of preferred nitrogen sources such as ammonium and amino acids, indicating that its expression is independent of NCR [61]. This latter finding is in striking contrast to *S. cerevisiae*, with its *GDH2* subject to tight NCR control (reviewed in [47]).

These clear differences between *C. albicans* and *S. cerevisiae* are not trivial, and clearly reflect divergent evolutionary trajectories and the need for *C. albicans* to rapidly respond to distinct host environments. For example, as *C. albicans* cells breach epithelial barriers and reach the blood, they are exposed to high concentrations of amino acids, a condition that likely represses NCR-controlled genes, including those required for the assimilation of nitrogen derived from the degradation of host proteins. The transcription factor *STP1* is NCR-controlled and under these conditions is not expressed, which limits the expression of the Stp1-dependent secreted protease Sap2 and oligopeptide transporters.

4. Extracellular Amino Acid Sensing and Uptake—The SPS Sensing System

The plasma-membrane-localized SPS (Ssy1-Ptr3-Ssy5) sensor of *C. albicans* has been characterized [36,38,59,62]. The SPS sensor enables cells to sense and respond to the presence of extracellular amino acids (Figure 4). Again, progress has largely been guided by ongoing studies using *S. cerevisiae* as a model. In *S. cerevisiae,* the SPS signaling pathway controls the expression of a distinct set of amino acid permease (AAP) genes encoding transporters catalyzing proton-driven amino acid uptake. Two homologous effector transcription factors, Stp1 and Stp2, are synthesized as inactive precursors that localize to the cytoplasm due to N-terminal regulatory domains. The regulatory domains possess cytoplasmic retention and nuclear degron motifs, both of which are required to maintain the "off-state" of SPS-sensor-regulated gene expression. The cytoplasmic retention motifs prevent these factors from efficiently entering the nucleus, and the degron motif targets the low levels of full-length Stp1 and Stp2 that escape cytoplasmic retention for degradation by means of a novel inner-nuclear-membrane-associated degradation (INMAD) pathway. The INMAD pathway is defined by the E3 ubiquitin ligase Asi complex (Asi1-Asi2-Asi3) [63–65]. Extracellular amino acids activate the SPS sensor by binding to the receptor component Ssy1, which undergoes a conformational change that activates the Ssy5 protease in a Ptr3-dependent manner: Ptr3 functions as a scaffold that mediates intracomplex protein–protein interactions. Activated Ssy5 cleaves the N-terminal regulatory domains of Stp1 and Stp2, a processing event that enables the cleaved factors, lacking cytoplasmic retention and degron motifs, to efficiently translocate to the nucleus and bind to upstream activating sequences (UASaa) in the promoters of AAP genes. AAPs are co-translationally inserted into the endoplasmic reticulum (ER) membrane, contiguous with the outer nuclear membrane. The movement of AAPs to the PM requires the ER membrane-localized chaperone Shr3, which facilitates their folding and packaging into ER-derived secretory vesicles, a requisite for their functional expression [66,67]. The SPS sensing system enables amino acids to induce their own uptake.

Orthologs of the SPS sensing system are present in *C. albicans* ([36–38,59,62]; reviewed in [35]) (Figure 4). There is, however, a major difference. In contrast to *S. cerevisiae,* Stp1 and Stp2 in *C. albicans* activate different sets of genes that express proteins facilitating the assimilation of distinct external nitrogen sources [36,62]. Stp1 regulates the expression of *SAP2,* encoding the major and broad-spectrum secreted aspartyl proteinase (Sap) and multiple oligopeptide transporters (Opts) [36]. *STP1* expression is subject to NCR and controlled by the GATA transcription factors Gln3 and Gat1 ([68], reviewed in [69]). Accordingly, *STP1* expression is repressed when preferred nitrogen sources, i.e., ammonium sulfate and amino acids, are available and is derepressed when these nitrogen sources become limiting or absent. *STP2* is constitutively expressed and functions analogously to Stp1/Stp2 in *S. cerevisiae* and derepresses the expression of multiple amino acid permeases. *C. albicans* strains lacking either Ssy1 or Csh3, the latter an ortholog of yeast Shr3, fail to efficiently respond to the presence of extracellular amino acids and have impaired an capacity to filament in amino acid-based media [37,38]. Not all amino acids activate the SPS sensor, as can be observed by monitoring the proteolytic processing of Stp2 [36,59]. The capacity to activate the sensor and induce Stp2 processing is limited to a subset of amino acids; the presence of glutamine and arginine leads to robust SPS sensor activation. Stp2 processing is observed 5 min post-induction, indicating that the SPS sensing system rapidly responds to the presence of extracellular amino acids.

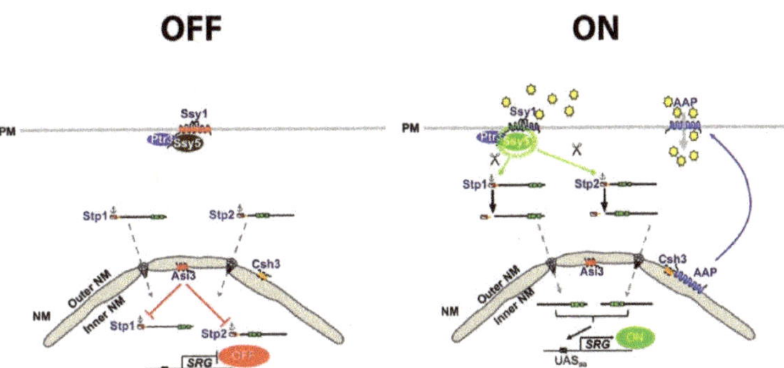

Figure 4. The SPS sensing system of *C. albicans*. Ssy1 is the primary amino acid sensor that functions with the scaffold protein Ptr3 and the protease Ssy5 as a multimeric receptor complex. Stp1 and Stp2 are the effector transcription factors of this pathway. ((**Left panel**), OFF state) In the absence of extracellular amino acids, Stp1 and Stp2 are produced as latent cytoplasmic precursors that are retained in the cytosol due to N-terminal regulatory domains that possess both a cytoplasmic retention motif and a nuclear degron, the latter recognized by the E3-ubiquitin ligase, Asi3. ((**Right panel**), ON state) In the presence of amino acids, Ssy1 is stabilized in a signaling conformation, which initiates downstream events, resulting in the activation of the Ssy5 protease. Activated Ssy5 endoproteolytically cleaves the N-terminal regulatory domains of Stp1 and Stp2. The shorter, cleaved forms efficiently translocate into the nucleus, where they bind upstream activating sequences (UASaa) and induce the expression of SPS-regulated genes (SRG). Importantly, Stp1 and Stp2 induce divergent subsets of genes, and *STP1* expression is under NCR control; *STP1* is repressed in cells grown in the presence of millimolar concentrations of amino acids, whereas Stp2 is constitutively expressed. Activated Stp2 induces the expression of amino acid permease (AAP) genes. AAPs are translated and initially inserted in the ER membrane, where they require the assistance of the ER-membrane-localized chaperone Csh3, the ortholog of yeast Shr3, to attain native structures. In the absence of Csh3, AAPs aggregate and are retained in the ER. Activated Stp1 derepresses the expression of *SAP2*, a secreted protease, and multiple oligopeptide transporter genes that facilitate peptide uptake. Stp1 triggers responses required for host protein utilization, whereas Stp2 induces amino acid utilization.

Additional sensing systems in *C. albicans*, capable of transmitting signals regarding extracellular amino acid availability, have been reported. The G-protein-coupled receptor (Gpr1) has been proposed to sense extracellular methionine [70] or even glucose, similarly to *S. cerevisiae* [71]. The idea that methionine is the primary activating ligand for Gpr1 stems from the fact that the addition of methionine could trigger the rapid internalization of Gpr1 in a manner consistent with ligand-mediated receptor internalization ([70]; reviewed in [72]). More recently, however, lactate has been proposed to be the primary activating ligand for Gpr1 [73]. Hence, the role of Gpr1 in amino-acid-induced morphogenesis remains to be defined. What is known is that ligand activation of Gpr1 stimulates GTP-GDP exchange in its effector Gα protein Gpa2; the active GTP-bound form of Gpa2 subsequently binds to the Gα-binding domain in the N-terminal region of the adenylate cyclase Cyr1, leading to enhanced cAMP production (reviewed in [74,75]). Null mutations of the *GPR1* or *GPA2* in *C. albicans* diminish filamentous growth on solid media, and consistently, filamentation can be restored via the addition of exogenous cAMP [71]. Interestingly, although Gpr1 and Gpa2 were initially characterized on the basis of increased cAMP synthesis in response to glucose, deletions of *GPR1* or *GPA2* do not affect glucose-induced cAMP signaling, and cells remain responsive to methionine and proline [39,70]. Furthermore, the levels of cAMP in *gpr1*-null mutants spike in response to serum or large amounts of glucose (100 mM, or 1.8%), suggesting that Cyr1 can be activated by Gpr1-independent processes [70].

5. Amino Acids from Proteins and Peptides

Although free amino acids are preferred, as they can be rapidly utilized as both carbon and nitrogen sources, the bulk of amino acids in hosts are typically fixed in proteins, e.g., the extracellular matrix proteins collagen and mucin. Consequently, extracellular proteolytic enzymes are required to cleave host proteins to release amino acids and peptide fragments that can subsequently be taken up by the cell. It is important to note that the breakdown of host proteins that occurs at sites of infection can be due to proteases secreted either from the fungal or host cells [76,77]. In vitro, *C. albicans* can acquire peptides and amino acids derived from extracellular proteins, e.g., albumin, collagen and mucin. This requires the expression of secreted aspartyl proteases (Saps) [78–80] or the activity of matrix metalloproteinases (MMPs) [81,82]. The host can also trigger the proteolytic degradation of tissues as observed in some pathological conditions such as cancer [83] or sarcopenia (muscle wasting) [84]. Once internalized, peptides are then degraded to amino acids through the activity of several intracellular proteases, e.g., metallopeptidase, dipeptidase, carboxypetidases and serine proteases, which liberate free amino acids [85–88]. The induced expression of some of these enzymes is complex and depends on the release from strict regulatory processes, including NCR [68]. Although the overall effect on virulence remains to be clarified, the discovery of the fungal toxin candidalysin is important as it can also trigger the release of free amino acids by contributing to the lysis of host cells [89–91].

6. Amino Acid Metabolism in *C. albicans*

Many amino acids, derived either from extracellular uptake or oligopeptide/protein degradation or released from the vacuolar compartment, are converted to glutamate in the cytosol via the catalytic activity of distinct aminotransferases (ATs) (Figure 5). Specifically, ATs transfer the α-amino group of an amino acid to α-ketoglutarate (α-KG; 2-oxoglutarate), resulting in the formation of glutamate. ATs are defined by the amino acid that serves as the amino group donor. ATs collectively contribute to the cytosolic glutamate pool (Glu_{cyto}). Examples include aspartate aminotransferase (Aat1, EC 2.6.1.1), which transfers the α-amino group of aspartate to α-KG, resulting in glutamate and oxaloacetate; ornithine transaminase (Car2; EC 2.6.1.13), which uses ornithine, forming glutamate and glutamyl-5-semialdehyde; alanine transaminase (Alt1; EC 2.6.1.2), which uses alanine, forming glutamate and pyruvate; and glutamate synthase (Glt1; EC 1.4.1.14), which uses glutamine, forming glutamate.

Glutamate is enzymatically converted to α-KG via oxidative deamination, catalyzed by NAD^+-dependent glutamate dehydrogenase (Gdh2; EC 1.4.1.2), yielding ammonia (NH_3) and reduced NADH [61,93]. At a physiologically relevant pH, ammonia is protonated to ammonium (NH_4^+). In mammalian cells, glutamate dehydrogenase is localized in the mitochondria, whereas in *S. cerevisiae*, there is a lack of consensus regarding its localization; Gdh2 has been reported to be a mitochondrial component [94,95] and alternatively a cytosolic component [96–98]. The bulk of ammonia produced by *C. albicans* during growth on amino acids as a sole nitrogen and carbon source comes from the reaction catalyzed by Gdh2 [61]. Contrary to the initial publication, now corrected [61], Gdh2 in *C. albicans* is clearly cytoplasmic. The correct assignment of Gdh2 as a cytoplasmic component is key to understanding its role in cellular energy production, as the reaction reduces NAD^+, forming NADH. The ammonium pool in the cytosol generated primarily from Gdh2 activity and possibly from the import of extracellular ammonium via the ammonium transporters Mep1 and Mep2 [57,99,100] can be assimilated by two key anabolic reactions catalyzed by the NADPH-dependent glutamate dehydrogenase (Gdh3, EC 1.4.1.4; Gdh1 in *S. cerevisiae*) and glutamine synthetase (Gln1; EC 6.3.1.2) (Figure 5). Gdh3 catalyzes the synthesis of glutamate from α-KG and ammonium [93], whereas Gln1 catalyzes the synthesis of glutamine from glutamate and ammonium in an ATP-dependent reaction [93]. In *S. cerevisiae*, cytosolic glutamate can be imported to the mitochondrial matrix via transporters localized at the inner mitochondrial membrane, such as Agc1 [101–104] or Ymc2 [104]. Putative orthologs of these proteins exist in *C. albicans* (see CGD, http://www.candidagenome.org (accessed on 19 December 2021); C1_13400C for Agc1 and C4_02080W for Ymc2). Furthermore, in *S. cerevisiae*, cytosolic α-KG can be imported into the mitochondria through oxodicarboxylate carriers that exists in two isoforms, Odc1 and Odc2; Odc1 is used during respiration and its expression is subject to glucose repression, whereas Odc2 is the predominant isoform under non-respiratory conditions [105,106]. The *C. albicans* genome has a putative ortholog for Odc1 (CR_05480W).

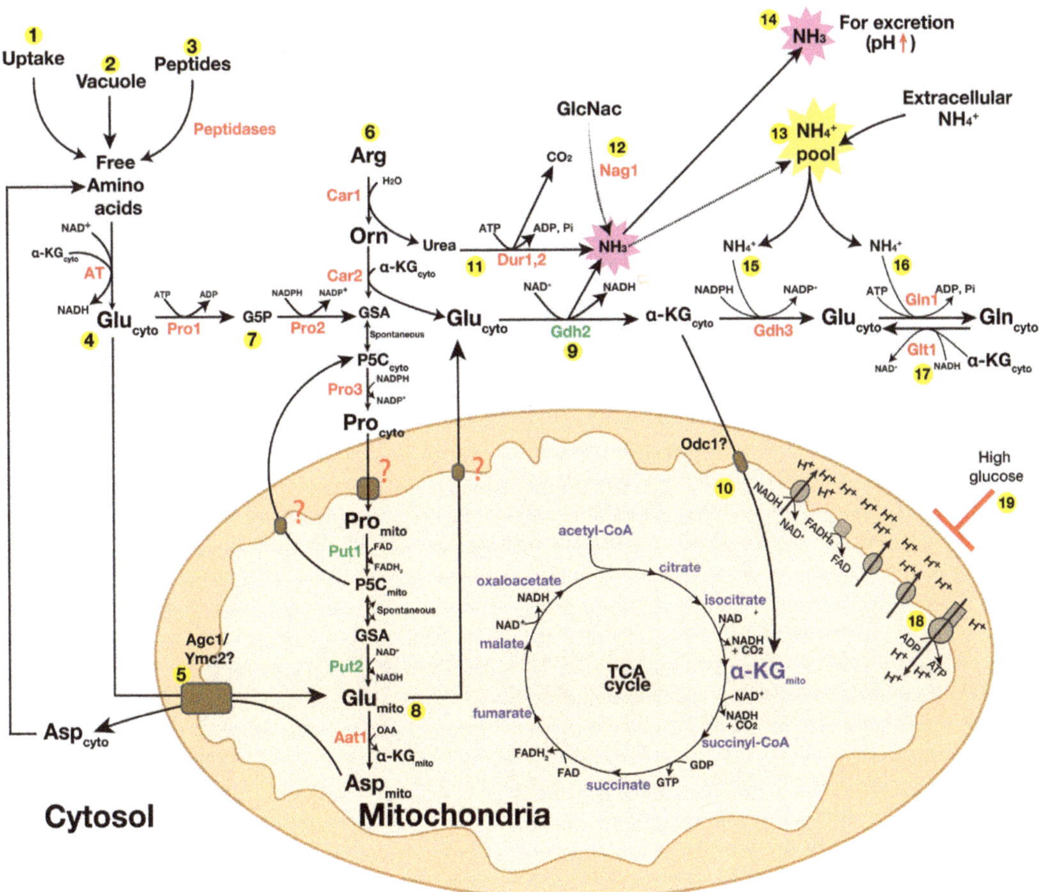

Figure 5. Nitrogen utilization in *C. albicans*. Most free amino acids, obtained from extracellular uptake (**1**), vacuolar release (**2**) or the degradation of small peptides by intracellular peptidases (**3**), are deaminated by specific aminotransferases (ATs) using α-ketoglutarate as the amino group acceptor, forming glutamate. Many ATs are found in the cytosol and there is an abundant supply of glutamate in the cytosol (Glu_cyto) (**4**). Glutamate can be imported into the mitochondria by transporters, such as Agc1 (**5**). Arginine and proline are converted to glutamate via the proline catabolic pathway (**6**). In the cytosolic portion of this pathway, arginine is converted to ornithine and urea by arginase (Car1). Ornithine is rapidly converted to glutamate semialdehyde (GSA) by ornithine transaminase (Car2), which is non-enzymatically converted to the cyclic Δ-1-pyrroline-5-carboxylate (P5C) and then reduced by P5C reductase (Pro3), generating proline (Pro_cyto). Pro_mito) via an unidentified transporter. Pro_mito is catabolized to glutamate (Glu_mito) via the concerted activities of proline dehydrogenase (Put1) and P5C dehydrogenase (Put2).

Proline enters the mitochondria (Glutamate produced in the mitochondria (Glu_{mito}) (**8**) is thought to exit the mitochondria and become part of Glu_{cyto}. Glu_{cyto} is used in the synthesis of proline (**7**); glutamate is first activated to produce glutamate-5-phosphate (G5P) by γ-glutamyl kinase (Pro1), followed by its conversion to GSA/P5C by γ-glutamyl phosphate reductase (Pro2) and is then reduced to proline by P5C reductase (Pro3). Cytosolic glutamate can be converted to α-ketoglutarate by the NAD^+-dependent glutamate dehydrogenase (Gdh2), which is critical for maintaining the α-ketoglutarate pool in the cytosol (α-KG_{cyto}). The reaction catalyzed by Gdh2 generates ammonia as a by-product (**9**). α-ketoglutarate can be transported in and out of mitochondria via putative oxodicarboxylate carriers (e.g., Odc1). The mitochondrial α-KG_{mito} pool is linked to the TCA cycle (**10**). Urea, derived either from arginine or from extracellular uptake, can be converted to ammonia and CO_2 via urea amidolyase (Dur1,2) (**11**). When grown in the presence of N-acetylglucosamine (GlcNac), ammonia is also produced when glucosamine-6-phosphate is converted to fructose-6-phosphate through glucosamine-6-phosphate isomerase (Nag1) (**12**). As the cytosolic pH is maintained near neutrality (pH~6.5), most ammonia is converted to its protonated form, ammonium (**13**). Free ammonia is membrane-permeable and can readily exit cells, where it contributes to the alkalization of the growth environment, a consequence of its conversion to ammonium (**14**). Ammonium in cells can be reassimilated to generate glutamate by the NADPH-dependent glutamate dehydrogenase (Gdh3), which uses α-ketoglutarate as a substrate (**15**); additionally, ammonium can be reassimilated via glutamine synthetase (Gln1), which catalyzes the conversion of glutamate to glutamine (**16**). Glutamate can also be generated from glutamine and α-ketoglutarate via the NADH-dependent glutamate synthase (Glt1) (**17**). In mitochondria, the $NADH/FADH_2$ generated by the TCA cycle and proline catabolism can be oxidized via the electron transport chain (ETC) to generate ATP (**18**). Mitochondrial function and multiple enzymatic activities are repressed in cells grown in the presence of high glucose ($\geq$0.2%, (**19**)). The localization of enzymes shown in green have been experimentally validated in *C. albicans*, whereas those shown in red are based on the localization of their corresponding orthologs in *S. cerevisiae* and the presence or absence of strong mitochondrial pre-sequences in the N-terminals of their respective protein sequences (Candida Genome Database (CGD, http://www.candidagenome.org (accessed on 18 December 2021)) analyzed using the MitoFate tool [92]. The following enzymes are present in the *C. albicans* genome—PRODH = *PUT1* (C5_02600W), P5CDH = *PUT2* (C5_04880C), GDH = *GDH2* (C2_07900W), P5CR = *PRO3* (C4_00240), OAT = *CAR2* (C4_00160C), ARG = *CAR1* (C5_04490C), GK = *PRO1* (CR_10580), GPR = *PRO2* (C3_07220C). *GLN1* (CR_05050W), *GLT1* (C1_06550W), *DUR1,2* (C1_04660W), NADPH-dependent GDH = *GDH3* (C4_06120W), *AAT1* (C2_05250C) and *AAT2* = *AAT21* (CR_07620W).

In addition to the cytoplasmic glutamate pool (Glu_{cyto}), a significant fraction of the intracellular glutamate is generated in the mitochondria (Glu_{mito}) via the proline catabolic pathway. In eukaryotes, the four-electron catabolic conversion of proline to glutamate is carried out through the successive actions of proline dehydrogenase (PRODH; EC 1.5.5.2) and Δ1-pyrroline-5-carboxylate (P5C) dehydrogenase (P5CDH; EC 1.2.1.88). PRODH and P5CDH are highly conserved enzymes throughout eukaryotes and bacteria (reviewed in [107–110]). In *C. albicans*, PRODH and P5CDH are called Put1 and Put2, respectively, and both are nuclear-encoded mitochondrial proteins [59,61]; in most eukaryotes, PRODH is associated with the inner mitochondrial membrane and is connected to complex II of the electron transport chain (ETC; reviewed in [110]). Cytosolic proline (Pro_{cyto}), derived from uptake, biosynthetic reactions or from the catabolism of arginine or ornithine [59], is imported into the mitochondria (Pro_{mito}) via a still-unidentified mitochondrial transporter. PRODH then transfers two electrons from proline to FAD to generate P5C and the reduced flavin cofactor ($FADH_2$). P5C tautomerizes spontaneously in a non-enzymatic reaction, forming glutamic-γ-semialdehyde (GSA). The prevailing pH strongly affects the equilibrium between P5C and GSA; P5C formation is favored when the pH is >6.5. P5CDH then catalyzes the oxidation of GSA to glutamate, reducing NAD^+ to NADH. When high levels of proline are available and catabolized, P5C can accumulate in the mitochondria and exert a toxic effect [111–113]. The reduced cofactors $FADH_2$ and NADH, generated by proline catabolism, are oxidized by the ETC of mitochondria to power ATP generation. Since Gdh2 catalyzes the conversion of glutamate to α-KG in the cytosol and that Gdh2-dependent alkalization is tightly linked to mitochondrial function [61], it is highly likely that glutamate resulting from proline catabolism (Glu$_{mito}$) is able to exit the mitochondria. To date, a dedicated glutamate transporter capable of exporting glutamate out of the mitochondria has yet to be identified. In yeast, mitochondrial glutamate is converted to aspartate by the mitochondrial aspartate aminotransferase (Aat1; EC 2.6.1.1), and aspartate exits the mitochondria via the Agc1 antiporter. The relevance of this transporter with respect to the export of glutamate is not clear, as the antiporter mechanism transports aspartate out and glutamate in. Aspartate in the cytosol can be converted back to glutamate via the cytosolic aspartate aminotransferase (Aat2; EC 2.6.1.1). The *C. albicans* genome has putative orthologs of Aat1 (*AAT1*; C2_05250C) and Aat2 (*AAT21*; CR_07620W). In yeast, cytosolic glutamate is used in the biosynthesis of several amino acids, including proline; 85% of the total cellular nitrogen is incorporated via the amino nitrogen of glutamate, and the remaining 15% is derived from the amide nitrogen of glutamine [114].

7. Mitochondrial Metabolism Is Sensitive to Glucose Availability in *C. albicans*

Amino acid metabolism can be directly or indirectly regulated by glucose. Direct control is exerted by Mig1 and Mig2, well-studied factors that bind promoters and repress transcription when glucose is abundant [115]. Indirectly, glucose can negatively and pleiotropically regulate amino acid metabolism by controlling the function of mitochondria. For example, we and others have shown that *C. albicans* mitochondrial activity can be down-regulated by glucose in a manner similar to *S. cerevisiae*, albeit to a lesser extent [59,116]. Our data indicate that the repressing effect of glucose is clearly evident in *C. albicans* at 0.2% or higher [59], and more sensitive transcriptomic studies have noted effects of 0.01% glucose, a very low level of glucose [116]. The more pronounced repressive effect of glucose on mitochondrial respiration in *S. cerevisiae* is likely due to the limited capacity to oxidize NADH when glycolytic flux is high [117]. This is expected to be similar in *C. albicans*; however, since *C. albicans* has a functional complex I with a higher capacity to oxidize NADH, the threshold level for glucose's repression of mitochondrial respiration is higher than that in *S. cerevisiae*. Consistently, we observed that the level of ATP is higher in 0.2% glucose than in 2% glucose [59]. Consistently with the model that ATP-dependent Ras1 activation drives filamentous growth [118], filamentation is more robust when glucose is <0.2% [39,59]. The lower ATP level observed for cells grown in the presence of 1% glycerol is likely due to the lower levels of reduced NADH (low NADH/NAD$^+$ ratio) that can be oxidized to generate the membrane potential needed to generate ATP [59].

The pleiotropic effect of glucose on mitochondrial activity and amino acid catabolism is nicely illustrated by arginine catabolism. Arginine catabolism occurs in a bifurcated manner that generates two products that are independently catabolized either in the cytosol (urea) or mitochondria (proline). When glucose is absent, the proline catabolic pathway becomes essential for arginine catabolism as cells lacking *PUT1* or *PUT2* failed to grow in the presence of arginine as the sole nitrogen and carbon source, whereas cells lacking *DUR1,2* grew [59]. However, when high glucose was added as the main carbon source, the *put1−/−* defect was rescued as the metabolism shifted from pure respiratory to mixed types of growth (respiratory/fermentative), shifting the metabolic burden of nitrogen assimilation to the cytosolic Dur1,2. Strikingly, the enzymes required to catabolize proline were still expressed; however, since the mitochondria were repressed by high glucose levels, they were unable to carry out their catabolic functions of converting proline to glutamate.

8. Ammonia Generation and Excretion

In the human body, amino acids are an abundant source of nitrogen for *C. albicans*. However, the utilization of amino acids in excess of the amount necessary to support growth and basic cellular functions must be controlled due to the accumulation of ammonia as a metabolic byproduct. Excess ammonia is toxic to cells. When cells are grown using amino acids as energy sources, excess ammonia exits into the extracellular medium, resulting in environmental alkalization (Figure 6A). Interestingly, this capacity to increase environmental pH via ammonia extrusion is believed to support the pathogenic growth of fungal pathogens such as *C. albicans* in certain acidic microenvironments, e.g., the phagosome of macrophages (reviewed in [119,120]).

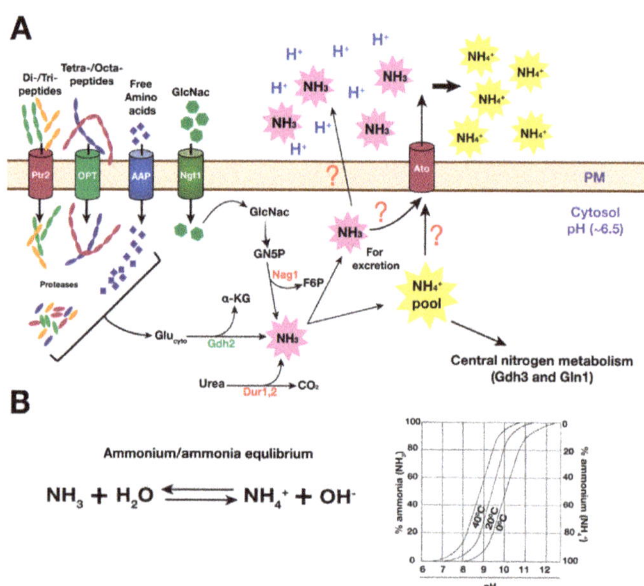

$$NH_3 + H_2O \rightleftharpoons NH_4^+ + OH^-$$

Figure 6. Ammonia extrusion in *C. albicans*. (**A**) When *C. albicans* utilizes amino acids or N-acetylglucosamine (GlcNac) as nitrogen sources for growth, ammonia is produced, promoting the alkalization of the extracellular pH. Amino acids can be either free or derived internally from the proteolytic degradation of oligopeptides, which are then converted to glutamate either in the cytosol (by specific aminotransferases) or mitochondria (proline catabolism), creating a glutamate pool in the cytosol (Glu_{cyto}). The NAD^+-dependent glutamate dehydrogenase (Gdh2) catalyzes the conversion of Glu_{cyto} to α-ketoglutarate, releasing ammonia in the process. Urea generated via arginine catabolism (or from the extracellular environment) can also be degraded to ammonia via the urea amidolyase (Dur1,2) enzyme. The deamination of GlcNac is dependent on glucosamine-6-phosphate isomerase (Nag1), which catalyzes the conversion of glucosamine-5-phosphate (GN5P) to fructose 6-phosphate (F6P). Excess ammonia produced in the cytosol must be removed in order to avoid its toxic effects; due to the cytosolic pH of around 6.5, most of the free ammonia (NH_3) is converted to ammonium (NH_4^+) (see graph below), which can be directly assimilated via central nitrogen metabolism (i.e., Gdh3 and Gln1), whereas a small fraction is released into the environment, where it could neutralize the acidic pH, producing ammonium (NH_4^+). Whether ammonia (or even ammonium) is exported through simple diffusion or via exporters (Ato) requires further study. (**B**) At the normal cytosolic pH of ≈6.5, ammonia (NH_3) is converted to ammonium (NH_4^+). We present a plot showing the relative concentrations of NH_3 and NH_4^+ in aqueous solution based on pH and temperature. Image adapted and redrawn from Huang, J; *Handbook of Environmental Engineering*) [121]. The ratio of NH_4^+ to NH_3 in this equilibrium is highly pH-dependent. At low acidic pH, the ammonium form (NH_4^+) dominates. As the pH increases, the ammonia (NH_3) form also increases, and the proportion becomes equal at the pKa value. Higher temperatures favor the NH_3 gas side of the equilibrium balance with NH_4^+.

It has been proposed that ammonia derived from amino acid catabolism enables *C. albicans* cells to neutralize the acidic luminal pH of the phagosome, reducing the activities of hydrolytic enzymes with low pH optima and inducing *C. albicans* to switch morphologies, resulting in hyphal growth, thus facilitating macrophage evasion [31]. This model was largely premised on studies using a strain lacking *STP2* (*stp2Δ/Δ*), the SPS transcription factor that positively regulates the expression of amino acid permeases required for amino acid uptake [31,32]. Strains carrying *stp2Δ/Δ* exhibit defects in both environmental alkalization and the capacity to escape the phagosome of the engulfing macrophage [31,32]. This model assigned the critical ammonia-generating event to the urea amidolyase (Dur1,2), which catalyzes the conversion of urea to ammonia and CO_2, as cells lacking this enzyme (*dur1,2Δ/Δ*) show alkalization defects when cells are grown in media with high glucose [32]. However, Dur1,2 has only been linked to ammonia generation in the presence of its substrate urea, and *DUR1,2* expression is under tight NCR control [44,56]. Thus, it is unlikely that Dur1,2 contributes to the alkalization of a growth medium with abundant preferred nitrogen sources such as amino acids and even ammonium sulfate.

There is mounting evidence that alkalization of the phagosomal compartment is not requisite for *C. albicans* cells to evade macrophages. Results obtained using dual-wavelength ratiometric fluorescence imaging to quantify the pH in the phagosome revealed that increased phagosomal pH is the consequence of elongating hyphal cells, physically stretching the phagosomal membrane, causing transient leaks [122]. The induction of hyphal growth was observed to precede alkalization. In addition, the proton-pumping activity of V-ATPase exceeds the rate of ammonia extrusion by several orders of magnitude [122]. Recently, we reported that Gdh2, the enzyme that catalyzes the conversion of glutamate to α-KG, is responsible for the bulk of ammonia produced from amino acid metabolism [61]; a *gdh2*-null strain is unable to alkalize a medium containing amino acids as the sole nitrogen and carbon source. Surprisingly, the capacity of *gdh2−/−* mutants to escape the macrophage phagosome or its virulence in murine systemic infection model was not affected, indicating that amino acid-dependent environmental alkalization is not essential for the virulence of *C. albicans*. Consistently with a previous report [122], we observed that viable wildtype cells pre-stained with a pH-sensitive dye (pHrodo), of which the fluorescence intensity varied inversely to pH, were retained in acidic phagosomes, and this observation is was even when a high MOI (more ammonia-extruding cells) was used [61].

In *C. albicans*, Gdh2 is a cytosolic component [61] and its expression is independent of NCR; Gdh2 is well expressed in cells grown in a medium with high levels of amino acids, even when supplemented with high levels of ammonium sulfate [61]. Consistently, a strain lacking *GLN3* and *GAT1*, which encode for the GATA transcription factors activating NCR-sensitive genes, remain alkalization-competent (our unpublished data). Despite its cytosolic localization, Gdh2-dependent alkalization is tightly linked to mitochondrial function as acute inhibition of the mitochondria with a sublethal dose of antimycin, a potent respiratory complex III inhibitor, virtually abolished alkalization in wildtype cells even when a very high starting cell density was used ($OD_{600} \approx 5$) [61]. Since proline catabolism is a major source of glutamate in the mitochondria and since this alkalization is partially dependent on proline catabolism [61], it is likely that pharmacological inhibition of the mitochondria pleiotropically prevented either the generation or export of mitochondrial glutamate. In a similar way, the inability of *C. albicans* to alkalinize the extracellular environment when grown in the presence of high glucose (2%) is likely due to the capacity of glucose to pleiotropically inhibit or downregulate mitochondrial function [59,116]. Gdh2 levels appear to be regulated by pH as the protein levels decrease as the pH of the growth medium approaches neutrality, which is consistent with the observed dependency of alkalization on the starting cell density [61]. In addition to amino acids, growth on N-acetylglucosamine (GlcNac) can also raise extracellular pH via ammonia extrusion. However, the origin of alkalinizing ammonia is distinct as it is catalyzed by the enzyme glucosamine-6-phosphate isomerase (Nag1), which deaminates glucosamine-5-phosphate (GN5P), converting it to fructose 6-phosphate (F6P) [123].

Regarding the fate of intracellular ammonia, in aqueous solution, ammonia can exist either as a gas (NH_3, ammonia) or as a cationic (NH_4^+, ammonium) species; the ratio (ammonia/ammonium) increases with pH (pKa = 9.25) (Figure 6B). Since the pH of the cytosol in actively growing *C. albicans* wildtype cells is maintained at ~6.5 [124], the protonated form NH_4^+ predominates and can be directly assimilated by the NADPH-dependent glutamate dehydrogenase (Gdh3) or glutamine synthetase (Gln1). Due to its being positively charged, ammonium cannot readily diffuse out of cells, but rather requires transporters or channels to traverse the phospholipid bilayer of biomembranes. A small fraction of the total ammonia species exists in the unprotonated form (NH_3) that can be released into the extracellular space, where it could exert its neutralizing effect by reacting to the hydrogen ions (H^+), generating ammonium (NH_4^+). How ammonia traverses the plasma membrane from the cytosol is unclear, as proposed earlier, because ammonia extrusion requires the ammonia transport outward (Ato) proteins, a family of plasma-membrane-bound proteins thought to facilitate ammonia export [32,125]. Strains lacking *ATO5* (*ato5Δ/Δ*) or a dominant point mutation in *ATO1* (*ATO1^{G53D}*) show strong alkalization defects, and consistently, the overexpression of *ATO* genes accelerates alkalization [32,125]. However, it is also known that ammonia (NH_3) is membrane-permeable and can easily diffuse out of yeast cells [126–128]. Whether ammonia (or even ammonium) is exported through simple diffusion or via exporters (Ato) remains to be clarified.

In addition to ammonia extrusion, yeast cells have an alternative mechanism to minimize the toxic effects of ammonia. *S. cerevisiae* can indirectly limit the production of ammonia by excreting cytosolic amino acids such as glutamate to the extracellular space via proteins that belong to the multidrug resistance transporter family that are thought to function as H^+ antiporters (e.g., Aqr1) [129]. Whether the same amino acid extrusion process, limiting intracellular ammonia production, operate in *C. albicans* is not yet known but a putative *AQR1* homolog has been identified in the *C. albicans* genome (*QDR2*/C3_05570W). Qdr2 may perform the same function, constituting a rudimentary ammonia detoxification mechanism in *C. albicans*.

9. Conclusions and Outlook

C. albicans is an opportunistic fungal pathogen that is intimately linked to its human hosts. Since *C. albicans* grows in symbiosis with humans, fungal cells must survive and propagate under identical physiological conditions as human cells. The capacity of *C. albicans* to establish persistent infections relies heavily on their capacity to assimilate nutrients in a competitive landscape where both hosts cells and even other members of the microbiome compete for nutrients. Amino acids are among the most versatile nutrients available in the hosts; they can be assimilated as both nitrogen and carbon precursors, transformed to key metabolic intermediates or utilized to modulate extracellular pH via ammonia formation. Although *S. cerevisiae* paved the way for most of our understanding of nutrient assimilation and metabolic processes in yeasts, there are clearly significant differences that exist in *C. albicans* that must be taken into account as they are crucial to our understanding of how this fungal pathogen assimilate nutrients in the host, especially in the context of infectious growth.

Some of the so-called poor or non-preferred nitrogen sources in *S. cerevisiae*, such as proline, are efficiently utilized by *C. albicans*. This observation is in alignment with recent findings that the enzymes required to utilize proline in *C. albicans* are independent of NCR, allowing the unrestricted utilization of proline regardless of whether other nitrogen sources are available [59,60]. Proline constitutes some of the most abundant proteins in humans (e.g., collagen, mucin); thus, given that *C. albicans* possesses a multi-subunit respiratory complex I (NADH dehydrogenase), similar to human cells, it is not surprising if *C. albicans* evolved to prefer this amino acid as an energy source for growth. Interestingly, proline has long been known as one of the most potent inducers of yeast-to-hyphal transitions, a key virulence factor in *C. albicans* [59,130–133]. We have shown that the induction of morphogenesis occurs via ATP-dependent Ras1 activation [59]. One molecule of proline can be completely oxidized to generate approximately 30 ATP equivalents [134,135], reinforcing the idea that proline is an important energy source for many types of cells, especially under nutrient-limited conditions. We have shown that *C. albicans* growth in the phagosome of the macrophage is dependent on proline catabolism to obtain energy to survive despite a multitude of environmental stresses [59]. The inadvertent replication of our previous data [59] in the corrected paper [61] highlights the idea that proline is sensed by *C. albicans* in the phagosome of macrophages. Our data are also consistent with recent transcriptomic data showing that proline induced the expression of *ICL1*, a gene encoding the key glyoxylate cycle enzyme isocitrate lyase 1 (Icl1), which is known to be derepressed in *C. albicans*, being engulfed by macrophages [60]. Consequently, strains lacking the capacity to utilize proline have defects in escaping the phagosome of macrophages [59]. In terms of environmental alkalization, proline catabolism plays a major role by virtue of glutamate production (Put2 product). In the presence of arginine as the sole nitrogen and carbon source, proline catabolism is essential as it is the primary catabolic route to generate glutamate, which can then be subsequently catabolized by Gdh2 to ammonia and α-KG, a key TCA cycle intermediate. However, in the presence of other amino acids, the proline catabolic pathway becomes less essential for alkalization as other amino acids can be transaminated to generate glutamate (Figure 5). Proline utilization in *C. albicans* provides a clear example of how evolution influences and fine-tunes metabolism, leading to unique capabilities, in this case to the utilization of nutrients in a manner not relevant for other related yeasts. Consequently, a thorough examination of other amino acid catabolic pathways in *C. albicans* is warranted, the premise being that many important regulatory differences may exist, and that these may be specifically linked to the evolution of *C. albicans* within mammalian hosts.

A major challenge to correctly interpret experimental results derived from studies examining nutrient sensing and assimilation in *C. albicans* is understanding how laboratory growth conditions influence the results. Many of the standard laboratory conditions do not reflect the mammalian micro-niches in which *C. albicans* resides. For example, many host–pathogen interaction experiments involving innate immune cells are carried out in cell culture medium (RPMI or DMEM) containing 5–10% fetal bovine serum. These media readily trigger filamentous growth in *C. albicans*, independently of host cell interactions (e.g., with macrophages), resulting in the false impression that certain genes are not important for the survival of *C. albicans* during co-culture with innate immune cells. This is especially crucial when looking at the role of specific genes that are required for nitrogen acquisition. For example, there is a possibility that the importance of certain genes under NCR control will be erroneously dismissed as they are not expressed under nitrogen-replete conditions such as those in cell culture media. In addition, it is common practice to use strains pre-grown in YPD, a complex medium that is high in glucose (2%) and rich in nitrogen (amino acids, peptides), prior to shifting cells to desired experimental test conditions. In humans, the level of blood glucose is maintained within homeostatic limits (0.05–0.1%; 3–5 mM glucose) that are well below the level used in YPD [116]. The dramatic reorientation of metabolism resulting from merely shifting conditions is likely to influence the response, and in many instances may provide a conflicting readout. For example, yeast cells grown in YPD build up an extensive reservoir of amino acids with vacuolar pools during growth in nitrogen-rich conditions [136,137]. This influences nutrient-based signals. Furthermore, many studies have relied on fixed-point microscopy coupled with differential staining to observe and deduce the role of specific mutations on filamentous growth. This approach relies heavily on observing obvious growth defects that may not be readily apparent on strains lacking genes relevant to nutrient acquisition. Although useful information has been obtained, many of these results often reflect "general" rather than "niche-specific" hyphal defects, highlighting the need to identify more suitable laboratory conditions that better mimic mammalian microenvironments.

Although a great deal of information regarding nutrient-induced processes in *C. albicans* is accumulating, there are major gaps in our knowledge with respect to the contribution of the host. The availability of assimilable nitrogen sources, i.e., the abundance of amino acids released as a result of host activities, is often overlooked. The contribution of host-derived activities to the degradation of the extracellular matrix (ECM) during stress due to the proteolytic activities of proteases secreted by different cell types is not fully understood. For example, in people of advanced age, who due to medical advances represent a growing population, often suffer from sarcopenia or muscle wasting. A hallmark of sarcopenia is that the amino acid proline is elevated in the blood, indicating the degradation of structural proteins rich in proline such as collagen [84]. Indeed, the elevation of free amino acids in the blood is linked to other pathological states in humans, including cancer [83,84]. It is likely that amino acid limitation influences the capacity of cancer cells to establish malignant forms of growth; cancer cells have been found to exhibit enhanced rates of amino acid uptake [138]. Furthermore, amino acid metabolism is an important factor during wasting in cancer patients (cachexia) and in aging individuals [139,140]. Clearly, illuminating the entire repertoire of regulatory mechanisms associated with amino acid signaling is crucial to understanding life processes in both healthy and disease states, and studies in *C. albicans* may provide important insights with clear therapeutic applications.

Author Contributions: Conceptualization, F.G.S.S. and P.O.L.; writing—original draft preparation, F.G.S.S.; writing—review and editing, F.G.S.S. and P.O.L.; visualization, F.G.S.S.; funding acquisition, P.O.L. All authors have read and agreed to the published version of the manuscript.

Funding: Original research in our laboratory is supported by the Swedish Research Council (P.O.L.) VR-M 2019-01547.

Acknowledgments: We would like to thank the members of the Ljungdahl laboratory for their patience and constructive comments throughout the course of this work. Many colleagues have contributed with strains and important insights and are collectively acknowledged. Furthermore, the vastness of the subject matter and space limitations have precluded the referencing of all relevant papers, and undoubtedly we have failed to cite some papers of equal or greater value than the ones cited; we apologize for the inadvertent omission of uncited work.

Conflicts of Interest: The authors declare no conflict of interest.

References

1. Poloni, J.A.T.; Rotta, L.N. Urine Sediment Findings and the Immune Response to Pathologies in Fungal Urinary Tract Infections Caused by *Candida* spp. *J. Fungi* **2020**, *6*, 245. [CrossRef]
2. Behzadi, P.; Behzadi, E.; Yazdanbod, H.; Aghapour, R.; Akbari Cheshmeh, M.; Salehian Omran, D. Urinary Tract Infections Associated with *Candida albicans*. *Maedica* **2010**, *5*, 277–279.
3. Rivett, A.G.; Perry, J.A.; Cohen, J. Urinary candidiasis: A prospective study in hospital patients. *Urol. Res.* **1986**, *14*, 183–186. [CrossRef]
4. Harris, A.D.; Castro, J.; Sheppard, D.C.; Carmeli, Y.; Samore, M.H. Risk factors for nosocomial candiduria due to *Candida glabrata* and *Candida albicans*. *Clin. Infect. Dis.* **1999**, *29*, 926–928. [CrossRef]
5. Lundstrom, T.; Sobel, J. Nosocomial candiduria: A review. *Clin. Infect. Dis.* **2001**, *32*, 1602–1607.
6. Kim, J.; Kim, D.S.; Lee, Y.S.; Choi, N.G. Fungal urinary tract infection in burn patients with long-term foley catheterization. *Korean J. Urol.* **2011**, *52*, 626–631. [CrossRef]
7. Jakubowska, A.; Kilis-Pstrusinska, K.; Pukajlo-Marczyk, A.; Samir, S.; Baglaj, M.; Zwolinska, D. Fungal urinary tract infection complicated by acute kidney injury in an infant with intestino-vesical fistula. *Postepy Hig. Med. Dosw.* **2013**, *67*, 719–721. [CrossRef]
8. Mazo, E.B.; Popov, S.V.; Shmel'kov, I. Fungal infections of the urinary tract. *Urologiia* **2007**, *5*, 67–70.
9. Fisher, J.F. Candida urinary tract infections—Epidemiology, pathogenesis, diagnosis, and treatment: Executive summary. *Clin. Infect. Dis.* **2011**, *52* (Suppl. 6), S429–S432. [CrossRef]
10. Kauffman, C.A.; Fisher, J.F.; Sobel, J.D.; Newman, C.A. Candida urinary tract infections—Diagnosis. *Clin. Infect. Dis.* **2011**, *52* (Suppl. 6), S452–S456. [CrossRef]
11. Veroux, M.; Corona, D.; Giuffrida, G.; Gagliano, M.; Tallarita, T.; Giaquinta, A.; Zerbo, D.; Cappellani, A.; Veroux, P.F. Acute renal failure due to ureteral obstruction in a kidney transplant recipient with *Candida albicans* contamination of preservation fluid. *Transpl. Infect. Dis.* **2009**, *11*, 266–268. [CrossRef]
12. Nishimoto, G.; Tsunoda, Y.; Nagata, M.; Yamaguchi, Y.; Yoshioka, T.; Ito, K. Acute renal failure associated with *Candida albicans* infection. *Pediatr. Nephrol.* **1995**, *9*, 480–482. [CrossRef]
13. Shimada, S.; Nakagawa, H.; Shintaku, I.; Saito, S.; Arai, Y. Acute renal failure as a result of bilateral ureteral obstruction by *Candida albicans* fungus balls. *Int. J. Urol.* **2006**, *13*, 1121–1122. [CrossRef] [PubMed]
14. Ekpanyapong, S.; Reddy, K.R. Fungal and Parasitic Infections of the Liver. *Gastroenterol. Clin. N. Am.* **2020**, *49*, 379–410. [CrossRef] [PubMed]

15. Fiore, M.; Cascella, M.; Bimonte, S.; Maraolo, A.E.; Gentile, I.; Schiavone, V.; Pace, M.C. Liver fungal infections: An overview of the etiology and epidemiology in patients affected or not affected by oncohematologic malignancies. *Infect. Drug Resist.* **2018**, *11*, 177–186. [CrossRef]

16. Gupta, A.; Bhowmik, D.M.; Dogra, P.M.; Mendonca, S.; Gupta, A. Candida lung abscesses in a renal transplant recipient. *Saudi J. Kidney Dis. Transpl.* **2013**, *24*, 315–317. [CrossRef]

17. Shweihat, Y.; Perry, J., 3rd; Shah, D. Isolated Candida infection of the lung. *Respir. Med. Case Rep.* **2015**, *16*, 18–19. [CrossRef]

18. Yokoyama, T.; Sasaki, J.; Matsumoto, K.; Koga, C.; Ito, Y.; Kaku, Y.; Tajiri, M.; Natori, H.; Hirokawa, M. A necrotic lung ball caused by co-infection with *Candida* and *Streptococcus pneumoniae*. *Infect. Drug Resist.* **2011**, *4*, 221–224. [CrossRef]

19. Schmidt, H.; Fischedick, A.R.; Peters, P.E.; von Lengerke, H.J. Candida abscesses in the liver and spleen. The sonographic and computed tomographic morphology. *Dtsch. Med. Wochenschr.* **1986**, *111*, 816–820. [CrossRef] [PubMed]

20. Raina, V.; Young, P.T.; Foulis, A.K.; Soukop, M. Hypersplenism due to fungal infection of spleen in a successfully treated patient with Hodgkin's disease. *Postgrad. Med. J.* **1989**, *65*, 83–85. [CrossRef] [PubMed]

21. Bezerra, L.S.; Silva, J.A.D.; Santos-Veloso, M.A.O.; Lima, S.G.; Chaves-Markman, A.V.; Juca, M.B. Antifungal Efficacy of Amphotericin B in *Candida albicans* Endocarditis Therapy: Systematic Review. *Braz. J. Cardiovasc. Surg.* **2020**, *35*, 789–796. [CrossRef]

22. Filizcan, U.; Cetemen, S.; Enc, Y.; Cakmak, M.; Goksel, O.; Eren, E. *Candida albicans* endocarditis and a review of fungal endocarditis: Case report. *Heart Surg. Forum* **2004**, *7*, E312–E314. [CrossRef]

23. Mugge, A.; Daniel, W.G.; Nonnast-Daniel, B.; Schroder, E.; Trotschel, H.; Lichtlen, P.R. Renal infarction with fatal bleeding–an unusual complication of *Candida albicans* endocarditis. *Klin. Wochenschr.* **1987**, *65*, 1169–1172. [CrossRef]

24. Oner, T.; Korun, O.; Celebi, A. Rare presentation of *Candida albicans*: Infective endocarditis and a pulmonary coin lesion. *Cardiol. Young* **2018**, *28*, 602–604. [CrossRef]

25. Prabhu, R.M.; Orenstein, R. Failure of caspofungin to treat brain abscesses secondary to *Candida albicans* prosthetic valve endocarditis. *Clin. Infect. Dis.* **2004**, *39*, 1253–1254. [CrossRef] [PubMed]

26. Sheikh, T.; Tomcho, J.C.; Awad, M.T.; Zaidi, S.R. *Candida albicans* endocarditis involving a normal native aortic valve in an immunocompetent patient. *BMJ Case Rep.* **2020**, *13*, e236902. [CrossRef]

27. Ancalle, I.M.; Rivera, J.A.; Garcia, I.; Garcia, L.; Valcarcel, M. *Candida albicans* meningitis and brain abscesses in a neonate: A case report. *Bol. Asoc. Med. Puerto Rico* **2010**, *102*, 45–48. [PubMed]

28. Black, J.T. Cerebral candidiasis: Case report of brain abscess secondary to *Candida albicans*, and review of literature. *J. Neurol. Neurosurg. Psychiatry* **1970**, *33*, 864–870. [CrossRef] [PubMed]

29. Tweddle, D.A.; Graham, J.C.; Shankland, G.S.; Kernahan, J. Cerebral candidiasis in a child 1 year after leukaemia. *Br. J. Haematol.* **1998**, *103*, 795–797. [CrossRef] [PubMed]

30. Zhang, S.C. Cerebral candidiasis in a 4-year-old boy after intestinal surgery. *J. Child Neurol.* **2015**, *30*, 391–393. [CrossRef]

31. Vylkova, S.; Lorenz, M.C. Modulation of phagosomal pH by *Candida albicans* promotes hyphal morphogenesis and requires Stp2p, a regulator of amino acid transport. *PLoS Pathog.* **2014**, *10*, e1003995. [CrossRef]

32. Vylkova, S.; Carman, A.J.; Danhof, H.A.; Collette, J.R.; Zhou, H.; Lorenz, M.C. The fungal pathogen *Candida albicans* autoinduces hyphal morphogenesis by raising extracellular pH. *mBio* **2011**, *2*, e00055-11. [CrossRef] [PubMed]

33. Bakker, B.M.; Overkamp, K.M.; van Maris, A.J.; Kotter, P.; Luttik, M.A.; van Dijken, J.P.; Pronk, J.T. Stoichiometry and compartmentation of NADH metabolism in *Saccharomyces cerevisiae*. *FEMS Microbiol. Rev.* **2001**, *25*, 15–37. [CrossRef] [PubMed]

34. Noble, S.M.; Gianetti, B.A.; Witchley, J.N. *Candida albicans* cell-type switching and functional plasticity in the mammalian host. *Nat. Rev. Microbiol.* **2017**, *15*, 96–108. [CrossRef] [PubMed]

35. Garbe, E.; Vylkova, S. Role of Amino Acid Metabolism in the Virulence of Human Pathogenic Fungi. *Curr. Clin. Microbiol. Rep.* **2019**, *6*, 108–119. [CrossRef]

36. Martinez, P.; Ljungdahl, P.O. Divergence of Stp1 and Stp2 transcription factors in *Candida albicans* places virulence factors required for proper nutrient acquisition under amino acid control. *Mol. Cell. Biol.* **2005**, *25*, 9435–9446. [CrossRef]

37. Martinez, P.; Ljungdahl, P.O. An ER packaging chaperone determines the amino acid uptake capacity and virulence of *Candida albicans*. *Mol. Microbiol.* **2004**, *51*, 371–384. [CrossRef] [PubMed]

38. Brega, E.; Zufferey, R.; Mamoun, C.B. Candida albicans Csy1p is a nutrient sensor important for activation of amino acid uptake and hyphal morphogenesis. *Eukaryot. Cell* **2004**, *3*, 135–143. [CrossRef]

39. Maidan, M.M.; Thevelein, J.M.; Van Dijck, P. Carbon source induced yeast-to-hypha transition in *Candida albicans* is dependent on the presence of amino acids and on the G-protein-coupled receptor Gpr1. *Biochem. Soc. Trans.* **2005**, *33*, 291–293. [CrossRef]

40. Kraidlova, L.; Schrevens, S.; Tournu, H.; Van Zeebroeck, G.; Sychrova, H.; Van Dijck, P. Characterization of the *Candida albicans* Amino Acid Permease Family: Gap2 Is the Only General Amino Acid Permease and Gap4 Is an *S*-Adenosylmethionine (SAM) Transporter Required for SAM-Induced Morphogenesis. *mSphere* **2016**, *1*, e00284-16. [CrossRef]

41. Dunkel, N.; Biswas, K.; Hiller, E.; Fellenberg, K.; Satheesh, S.V.; Rupp, S.; Morschhauser, J. Control of morphogenesis, protease secretion and gene expression in *Candida albicans* by the preferred nitrogen source ammonium. *Microbiology* **2014**, *160*, 1599–1608. [CrossRef]

42. Du, H.; Ennis, C.L.; Hernday, A.D.; Nobile, C.J.; Huang, G. N-Acetylglucosamine (GlcNAc) Sensing, Utilization, and Functions in *Candida albicans*. *J. Fungi* **2020**, *6*, 129. [CrossRef]

43. Alvarez, F.J.; Konopka, J.B. Identification of an N-acetylglucosamine transporter that mediates hyphal induction in *Candida albicans*. *Mol. Biol. Cell* **2007**, *18*, 965–975. [CrossRef]

44. Navarathna, D.H.; Das, A.; Morschhauser, J.; Nickerson, K.W.; Roberts, D.D. Dur3 is the major urea transporter in *Candida albicans* and is co-regulated with the urea amidolyase Dur1,2. *Microbiology* **2011**, *157*, 270–279. [CrossRef]

45. Dunkel, N.; Hertlein, T.; Franz, R.; Reuss, O.; Sasse, C.; Schafer, T.; Ohlsen, K.; Morschhauser, J. Roles of different peptide transporters in nutrient acquisition in *Candida albicans*. *Eukaryot. Cell* **2013**, *12*, 520–528. [CrossRef] [PubMed]

46. Reuss, O.; Morschhauser, J. A family of oligopeptide transporters is required for growth of *Candida albicans* on proteins. *Mol. Microbiol.* **2006**, *60*, 795–812. [CrossRef] [PubMed]

47. Ljungdahl, P.O.; Daignan-Fornier, B. Regulation of amino acid, nucleotide, and phosphate metabolism in *Saccharomyces cerevisiae*. *Genetics* **2012**, *190*, 885–929. [CrossRef] [PubMed]

48. Georis, I.; Feller, A.; Vierendeels, F.; Dubois, E. The yeast GATA factor Gat1 occupies a central position in nitrogen catabolite repression-sensitive gene activation. *Mol. Cell. Biol.* **2009**, *29*, 3803–3815. [CrossRef] [PubMed]

49. Georis, I.; Tate, J.J.; Feller, A.; Cooper, T.G.; Dubois, E. Intranuclear function for protein phosphatase 2A: Pph21 and Pph22 are required for rapamycin-induced GATA factor binding to the *DAL5* promoter in yeast. *Mol. Cell. Biol.* **2011**, *31*, 92–104. [CrossRef]

50. Cunningham, T.S.; Cooper, T.G. Expression of the *DAL80* gene, whose product is homologous to the GATA factors and is a negative regulator of multiple nitrogen catabolic genes in *Saccharomyces cerevisiae*, is sensitive to nitrogen catabolite repression. *Mol. Cell. Biol.* **1991**, *11*, 6205–6215. [PubMed]

51. Coffman, J.A.; Rai, R.; Cunningham, T.; Svetlov, V.; Cooper, T.G. Gat1p, a GATA family protein whose production is sensitive to nitrogen catabolite repression, participates in transcriptional activation of nitrogen-catabolic genes in *Saccharomyces cerevisiae*. *Mol. Cell. Biol.* **1996**, *16*, 847–858. [CrossRef] [PubMed]

52. Soussi-Boudekou, S.; Vissers, S.; Urrestarazu, A.; Jauniaux, J.C.; André, B. Gzf3p, a fourth GATA factor involved in nitrogen-regulated transcription in *Saccharomyces cerevisiae*. *Mol. Microbiol.* **1997**, *23*, 1157–1168. [CrossRef] [PubMed]

53. Rowen, D.W.; Esiobu, N.; Magasanik, B. Role of GATA factor Nil2p in nitrogen regulation of gene expression in *Saccharomyces cerevisiae*. *J. Bacteriol.* **1997**, *179*, 3761–3766. [CrossRef] [PubMed]

54. Coffman, J.A.; Rai, R.; Loprete, D.M.; Cunningham, T.; Svetlov, V.; Cooper, T.G. Cross regulation of four GATA factors that control nitrogen catabolic gene expression in *Saccharomyces cerevisiae*. *J. Bacteriol.* **1997**, *179*, 3416–3429. [CrossRef]

55. Limjindaporn, T.; Khalaf, R.A.; Fonzi, W.A. Nitrogen metabolism and virulence of *Candida albicans* require the GATA-type transcriptional activator encoded by *GAT1*. *Mol. Microbiol.* **2003**, *50*, 993–1004. [CrossRef]

56. Liao, W.L.; Ramon, A.M.; Fonzi, W.A. *GLN3* encodes a global regulator of nitrogen metabolism and virulence of *C. albicans*. *Fungal Genet. Biol.* **2008**, *45*, 514–526. [CrossRef]

57. Dabas, N.; Morschhauser, J. Control of ammonium permease expression and filamentous growth by the GATA transcription factors GLN3 and GAT1 in *Candida albicans*. *Eukaryot. Cell* **2007**, *6*, 875–888. [CrossRef]

58. Magasanik, B.; Kaiser, C.A. Nitrogen regulation in *Saccharomyces cerevisiae*. *Gene* **2002**, *290*, 1–18. [CrossRef]

59. Silao, F.G.S.; Ward, M.; Ryman, K.; Wallstrom, A.; Brindefalk, B.; Udekwu, K.; Ljungdahl, P.O. Mitochondrial proline catabolism activates Ras1/cAMP/PKA-induced filamentation in *Candida albicans*. *PLoS Genet.* **2019**, *15*, e1007976. [CrossRef]

60. Tebung, W.A.; Omran, R.P.; Fulton, D.L.; Morschhauser, J.; Whiteway, M. Put3 Positively Regulates Proline Utilization in *Candida albicans*. *mSphere* **2017**, *2*, e00354-17. [CrossRef]

61. Silao, F.G.S.; Ryman, K.; Jiang, T.; Ward, M.; Hansmann, N.; Molenaar, C.; Liu, N.N.; Chen, C.; Ljungdahl, P.O. Glutamate dehydrogenase (Gdh2)-dependent alkalization is dispensable for escape from macrophages and virulence of *Candida albicans*. *PLoS Pathog.* **2020**, *16*, e1008328, Correction in *PLoS Pathog.* **2021**, *17*, e1009877. [CrossRef]

62. Miramon, P.; Lorenz, M.C. The SPS amino acid sensor mediates nutrient acquisition and immune evasion in *Candida albicans*. *Cell. Microbiol.* **2016**, *18*, 1611–1624. [CrossRef] [PubMed]

63. Boban, M.; Zargari, A.; Andreasson, C.; Heessen, S.; Thyberg, J.; Ljungdahl, P.O. Asi1 is an inner nuclear membrane protein that restricts promoter access of two latent transcription factors. *J. Cell Biol.* **2006**, *173*, 695–707. [CrossRef] [PubMed]

64. Zargari, A.; Boban, M.; Heessen, S.; Andreasson, C.; Thyberg, J.; Ljungdahl, P.O. Inner nuclear membrane proteins Asi1, Asi2, and Asi3 function in concert to maintain the latent properties of transcription factors Stp1 and Stp2. *J. Biol. Chem.* **2007**, *282*, 594–605. [CrossRef] [PubMed]

65. Khmelinskii, A.; Blaszczak, E.; Pantazopoulou, M.; Fischer, B.; Omnus, D.J.; Le Dez, G.; Brossard, A.; Gunnarsson, A.; Barry, J.D.; Meurer, M.; et al. Protein quality control at the inner nuclear membrane. *Nature* **2014**, *516*, 410–413. [CrossRef] [PubMed]

66. Kota, J.; Gilstring, C.F.; Ljungdahl, P.O. Membrane chaperone Shr3 assists in folding amino acid permeases preventing precocious ERAD. *J. Cell Biol.* **2007**, *176*, 617–628. [CrossRef] [PubMed]

67. Ljungdahl, P.O.; Gimeno, C.J.; Styles, C.A.; Fink, G.R. SHR3: A novel component of the secretory pathway specifically required for localization of amino acid permeases in yeast. *Cell* **1992**, *71*, 463–478. [CrossRef]

68. Dabas, N.; Morschhauser, J. A transcription factor regulatory cascade controls secreted aspartic protease expression in *Candida albicans*. *Mol. Microbiol.* **2008**, *69*, 586–602. [CrossRef]

69. Morschhauser, J. Nitrogen regulation of morphogenesis and protease secretion in *Candida albicans*. *Int. J. Med. Microbiol.* **2011**, *301*, 390–394. [CrossRef]

70. Maidan, M.M.; De Rop, L.; Serneels, J.; Exler, S.; Rupp, S.; Tournu, H.; Thevelein, J.M.; Van Dijck, P. The G protein-coupled receptor Gpr1 and the Galpha protein Gpa2 act through the cAMP-protein kinase A pathway to induce morphogenesis in *Candida albicans*. *Mol. Biol. Cell* **2005**, *16*, 1971–1986. [CrossRef]

71. Miwa, T.; Takagi, Y.; Shinozaki, M.; Yun, C.W.; Schell, W.A.; Perfect, J.R.; Kumagai, H.; Tamaki, H. Gpr1, a putative G-protein-coupled receptor, regulates morphogenesis and hypha formation in the pathogenic fungus *Candida albicans*. *Eukaryot. Cell* **2004**, *3*, 919–931. [CrossRef]

72. Van Ende, M.; Wijnants, S.; Van Dijck, P. Sugar Sensing and Signaling in *Candida albicans* and *Candida glabrata*. *Front. Microbiol.* **2019**, *10*, 99. [CrossRef] [PubMed]

73. Ballou, E.R.; Avelar, G.M.; Childers, D.S.; Mackie, J.; Bain, J.M.; Wagener, J.; Kastora, S.L.; Panea, M.D.; Hardison, S.E.; Walker, L.A.; et al. Lactate signalling regulates fungal beta-glucan masking and immune evasion. *Nat. Microbiol.* **2016**, *2*, 16238. [CrossRef]

74. Hogan, D.A.; Sundstrom, P. The Ras/cAMP/PKA signaling pathway and virulence in *Candida albicans*. *Future Microbiol.* **2009**, *4*, 1263–1270. [CrossRef] [PubMed]

75. Wang, Y. Fungal adenylyl cyclase acts as a signal sensor and integrator and plays a central role in interaction with bacteria. *PLoS Pathog.* **2013**, *9*, e1003612. [CrossRef]

76. Lu, P.; Takai, K.; Weaver, V.M.; Werb, Z. Extracellular matrix degradation and remodeling in development and disease. *Cold Spring Harb. Perspect. Biol.* **2011**, *3*, a005058. [CrossRef] [PubMed]

77. Pandhare, J.; Donald, S.P.; Cooper, S.K.; Phang, J.M. Regulation and function of proline oxidase under nutrient stress. *J. Cell. Biochem.* **2009**, *107*, 759–768. [CrossRef]

78. Taylor, B.N.; Staib, P.; Binder, A.; Biesemeier, A.; Sehnal, M.; Rollinghoff, M.; Morschhauser, J.; Schroppel, K. Profile of *Candida albicans*-secreted aspartic proteinase elicited during vaginal infection. *Infect. Immun.* **2005**, *73*, 1828–1835. [CrossRef]

79. Staib, P.; Kretschmar, M.; Nichterlein, T.; Hof, H.; Morschhauser, J. Differential activation of a *Candida albicans* virulence gene family during infection. *Proc. Natl. Acad. Sci. USA* **2000**, *97*, 6102–6107. [CrossRef]

80. Kretschmar, M.; Felk, A.; Staib, P.; Schaller, M.; Hess, D.; Callapina, M.; Morschhauser, J.; Schafer, W.; Korting, H.C.; Hof, H.; et al. Individual acid aspartic proteinases (Saps) 1-6 of *Candida albicans* are not essential for invasion and colonization of the gastrointestinal tract in mice. *Microb. Pathog.* **2002**, *32*, 61–70. [CrossRef] [PubMed]

81. Imbert, C.; Kauffmann-Lacroix, C.; Daniault, G.; Jacquemin, J.L.; Rodier, M.H. Effect of matrix metalloprotease inhibitors on the 95 kDa metallopeptidase of *Candida albicans*. *J. Antimicrob. Chemother.* **2002**, *49*, 1007–1010. [CrossRef]

82. Yuan, X.; Mitchell, B.M.; Wilhelmus, K.R. Expression of matrix metalloproteinases during experimental *Candida albicans* keratitis. *Investig. Ophthalmol. Vis. Sci.* **2009**, *50*, 737–742. [CrossRef]

83. Bi, X.; Henry, C.J. Plasma-free amino acid profiles are predictors of cancer and diabetes development. *Nutr. Diabetes* **2017**, *7*, e249. [CrossRef]

84. Toyoshima, K.; Nakamura, M.; Adachi, Y.; Imaizumi, A.; Hakamada, T.; Abe, Y.; Kaneko, E.; Takahashi, S.; Shimokado, K. Increased plasma proline concentrations are associated with sarcopenia in the elderly. *PLoS ONE* **2017**, *12*, e0185206. [CrossRef] [PubMed]

85. Logan, D.A. Partial purification and characterization of intracellular carboxypeptidase of *Candida albicans*. *Exp. Mycol.* **1987**, *11*, 115–121. [CrossRef]

86. Logan, D.A.; Naider, F.; Becker, J.M. Peptidases of Yeast and Filamentous Forms of *Candida albicans*. *Exp. Mycol.* **1983**, *7*, 116–126. [CrossRef]

87. El Moudni, B.; Rodier, M.H.; Barrault, C.; Ghazali, M.; Jacquemin, J.L. Purification and characterisation of a metallopeptidase of *Candida albicans*. *J. Med. Microbiol.* **1995**, *43*, 282–288. [CrossRef]

88. Rodier, M.-H.; El Moudni, B.; Ghazali, M.; Lacroix, C.; Jacquemin, J.-L. Electrophoretic Detection of Cytoplasmic Serine Proteinases (Gelatinases) in *Candida albicans*. *Exp. Mycol.* **1994**, *18*, 267–270. [CrossRef]

89. Chu, H.; Duan, Y.; Lang, S.; Jiang, L.; Wang, Y.; Llorente, C.; Liu, J.; Mogavero, S.; Bosques-Padilla, F.; Abraldes, J.G.; et al. The *Candida albicans* exotoxin candidalysin promotes alcohol-associated liver disease. *J. Hepatol.* **2020**, *72*, 391–400. [CrossRef] [PubMed]

90. Kasper, L.; Konig, A.; Koenig, P.A.; Gresnigt, M.S.; Westman, J.; Drummond, R.A.; Lionakis, M.S.; Gross, O.; Ruland, J.; Naglik, J.R.; et al. The fungal peptide toxin Candidalysin activates the NLRP3 inflammasome and causes cytolysis in mononuclear phagocytes. *Nat. Commun.* **2018**, *9*, 4260. [CrossRef] [PubMed]

91. Moyes, D.L.; Wilson, D.; Richardson, J.P.; Mogavero, S.; Tang, S.X.; Wernecke, J.; Hofs, S.; Gratacap, R.L.; Robbins, J.; Runglall, M.; et al. Candidalysin is a fungal peptide toxin critical for mucosal infection. *Nature* **2016**, *532*, 64–68. [CrossRef]

92. Fukasawa, Y.; Tsuji, J.; Fu, S.C.; Tomii, K.; Horton, P.; Imai, K. MitoFates: Improved prediction of mitochondrial targeting sequences and their cleavage sites. *Mol. Cell. Proteom.* **2015**, *14*, 1113–1126. [CrossRef]

93. Han, T.L.; Cannon, R.D.; Gallo, S.M.; Villas-Bôas, S.G. A metabolomic study of the effect of *Candida albicans* glutamate dehydrogenase deletion on growth and morphogenesis. *NPJ Biofilms Microbiomes* **2019**, *5*, 13. [CrossRef]

94. Sickmann, A.; Reinders, J.; Wagner, Y.; Joppich, C.; Zahedi, R.; Meyer, H.E.; Schonfisch, B.; Perschil, I.; Chacinska, A.; Guiard, B.; et al. The proteome of *Saccharomyces cerevisiae* mitochondria. *Proc. Natl. Acad. Sci. USA* **2003**, *100*, 13207–13212. [CrossRef]

95. Mara, P.; Fragiadakis, G.S.; Gkountromichos, F.; Alexandraki, D. The pleiotropic effects of the glutamate dehydrogenase (GDH) pathway in *Saccharomyces cerevisiae*. *Microb. Cell Factories* **2018**, *17*, 170. [CrossRef]

96. Schwencke, J.; Canut, H.; Flores, A. Simultaneous isolation of the yeast cytosol and well-preserved mitochondria with negligible contamination by vacuolar proteinases. *FEBS Lett.* **1983**, *156*, 274–280. [CrossRef]

97. Hollenberg, C.P.; Riks, W.F.; Borst, P. The glutamate dehydrogenases of yeast: Extra-mitochondrial enzymes. *Biochim. Biophys. Acta* **1970**, *201*, 13–19. [CrossRef]

98. Huh, W.K.; Falvo, J.V.; Gerke, L.C.; Carroll, A.S.; Howson, R.W.; Weissman, J.S.; O'Shea, E.K. Global analysis of protein localization in budding yeast. *Nature* **2003**, *425*, 686–691. [CrossRef]

99. Biswas, K.; Morschhauser, J. The Mep2p ammonium permease controls nitrogen starvation-induced filamentous growth in *Candida albicans*. *Mol. Microbiol.* **2005**, *56*, 649–669. [CrossRef]

100. Dabas, N.; Schneider, S.; Morschhauser, J. Mutational analysis of the *Candida albicans* ammonium permease Mep2p reveals residues required for ammonium transport and signaling. *Eukaryot. Cell* **2009**, *8*, 147–160. [CrossRef]

101. Palmieri, L.; Runswick, M.J.; Fiermonte, G.; Walker, J.E.; Palmieri, F. Yeast mitochondrial carriers: Bacterial expression, biochemical identification and metabolic significance. *J. Bioenerg. Biomembr.* **2000**, *32*, 67–77. [CrossRef]

102. Cavero, S.; Vozza, A.; del Arco, A.; Palmieri, L.; Villa, A.; Blanco, E.; Runswick, M.J.; Walker, J.E.; Cerdan, S.; Palmieri, F.; et al. Identification and metabolic role of the mitochondrial aspartate-glutamate transporter in *Saccharomyces cerevisiae*. *Mol. Microbiol.* **2003**, *50*, 1257–1269. [CrossRef]

103. Amoedo, N.D.; Punzi, G.; Obre, E.; Lacombe, D.; De Grassi, A.; Pierri, C.L.; Rossignol, R. AGC1/2, the mitochondrial aspartate-glutamate carriers. *Biochim. Biophys. Acta* **2016**, *1863*, 2394–2412. [CrossRef]

104. Porcelli, V.; Vozza, A.; Calcagnile, V.; Gorgoglione, R.; Arrigoni, R.; Fontanesi, F.; Marobbio, C.M.T.; Castegna, A.; Palmieri, F.; Palmieri, L. Molecular identification and functional characterization of a novel glutamate transporter in yeast and plant mitochondria. *Biochim. Biophys. Acta Bioenerg.* **2018**, *1859*, 1249–1258. [CrossRef]

105. Palmieri, L.; Agrimi, G.; Runswick, M.J.; Fearnley, I.M.; Palmieri, F.; Walker, J.E. Identification in *Saccharomyces cerevisiae* of two isoforms of a novel mitochondrial transporter for 2-oxoadipate and 2-oxoglutarate. *J. Biol. Chem.* **2001**, *276*, 1916–1922. [CrossRef]

106. Palmieri, F.; Agrimi, G.; Blanco, E.; Castegna, A.; Di Noia, M.A.; Iacobazzi, V.; Lasorsa, F.M.; Marobbio, C.M.; Palmieri, L.; Scarcia, P.; et al. Identification of mitochondrial carriers in *Saccharomyces cerevisiae* by transport assay of reconstituted recombinant proteins. *Biochim. Biophys. Acta* **2006**, *1757*, 1249–1262. [CrossRef]

107. Tanner, J.J. Structural Biology of Proline Catabolic Enzymes. *Antioxid. Redox Signal.* **2019**, *30*, 650–673. [CrossRef]

108. Liu, L.K.; Becker, D.F.; Tanner, J.J. Structure, function, and mechanism of proline utilization A (PutA). *Arch. Biochem. Biophys.* **2017**, *632*, 142–157. [CrossRef]

109. Christgen, S.L.; Becker, D.F. Role of Proline in Pathogen and Host Interactions. *Antioxid. Redox Signal.* **2018**, *30*, 683–709. [CrossRef]

110. Phang, J.M. Proline Metabolism in Cell Regulation and Cancer Biology: Recent Advances and Hypotheses. *Antioxid. Redox Signal.* **2019**, *30*, 635–649. [CrossRef]

111. Nishimura, A.; Nasuno, R.; Takagi, H. The proline metabolism intermediate Delta1-pyrroline-5-carboxylate directly inhibits the mitochondrial respiration in budding yeast. *FEBS Lett.* **2012**, *586*, 2411–2416. [CrossRef]

112. Deuschle, K.; Funck, D.; Forlani, G.; Stransky, H.; Biehl, A.; Leister, D.; van der Graaff, E.; Kunze, R.; Frommer, W.B. The role of [Delta]1-pyrroline-5-carboxylate dehydrogenase in proline degradation. *Plant Cell* **2004**, *16*, 3413–3425. [CrossRef]

113. Miller, G.; Honig, A.; Stein, H.; Suzuki, N.; Mittler, R.; Zilberstein, A. Unraveling delta1-pyrroline-5-carboxylate-proline cycle in plants by uncoupled expression of proline oxidation enzymes. *J. Biol. Chem.* **2009**, *284*, 26482–26492. [CrossRef]

114. Cooper, T. Nitrogen metabolism in *Saccharomyces cerevisiae*. In *The Molecular Biology of the Yeast Saccharomyces: Metabolism and Gene Expression*; Strathern, J.N., Jones, E.W., Broach, J.R., Eds.; Cold Spring Harbor Laboratory: Cold Spring Harbor, NY, USA, 1982; Volume 2, pp. 39–100.

115. Lagree, K.; Woolford, C.A.; Huang, M.Y.; May, G.; McManus, C.J.; Solis, N.V.; Filler, S.G.; Mitchell, A.P. Roles of *Candida albicans* Mig1 and Mig2 in glucose repression, pathogenicity traits, and SNF1 essentiality. *PLoS Genet.* **2020**, *16*, e1008582. [CrossRef]

116. Rodaki, A.; Bohovych, I.M.; Enjalbert, B.; Young, T.; Odds, F.C.; Gow, N.A.; Brown, A.J. Glucose promotes stress resistance in the fungal pathogen *Candida albicans*. *Mol. Biol. Cell* **2009**, *20*, 4845–4855. [CrossRef]

117. Vemuri, G.N.; Eiteman, M.A.; McEwen, J.E.; Olsson, L.; Nielsen, J. Increasing NADH oxidation reduces overflow metabolism in *Saccharomyces cerevisiae*. *Proc. Natl. Acad. Sci. USA* **2007**, *104*, 2402–2407. [CrossRef]

118. Grahl, N.; Demers, E.G.; Lindsay, A.K.; Harty, C.E.; Willger, S.D.; Piispanen, A.E.; Hogan, D.A. Mitochondrial Activity and Cyr1 Are Key Regulators of Ras1 Activation of *C. albicans* Virulence Pathways. *PLoS Pathog.* **2015**, *11*, e1005133. [CrossRef]

119. Fernandes, T.R.; Segorbe, D.; Prusky, D.; Di Pietro, A. How alkalinization drives fungal pathogenicity. *PLoS Pathog.* **2017**, *13*, e1006621. [CrossRef]

120. Vylkova, S. Environmental pH modulation by pathogenic fungi as a strategy to conquer the host. *PLoS Pathog.* **2017**, *13*, e1006149. [CrossRef]

121. Huang, J.-C.; Shang, C. Air Stripping. In *Advanced Physicochemical Treatment Processes*; Humana Press: Totowa, NJ, USA, 2006; pp. 47–79. [CrossRef]

122. Westman, J.; Moran, G.; Mogavero, S.; Hube, B.; Grinstein, S. *Candida albicans* Hyphal Expansion Causes Phagosomal Membrane Damage and Luminal Alkalinization. *mBio* **2018**, *9*, e01226-18. [CrossRef]

123. Naseem, S.; Araya, E.; Konopka, J.B. Hyphal growth in *Candida albicans* does not require induction of hyphal-specific gene expression. *Mol. Biol. Cell* **2015**, *26*, 1174–1187. [CrossRef] [PubMed]

124. Rane, H.S.; Hayek, S.R.; Frye, J.E.; Abeyta, E.L.; Bernardo, S.M.; Parra, K.J.; Lee, S.A. *Candida albicans* Pma1p Contributes to Growth, pH Homeostasis, and Hyphal Formation. *Front. Microbiol.* **2019**, *10*, 1012. [CrossRef] [PubMed]
125. Danhof, H.A.; Lorenz, M.C. The *Candida albicans* ATO Gene Family Promotes Neutralization of the Macrophage Phagolysosome. *Infect. Immun.* **2015**, *83*, 4416–4426. [CrossRef]
126. Cueto-Rojas, H.F.; Milne, N.; van Helmond, W.; Pieterse, M.M.; van Maris, A.J.A.; Daran, J.M.; Wahl, S.A. Membrane potential independent transport of NH3 in the absence of ammonium permeases in *Saccharomyces cerevisiae*. *BMC Syst. Biol.* **2017**, *11*, 49. [CrossRef]
127. Ip, Y.K.; Chew, S.F. Ammonia production, excretion, toxicity, and defense in fish: A review. *Front. Physiol.* **2010**, *1*, 134. [CrossRef]
128. Antonenko, Y.N.; Pohl, P.; Denisov, G.A. Permeation of ammonia across bilayer lipid membranes studied by ammonium ion selective microelectrodes. *Biophys. J.* **1997**, *72*, 2187–2195. [CrossRef]
129. Velasco, I.; Tenreiro, S.; Calderon, I.L.; Andre, B. *Saccharomyces cerevisiae* Aqr1 Is an Internal-Membrane Transporter Involved in Excretion of Amino Acids. *Eukaryot. Cell* **2004**, *3*, 1492–1503. [CrossRef]
130. Land, G.A.; McDonald, W.C.; Stjernholm, R.L.; Friedman, T.L. Factors affecting filamentation in *Candida albicans*: Relationship of the uptake and distribution of proline to morphogenesis. *Infect. Immun.* **1975**, *11*, 1014–1023. [CrossRef]
131. Dabrowa, N.; Taxer, S.S.; Howard, D.H. Germination of *Candida albicans* induced by proline. *Infect. Immun.* **1976**, *13*, 830–835. [CrossRef] [PubMed]
132. Dabrowa, N.; Howard, D.H. Proline uptake in *Candida albicans*. *Microbiology* **1981**, *127*, 391–397. [CrossRef]
133. Holmes, A.R.; Shepherd, M.G. Proline-induced germ-tube formation in *Candida albicans*: Role of proline uptake and nitrogen metabolism. *Microbiology* **1987**, *133*, 3219–3228. [CrossRef]
134. Zhang, L.; Becker, D.F. Connecting proline metabolism and signaling pathways in plant senescence. *Front. Plant Sci.* **2015**, *6*, 552. [CrossRef] [PubMed]
135. Liang, X.; Zhang, L.; Natarajan, S.K.; Becker, D.F. Proline mechanisms of stress survival. *Antioxid. Redox Signal.* **2013**, *19*, 998–1011. [CrossRef] [PubMed]
136. Ohsumi, Y.; Kitamoto, K.; Anraku, Y. Changes induced in the permeability barrier of the yeast plasma membrane by cupric ion. *J. Bacteriol.* **1988**, *170*, 2676–2682. [CrossRef]
137. Kitamoto, K.; Yoshizawa, K.; Ohsumi, Y.; Anraku, Y. Dynamic aspects of vacuolar and cytosolic amino acid pools of *Saccharomyces cerevisiae*. *J. Bacteriol.* **1988**, *170*, 2683–2686. [CrossRef]
138. Singh, R.K.; Rinehart, C.A.; Kim, J.P.; Tolleson-Rinehart, S.; Lawing, L.F.; Kaufman, D.G.; Siegal, G.P. Tumor cell invasion of basement membrane in vitro is regulated by amino acids. *Cancer Investig.* **1996**, *14*, 6–18. [CrossRef]
139. Pasini, E.; Aquilani, R.; Dioguardi, F.S. Amino acids: Chemistry and metabolism in normal and hypercatabolic states. *Am. J. Cardiol.* **2004**, *93*, 3A–5A. [CrossRef] [PubMed]
140. Dioguardi, F.S. Wasting and the substrate-to-energy controlled pathway: A role for insulin resistance and amino acids. *Am. J. Cardiol.* **2004**, *93*, 6A–12A. [CrossRef]

pathogens

Review

Role of Cellular Metabolism during *Candida*-Host Interactions

Aize Pellon [1,*], Neelu Begum [1], Shervin Dokht Sadeghi Nasab [1], Azadeh Harzandi [1], Saeed Shoaie [1,2] and David L. Moyes [1,*]

[1] Centre for Host-Microbiome Interactions, Faculty of Dentistry, Oral & Craniofacial Sciences, King's College London, London SE1 9RT, UK; neelu.begum@kcl.ac.uk (N.B.); shervin_dokht.sadeghi_nasab@kcl.ac.uk (S.D.S.N.); azadeh.1.harzandi@kcl.ac.uk (A.H.); saeed.shoaie@kcl.ac.uk (S.S.)

[2] Science for Life Laboratory, KTH—Royal Institute of Technology, SE-171 21 Stockholm, Sweden

* Correspondence: aize.pellon@kcl.ac.uk (A.P.); david.moyes@kcl.ac.uk (D.L.M.)

Abstract: Microscopic fungi are widely present in the environment and, more importantly, are also an essential part of the human healthy mycobiota. However, many species can become pathogenic under certain circumstances, with *Candida* spp. being the most clinically relevant fungi. In recent years, the importance of metabolism and nutrient availability for fungi-host interactions have been highlighted. Upon activation, immune and other host cells reshape their metabolism to fulfil the energy-demanding process of generating an immune response. This includes macrophage upregulation of glucose uptake and processing via aerobic glycolysis. On the other side, *Candida* modulates its metabolic pathways to adapt to the usually hostile environment in the host, such as the lumen of phagolysosomes. Further understanding on metabolic interactions between host and fungal cells would potentially lead to novel/enhanced antifungal therapies to fight these infections. Therefore, this review paper focuses on how cellular metabolism, of both host cells and *Candida*, and the nutritional environment impact on the interplay between host and fungal cells.

Keywords: immunometabolism; metabolism; macrophages; epithelial cells; glycolysis; glucose; moonlighting proteins; *Candida albicans*

Citation: Pellon, A.; Begum, N.; Sadeghi Nasab, S.D.; Harzandi, A.; Shoaie, S.; Moyes, D.L. Role of Cellular Metabolism during *Candida*-Host Interactions. *Pathogens* **2022**, *11*, 184. https://doi.org/10.3390/pathogens11020184

Academic Editor: Jeniel Nett

Received: 12 October 2021
Accepted: 26 January 2022
Published: 28 January 2022

Publisher's Note: MDPI stays neutral with regard to jurisdictional claims in published maps and institutional affiliations.

1. Introduction

Fungal microorganisms inhabiting the human body, namely the mycobiota, constitute an essential part of the microbiota, despite their relatively low number compared to their bacterial counterparts [1,2]. Commensal fungi, either being permanent or transient colonisers, populate the skin and mucosae covering the oral cavity and the respiratory, gastrointestinal, and genitourinary tracts. Unsurprisingly, different genera governing each body site, including *Candida* (oral cavity and gut), *Malassezia* (skin), *Saccharomyces* (gut) or *Eremothecium* (lung) [3–5]. Remarkably, many of these genera, as well as other species present in our environment, are pathobionts, capable of becoming pathogenic when host immunity or tissue microenvironment changes.

Among fungal pathogens, *Candida* spp., and specifically *C. albicans*, remain the most clinically relevant fungi, causing a wide range of infections in humans from superficial to systemic candidiasis [6]. The emergence of antifungal drug resistance in *C. albicans*, as well as the increasing prevalence of infections by other *Candida* species that are intrinsically resistant to available drugs (e.g., *Candida auris*) [7], highlights the importance of finding novel therapeutic strategies to deal with these infections.

In the last couple of decades, the importance of the nutritional environment and metabolism of both host and pathogens during infectious processes has been highlighted [8]. The presence or abundance of certain metabolites, including simple carbohydrates such as glucose or galactose, modulates cellular responses of both pathogen and host, therefore being essential factors during their interactions. The stress derived from the interaction with the other organisms often leads to metabolic reprogramming that supports immune responses on one side and pathogenic/commensal growth on the other.

This review paper aims to explore the current knowledge regarding the role of host metabolism in the control of innate immune responses to fungal microbes on one side, and the importance of *C. albicans* metabolism for commensalism and virulence on the other. We will also highlight how metabolism modulates the biology of both host and fungal cells during their interactions, and the emerging strategies to develop novel therapeutic tools.

2. Immunometabolism: Feeding Immune Responses in the Host

During the last two decades, an increasing body of evidence has identified the key role of cellular metabolism in developing immune responses, either enhancing (contributing to pathogen clearance) or diminishing (contributing to tolerogenic states) them. Thus, a new research field termed immunometabolism developed to delve into the control of immunity driven by metabolic processes [9]. Metabolic regulation of immunity has been described in both adaptive (e.g., T cells) [10] and innate (e.g., macrophages) [11] cells. Both types of cells show a wide spectrum of metabolic profiles upon activation with different stimuli. Since host immunometabolism has been extensively reviewed in recent years, we will focus on innate immunity, giving a general overview of how metabolic reprograming occurs and the modulation of immune responses by metabolites and metabolic enzymes.

2.1. Metabolic Reprogramming in Immune and Non-Immune Cells

Interaction of innate immune cells, such as macrophages and monocytes, with different microorganisms leads to metabolic shifts on which their responses rely. These responses are either boosted or decreased to promote infection clearance or microbial tolerance, respectively. Since there is a great diversity of microbial structures (e.g., pathogen-associated molecular patterns (PAMPs)) and of host receptors (pattern recognition receptors (PRRs)) involved in their detection, the metabolic profiles of these differently stimulated cells, along with their derived immune responses, are also very diverse [12].

Alterations in glucose uptake and metabolism are the main hallmark of metabolic shifts in innate immune cells (Figure 1A). For instance, when macrophages are challenged with bacterial lipopolysaccharide (LPS), glucose uptake and processing via aerobic glycolysis increases, that is glycolysis coupled with lactate dehydrogenase activity leading to lactic acid production in normoxic conditions. Conversely, there is decreased activity in the tricarboxylic acid (TCA) cycle and oxidative phosphorylation (OxPhos) [13]. In contrast, cells stimulated with fungal β-glucan show an increase in both aerobic glycolysis and OxPhos [14]. Shifts in cellular metabolism towards aerobic glycolysis provide cells with more rapid energy and building blocks generation, and leading to increased cytokine release, etc. [15]. In contrast, anti-inflammatory macrophages rely on aerobic respiration, completely oxidising glucose through glycolysis, the TCA cycle and OxPhos [9].

Besides glucose metabolism, other pathways are involved during metabolic reprogramming of innate immune cells [9,16]. These pathways provide energy or redox molecules, or intermediate metabolites serving as building blocks or having regulatory functions, as explained below. The pentose phosphate pathway (PPP) provides proliferative cells with metabolites needed for nucleotide synthesis, but also contributes to NADPH production. Notably, this pathway is upregulated after LPS activation of macrophages [17], which has been related to the increased reactive oxygen species (ROS) generation in these cells via NADPH oxidase [18,19]. Fatty acid synthesis (FAS) or oxidation (FAO), alongside other lipid metabolism pathways, are also differentially regulated in activated macrophages [20]. Pro-inflammatory cells use FAS and citrate accumulated due to the TCA cycle shut down to synthesise fatty acids, prostaglandins, and leukotrienes, essential molecules for signalling events, inflammation, etc. [20,21]. In contrast, since anti-inflammatory macrophages keep their TCA and OxPhos intact but lower glycolytic levels, they rely on FAO and fatty acid uptake to feed those pathways. Finally, amino acid metabolism, such as glutamine or arginine, is key for innate immune responses, including nitric oxide production or cytokine production [9].

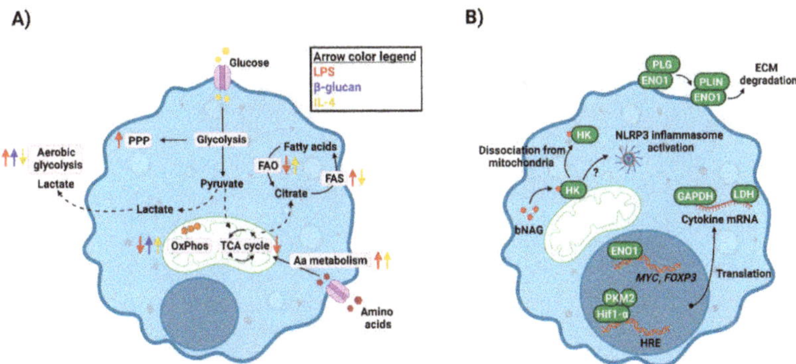

Figure 1. Immunometabolism in innate immune cells. (**A**) Challenging innate immune cells with either exogenous (LPS, β-glucan) or endogenous (IL-4) stimuli leads to metabolic reprogramming, which involves changes in pathways as glycolysis, FAO/FAS, OxPhos, etc. These shifts in metabolism provide cells with energy and building blocks to develop their functions. (**B**) Metabolic enzymes regulate immune responses at many levels, including their moonlighting functions. They can act as transcription/translation facilitators (ENO1, PKM2, GAPDH, LDH), immune receptors or activators (HK) or facilitators of immune cell migration (ENO1). bNAG, bacterial N-acetylglucosamine; ECM, extracellular matrix; ENO1, enolase 1; FAO, fatty acid oxidation; FAS, fatty acid synthesis; GAPDH, glyceraldehyde dehydrogenase; HK, hexokinase; LDH, lactate dehydrogenase; OxPhos, oxidative phosphorylation; PKM2, pyruvate kinase M2; PLG, plasminogen; PLIN, plasmin; PPP, pentose phosphate pathway; TCA, tricarboxylic acid. Created with BioRender.com (last accessed 12 January 2022).

A number of signalling pathways govern metabolic shifts in innate immune cells [22]. Among them, the activation of the transcription factor hypoxia-inducible 1α (HIF-1α) is involved in the increase in glycolytic activity observed in LPS-activated macrophages and is responsible for the expression of several immunity-related genes [23,24]. Similarly, signalling via mechanistic target of rapamycin (mTOR) is involved in promoting cholesterol and FAS, as well as sensing amino acid and glucose availability [25].

In recent years, the ability of innate immune cells to develop long-term responses has been described, adding more complexity to the biology of these types of cells [26,27]. Essentially, the term "innate immune memory" involves a wide range of phenotypes mainly observed in monocytes/macrophages that renders them more tolerogenic or reactive against a second encounter [28]. Notably, these events are intimately linked to epigenetic and metabolic reprogramming of cells [29,30], both of which are the consequence of signalling promoted by the first encounter with the microbial challenge.

Since the discovery by Otto Warburg of the metabolic shift towards aerobic glycolysis undergone by some cancer cells [31], metabolic reprogramming has been observed in many cell types, especially in a pathologic context such as cancer. Although much of the work on these shifts driven by microbes has been carried out on immune cells, non-immune cells playing paramount roles during the infectious processes undergo similar shifts. Viral infections have been shown to modulate metabolic profiles of airway epithelial cells [32] and endothelial cells [33]. Similarly, the murine bacterial pathogen *Citrobacter rodentium* promotes a decrease in carbohydrate metabolism in intestinal epithelial cells [34], whilst skin keratinocytes increase their aerobic glycolytic metabolism in response to *Staphylococcus aureus* [35]. Despite this, the consequences of these metabolic shifts on immune responses developed by these cell types are yet to be further explored.

2.2. Beyond Metabolic Reprogramming: Immune Regulatory Roles of Metabolic Enzymes and Metabolites

Besides the direct impact of metabolic reprogramming on immune cell activity (e.g., energy and redox balance, metabolite catabolism/anabolism, etc.), there are other levels of regulation of immune responses in which metabolic enzymes or metabolites play a role (Figure 1B).

Some metabolic enzymes have been observed to display regulatory functions distinct from their metabolic activities. Therefore, these proteins have been termed as "moonlighting proteins". These alternative functions of metabolic enzymes can be found among diverse biological organisms and were firstly observed in microorganisms, including bacteria and fungi, such as *Candida* spp., in which they have roles in microbial cell adhesion, pathogenicity, etc. [36,37]. Notably, the capacity of these proteins to develop moonlighting functions has been conserved in mammalian cells [38].

Glycolytic enzymes, such as hexokinase, glyceraldehyde-3-phosphate dehydrogenase (GAPDH), or enolase, display these moonlighting functions in very different ways. Hexokinase, the enzyme catalysing the first step in glycolysis, is one of the main proteins upregulated upon cell activation, and its inhibition by 2-deoxyglucose (2-DG) leads to a significant reduction in pro-inflammatory marker release [17]. However, hexokinase was recently described as a new intracellular pattern recognition receptor able to bind to the bacterial peptidoglycan component N-acetylglucosamine. This binding leads to hexokinase separation from mitochondria and drives NLRP3 inflammasome activation [39]. GAPDH regulates cytokine release in both T cells [40] and monocytes [41] by directly binding to cytokine mRNA. Enolase, involved in one of the final steps of glycolysis, has been associated with monocyte binding to plasminogen, facilitating their migration [42]. Moreover, this enzyme can modulate gene expression, including *MYC* [43] and *FOXP3* [44], by directly binding to gene regulatory elements. Similarly, PKM2 (pyruvate kinase isoform M2) can act as a co-activator of Hif-1α and regulates IL-1β expression through the activation of NLRP3 and AIM2 inflammasome [45,46]. Finally, lactate dehydrogenase (LDH), the last enzyme in the aerobic glycolytic pathway converting pyruvate in lactate, is able to bind to cytokine transcripts to modulate their translation [47,48].

It is not just proteins/enzymes involved in metabolic processes that can have these alternative immune functions. Metabolites derived from central metabolic pathways, both intermediates and final products, have been shown to modify protein function/structure and in that way modulate immune cell biology. The best-described process by which metabolites regulate immune responses is via protein post-translational modifications (PTMs). These modifications are of special relevance in the case of histones as they lead to changes in the expression of a wide range of genes—the field of epigenetics. These histones PTMs are manifold, including acetylation, phosphorylation, deamination, and methylation among others. Of these, lysine acetylation is one the clearest examples of the link between metabolism and cell functions. Acetyl-CoA is a key metabolite used by lysine acetyltransferases as a donor to acetylate proteins, although this process can occur non-enzymatically [49]. Moreover, acetyl-CoA intracellular levels correlate with protein acetylation rates and thus, changes in the nutritional environment of cells or tissues are associated with changes in acetylation levels [50].

Besides acetylation, a great variety of histone PTMs associated with metabolism has been described to date, most of them involving short-chain fatty acids (SCFAs) such as propionate, butyrate, crotonate or succinate [51]. This process is thus tightly regulated by cellular metabolism and the nutritional environment since the level of each histone acylation depends on the concentration of their respective acyl-CoA [52]. Notably, many of these PTMs have been discovered very recently and novel forms are predicted to be found in the near future. In fact, histone lysine lactylation was recently described in both human and mouse cells [53]. The event was regulated by exogenous glucose, hypoxia, and glycolytic activity levels, all three being positively correlated with intracellular lactate levels. Specifically looking at macrophages, the authors showed that stimulation of M1 polarisation using an acute LPS and interferon-γ challenge led to higher lactate production because of the expected shift towards aerobic glycolysis. Coupled RNA-seq and lactylation-specific ChIP-seq analyses of activated macrophages showed the modulation of gene expression by these PTMs, with pro-inflammatory genes being regulated at early timepoints whilst M2 profile-related gene expression was modulated during a later phase. This suggests that histone lactylation sets a gene expression "timer" that leads to homeostasis after the inflammatory burst [53].

Metabolic intermediates can also act as intra- or extracellular signals to modulate immune responses via mechanisms beyond epigenetic modifications [54,55]. The proven existence of a wide range of metabolite transporters [56] and receptors [57] has shown the potential impact of their availability on immune cell biology. Metabolite transporters facilitate metabolite uptake and secretion, highlighting the paramount relevance of the nutritional microenvironments created during, for example, inflammatory processes. The second, metabolite receptors, are usually G-protein-coupled receptors sensing metabolites and triggering intracellular signalling events, which has led to the hypothesis of some metabolites having cytokine/chemokine-like functions [54]. Moreover, some of these metabolites, such as lactate [58] or succinate [17], have been associated with functional stabilisation of such relevant proteins as HIF-1α, the master regulator linking metabolism to immunity.

3. Candida Metabolism: The Significance of Being Adaptable

As commensals and opportunistic pathogens, *Candida* spp. have developed high degree of phenotypic plasticity to adapt to diverse and changing environments. Therefore, metabolism is an essential part of *Candida* survival for nutrient assimilation and pathogenicity. The virulence of *C. albicans* is related to gene expression and host immune status [59]. *Candida* genes encoding metabolic enzymes directly interact with the host mediating fungal virulence. These virulence mechanisms include yeast-hyphal morphogenesis, phenotypic switching in the opacity of cells, adhesion, secreted hydrolases, and moonlighting proteins. *Candida* metabolic flexibility and evolution emphasises the challenges in investigating metabolic divergency with particular attention to clinical and therapeutic intervention [60,61].

Carbon assimilation and its accompanying metabolic pathway plasticity has been widely explored in *C. albicans* [62]. The carbon metabolic framework, including glycolysis, the TCA cycle and gluconeogenesis, is controlled by regulatory networks based on local nutrient availability. Metabolic plasticity allows *C. albicans* to assimilate glucose and other carbon sources simultaneously, unlike *S. cerevisiae* that switches to fermentative pathway in the presence of glucose [63]. This confers fitness in survival and adaptation to *Candida* in different host niches. General control of amino acid metabolism (GCN response) has also been linked to pathogenicity and virulence attributes of *Candida* species [61,64].

3.1. Impact of Metabolism on Fungal Biology: From Morphogenesis to Cell Wall Synthesis

C. albicans displays a remarkable metabolic plasticity, being able to grow in the presence of different carbon sources, such as glucose, fructose, or galactose (Figure 2A). However, it shows preference towards the first one and in fact, growing on glucose as the only carbon source allows the fungus to thrive in the presence of a wide range of nutritional and stress conditions [65]. The transcriptional regulators Tye7 and Gal4 are key for the catabolism of glucose and other hexoses by *C. albicans*, controlling the expression of genes involved in glycolysis, fermentation, pyruvate dehydrogenase complex (Gal4 only), or trehalose metabolism (Tye7 only) [66]. Furthermore, Tye7 assists in cohesiveness and hyphal formation in biofilms although its absence does not impact on hyphal growth in planktonic conditions [67]. Defects in Tye7 function do not have a great impact on systemic candidiasis but have a significant effect on *C. albicans* ability to colonise the gut [68]. Gal4 regulates a unique set of carbohydrate genes initiated in hypoxic conditions that are essential for pathogenicity. Fermentable carbon sources such as galactose enhance the glycolytic pathway and minimise dependency on fermentation [66]. Interestingly, two Gal4 analogues, Rtg1 and Rtg3, have a great impact during both systemic infections and gut colonisation, although they are involved in the regulation of a broader range of cellular processes [68].

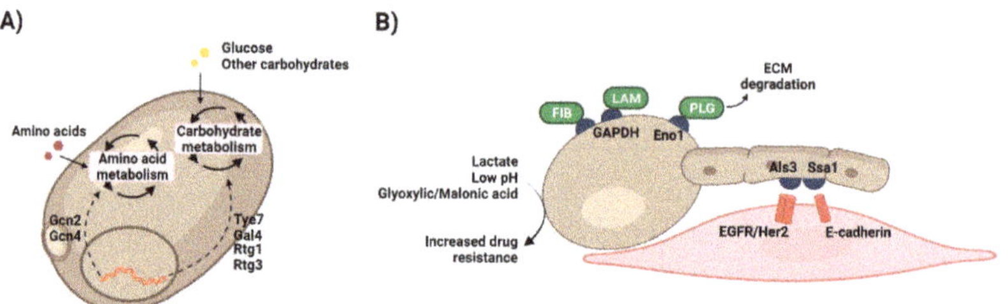

Figure 2. Metabolic plasticity in *Candida*. (**A**) *Candida* species are able to grow on a wide range of compounds, giving them the chance to thrive in very different environments. This metabolic plasticity is tightly regulated by a network of transcription factors that are activated depending on the nutritional requirements of the fungus. (**B**) Growing on different compounds leads to changes in *Candida*, for example cell wall structure or composition. This is of special importance when physiologically relevant nutrients, such as lactate, are present. Utilization of these metabolites by *C. albicans* remodel its cells wall increasing antifungal drug resistance. Moreover, like host cells metabolic enzymes in *Candida* display moonlighting functions associated with, for instance, cell adhesion to the ECM (GAPDH, Eno1) or host cells (Ssa1). ECM, extracellular matrix; Eno1, enolase 1; GAPDH, glyceraldehyde dehydrogenase; FIB, fibronectin; LAM, laminin; PLG, plasminogen. Created with BioRender.com (last accessed 12 January 2022).

Moreover, carbohydrate metabolism is intimately linked to *C. albicans* morphogenesis, with nutrient starvation or serum presence being among the factors inducing the yeast-to-hypha transition [69,70]. Metabolic genes are regulated during hyphal growth, including Adh1, Pgk1, and Gpm1 [71]. Similarly, white and opaque *Candida* cells display different metabolic profiles, with the white phenotype being more fermentative and the opaque being more oxidative and using FAO [65]. In fact, metabolic genes including Eno1, Fba1, Pyk1, Tpi1 and Pgi1 [72], are regulated by the central morphogenetic regulator Efg1, a transcription factor related to the white-opaque transition. Efg1 expression appears to be mechanistically connected to carbon metabolism in *Candida*. In general, Efg1 is downregulated in fermentative metabolism and upregulated in oxidative metabolism involved in morphogenesis [59,73,74]. Moreover, Efg1 stimulates fermentation and suppression of respiratory metabolism, demonstrating the importance of glycolytic metabolism in controlling virulence attributes [73]. This ability allows *Candida* species to switch between opaque and white cells (fermentative metabolism) depending on the nutritional environment [74].

As well as glycolysis, other metabolic pathways have an impact on *Candida* virulence. Knockout of FAO, for example, does not prevent candidiasis but assists in systemic virulence [75,76]. Conserved GCN networks, including *GCN4* and *GCN2* genes, are vital regulators activated during amino acid starvation. They act to reduce protein translation rates and induce cellular morphogenesis in *C. albicans* [77]. *GCN4* is a master regulator that activates morphogenesis via the Ras-cAMP signalling pathway to form pseudo-hyphae and activating amino acid biosynthetic genes [77]. In addition, GCN, particularly upregulation of *GCN4* gene, is further required for efficient biofilm formation in *C. albicans* [78]. On the other hand, the arginine pathway, meanwhile, appears to be essential in *C. albicans* as mutations in this pathway caused a defect in germ tube and hyphal formation [79]. Finally, the amino sugar N-acetylglucosamine (GlcNAc), which is the main component in chitin within the fungal cell wall, stimulates cellular responses mediating virulence, comprising of yeast-hyphae transition and stress responses [80].

The cell wall protects fungal cells from the environmental stress, controls cell morphogenesis, allows for immune recognition and is essential for cell growth of *Candida* species [81]. *C. albicans* uses sugars such as glucose, mannose, and galactose to provide energy to synthesise the cell wall. Thus, metabolic regulation is important in cell wall remodelling with the main constituents being β-glucan, chitin, and an outer layer consisting of mannoproteins (mannosylated proteins) [82]. The generation of these cell wall components requires glucose via both glycolysis catabolic and gluconeogenesis biosynthetic pathways [83]. The relative proportions of these components in the cell wall changes depending on the cells' environment. For example, β-glucan in cells within biofilms is elevated compared to non-biofilms [84]. The use of carbon sources alternative to glucose have been attributed to differences in cell wall architecture, adherence, biofilm formation, resistance to antifungal drugs and responses to stress [59,85,86]. For instance, the growth of *C. albicans* on lactate led to cell wall restructuring leading to increased resistance to azoles and oxidative stress. However, fungal cells grown on lactate media showed increased pores, higher hydrophobicity, and less elastic cell walls with reduced thickness of β-glucan and chitin [85].

3.2. Impact of Metabolism on Candida Pathogenic Potential

As discussed above, *Candida* species have a robust metabolism that contributes to virulence factors (Figure 2B). Like host cells, *Candida* invasion strategies include moonlighting proteins with distinct functions. These multifunctional proteins perform additional actions to their canonical biochemical function [87,88]. Owing to evolution, some moonlighting proteins can display their different functions simultaneously, whilst others alter their activity or cellular location in response to environmental changes and cell survival needs [88]. To survive within different environments in the host organism during disease progression, microbes need to use adaptable mechanisms other than common virulence features, such as adhesion molecules and hydrolytic enzymes. In *Candida* species, different moonlighting proteins can be found attached to the cell wall, and they enable microbial cells to be more flexible and adaptable in a dynamic host environment during colonisation and invasion [89]. GAPDH, usually present in the cytoplasm, may be localised in the cell surface of *C. albicans* where it facilitates cell adhesion to fibronectin and laminin, hence helping the fungal attachment to the host and initiation of candidiasis [90]. Similarly, enolase has been identified in the surface of several clinically relevant fungi, with this enzyme being involved in fungal cell adhesion via plasminogen binding (as with macrophages) and in the degradation of the extracellular matrix (ECM) [91,92]. Moreover, the intracellular chaperone Ssa1, a member of the heat shock protein 70 family, has been shown as another atypical protein with localisation in *C. albicans* cell wall. This moonlighting protein also plays a key role during colonisation of host cells as an adhesin, acting jointly with Als3 to bind to EGFR/Her2 and E-cadherin [93–95].

While for decades there have been well-known classes of anti-fungal drugs, some of them do not specifically target fungi, hence showing toxicity for mammalian cells. Therefore, there is an urgent need to develop novel drug strategies [96]. As mentioned previously, *Candida* cell functions, such as cell wall construction and adaptation to environmental stress, significantly rely on nutrient availability and the type of carbon source. Equally, antifungal drug resistance can be also modulated by the nutritional environment. Deficit of glucose as the main carbon source force *C. albicans* cells to find an alternative source and therefore changes in the downstream machinery pathways, which could result in adaptation of the cell against different stress. Previously, it has been observed the *C. albicans* growth in presence of fermentable substrate, glucose, and non-fermentable, lactate, can change cell secretome, as well as alter the cell wall structure and proteome [97,98]. These modifications affect resistance to antifungal drugs and susceptibility to stress. In fact, *C. albicans* grown in the presence of lactate was more resistant to amphotericin B, caspofungin, and tunicamycin, whilst it showed increased susceptibility to miconazole [85]. In addition to alternative carbohydrate sources, the acidity of the environment can also make *Candida* susceptible to antifungal drugs. Growing *C. albicans* under vaginal simulated media and in the presence of acetic acid rendered it more susceptible to fluconazole. However, *C. albicans* susceptibility to fluconazole remained unchanged when some other organic acids, such as glyoxylic acid and malonic acid, were present [99]. Similarly, *C. glabrata* shows higher susceptibility in the presence of acetic acid compared to when it is just grown on glucose [100]. These examples show the importance of carbon source availability and elucidation of the role of different nutrient in the *Candida* pathogenicity and antifungal resistance, and the need for more in depth and targeted metabolic analysis on the drug efficacy to tackle the resistome problem in fungal infections.

4. The Role of Metabolism during Host-*Candida* Interactions

Interactions of *C. albicans* with innate immune and epithelial cells have been extensively studied in the past [101,102]. Great strides have been made in our understanding of how host cells recognize this fungus via PRRs, although the relevance of each receptor varies depending on the infection context, either being systemic [103] or at the mucosal barriers [104]. Phagocytosis by immune cells [105], or attachment to and invasion of epithelia [29], are the next steps in the infectious process and are essential to promote immune responses in these cell types. Secretion of the peptide toxin candidalysin contributes to cell damage and activation, especially in the case of epithelial cells [106,107].

However, there are still a lot of gaps in our knowledge of how all these responses are regulated. As explained above, host cells undergo metabolic reprogramming upon interacting with microbes or microbial components, modulating how they respond to infections and competing over nutrients. In this section, we will discuss the current knowledge regarding the role of (immuno)metabolism during fungal interactions with epithelial and innate immune cells (Figure 3).

4.1. Impact of Metabolism during C. albicans Interactions with Immune Cells

Following their first contact with *C. albicans* (i.e., recognition, phagocytosis, etc.), activated immune cells reprogram their metabolism to mount an effective response. Transcriptomics-based analysis of peripheral blood mononuclear cells (PBMCs) stimulated with *C. albicans* shows a consistent upregulation of glycolysis, whilst no change (TCA cycle) or even downregulation (PPP) was observed for other pathways [108]. Specific stimulation of monocytes by heat-killed yeast or hyphae drives upregulation of several glycolytic enzymes, along with increased lactate production and glucose consumption, suggesting a shift towards aerobic glycolysis. Like β-glucan-stimulated cells [14], heat-killed cells promoted both higher ECAR (extracellular acidification rate) and OCR (oxygen consumption rate) levels, showing that OxPhos is also upregulated. This increased glycolysis plays a key role in immune responses as inhibiting glycolysis (2-DG and dichloroacetate, DCA) and mTOR pathway signalling (Torin1) significantly downregulates cytokine production post-fungal challenge [108].

The induced shift in metabolic pathways of infected monocytes with *C. albicans* differs between yeast and hyphal stimulation and is mediated by C-type lectins (CLR) but not by Toll-like receptors (TLR), showing the heterogenicity of host receptors in fungal recognition and responses. The responses generated to different *Candida* morphotypes is also varied. Monocytes infected with yeast cells activate glycolysis, oxidative phosphorylation, and glutaminolysis, whilst those infected with hyphae activate only glycolysis. Thus, we can see that the mechanisms of glucose metabolism are central players in regulating anti-*C. albicans* immunity and cytokine production [108]. Similarly, *A. fumigatus* induces an increase in aerobic glycolysis that is involved in macrophage responses to this filamentous fungus [109]. Of note, induction of metabolic reprogramming is mediated by the phagosomal removal of *A. fumigatus* melanin and its detection by the recently discovered melanin receptor MelLec [110]. This recognition is involved in HIF-1α mobilisation and subsequent cytokine release [109].

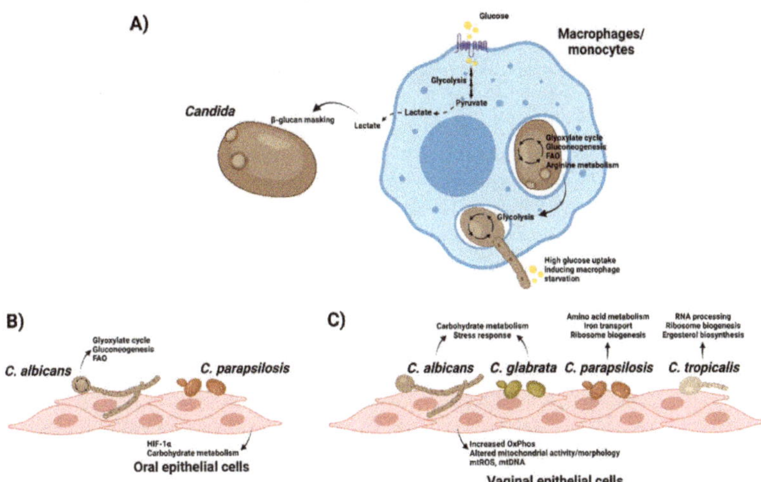

Figure 3. Interplay between *Candida* and host cells lead to metabolic reprogramming in both organisms. (**A**) Recognition of *C. albicans* by macrophages/monocytes drive changes in metabolism towards aerobic glycolysis, leading to the production of lactate that can be used by the fungus to enhance its β-glucan masking to evade immune responses. In turn, phagocytosed fungal cells use their metabolic plasticity to adapt to the nutrient-poor environment inside phagolysosomes. After piercing cell membranes using hyphae, *C. albicans* switches to glycolysis and depletes glucose from the medium, leading to macrophage cell death. (**B**) Oral epithelial cells activate HIF-1α when challenged with *C. parapsilosis*, whilst *C. albicans* adapts its metabolism upon interaction with these cells. (**C**) Variable responses are observed in different *Candida* species after interacting with vaginal epithelial cells. However, host cells exert a common early response to all of them mediated by changes in mitochondrial activity and morphology, with higher release of mitochondrial reactive oxygen species (mtROS) and DNA (mtDNA). Created with BioRender.com (last accessed 12 January 2022).

The main interface of host-fungal interaction is the fungal cell wall, a highly flexible structure with ability to remodel itself in the presence of a variety of environmental pressures, such as antifungal drugs [111]. Immune cells activated by *C. albicans* infection generate metabolites that can be sensed by fungi, which then remodel their cell wall in response to improve their immune evasion/protection. The major component of fungal cell walls, β-glucan, is a major fungal PAMP involved in the activation of many of the host antifungal responses [112]. The increased lactate levels associated with the shift to aerobic glycolysis may lead to β-glucan masking, preventing recognition of this key PAMP [113,114]. This phenotype (observed in multiple pathogenic *Candida* species) is only activated by appropriate lactate concentrations, and not by other metabolites such as proline, acetate, and methionine. This phenomenon is facilitated by the activation of the Crz1 transcription factor by the G protein-coupled receptor, Gpr1. Crz1 modulates the expression of genes involved in lactate-induced β-glucan masking. The outcome of this masking is significantly reduced visibility of *Candida* cells in terms of immune responses and thus diminished levels of tumour necrosis factor-alpha (TNFα) release and neutrophils [113]. Thus, *Candida* can successfully escape from macrophage uptake by taking advantage of the carbon sources released during metabolic rewiring of host cells in response to infection.

Upon *C. albicans* infection, macrophages are recruited to the site of infection and engulf fungal cells to try to destroy them or inhibit their growth in the phagolysosome through oxidative and nitrosative mechanisms [115]. *C. albicans*, however, has developed mechanisms to survive inside macrophages through metabolism manipulation. Two successive reprograming events of macrophages in response to *Candida* have been identified as follows, using whole-genome arrays: the early and late responses [116]. Their transcriptional profiles show the enhancement of gluconeogenesis and the glyoxylate cycle by upregulation of all genes involved in the conversion of fatty acid to glucose and a massive down-regulation of translation-related genes during the early response, suggesting a switch from glycolysis to gluconeogenesis, the glyoxylate cycle, and FAO during this early response. In addition, *Candida* cells phagocytosed by either macrophages [116] or neutrophils [117] upregulate arginine biosynthetic genes in response to ROS, rather than nutrient starvation, with these genes being important for germ tube and hyphal formation [118]. In contrast, the late response includes the reactivation of protein translation machinery and glycolysis. The metabolic shift in *C. albicans* cells following interaction with macrophages is assumed to be driven by the poor nutrient availability inside macrophage phagolysosomes, in which glucose concentration for instance is extremely low. This metabolic remodelling is dependent on the pathogenicity of *C. albicans* and *C. glabrata* since the non-pathogenic fungus *Saccharomyces cerevisiae* fails to demonstrate this response [76,116,119–121]. Moreover, similar events occur in *C. albicans* when it is exposed to neutrophils or whole human blood [117,120,122].

As stated earlier, during infection, activated macrophages shift their metabolism to aerobic glycolysis to activate antimicrobial inflammation and host defences. This means that for their survival *Candida*-activated macrophages rely specifically on glucose as their carbon source and additionally cannot reactivate mitochondrial oxidative phosphorylation. At the same time, ingested *C. albicans* cells similarly switch to aerobic glycolysis in the later phase of infection. As a result, macrophages and their ingested *C. albicans* compete for the available glucose [123]. During this nutrient war, the combatants rapidly consume the local glucose, leading to glucose depletion and triggering the "starvation" death of macrophages. Unlike macrophages, *C. albicans* cells have enough metabolic plasticity to switch their carbon source to alternatives such as the glyoxylate pathway, and in doing so survive the loss of glucose. As described earlier, these events are regulated by Tye7 and Gal4 *C. albicans* transcription factors. In *tye7Δ/Δgal4Δ/Δ* mutant strains, glycolysis and glucose consumption occurs at a far lower rate and, therefore, induction of macrophage starvation and cell death is lower. Using metformin to shut down the mitochondrial respiratory chain and drive faster glucose consumption ramps up the rate of death of activated macrophages by *C. albicans* with the knock-on effect of increasing mortality. In contrast, boosting local glucose levels by continuous administration of glucose improved these outcomes [123].

Glucose depletion not only leads to the rapid death of activated macrophages but also causes inflammasome activation by activating NLRP3, due to increased fungal burden [124]. NLRP3 has a protective role during infection, being a PRR that triggers processing and secretion of IL-1α. Therefore, the regulation of NLRP3 is crucial during *C. albicans* infection [125]. In a recent study, the mechanism behind NLRP3 activation during infection of macrophages was investigated, showing that inflammasome activation was broadly uniform among multiple clinical isolates of *C. albicans*, and rather than being dependent on hyphal formation, was purely down to glucose competition. Notably, reducing fungal ability to consume glucose (by using the *tye7Δ/Δgal4Δ/Δ* mutant strains) or increasing the glucose levels both reduce NLRP3 activation and IL-1β production [124].

It was believed for a long time that hyphae are essential for the pathogenicity of *C. albicans* during infections. This hypothesis, however, was challenged with the discovery that metabolic adaptation during systemic infections can be as important as morphological plasticity [126]. In a murine model of systemic candidiasis using the yeast-locked *eed1Δ/Δ* mutant, virulence was retained, leading to rapid yeast proliferation, and higher fungal loads in organs such as the kidneys or liver. Phenotypic analyses of the mutant strain showed enhanced growth rates in physiologically relevant carbon sources, including lactate, acetate, and citrate. A few genes involved in carboxylic acid and citrate metabolism were upregulated, alongside with *GAT1* that promotes proliferation in casamino acid rich environments. Therefore, the metabolic flexibility of *C. albicans* yeast-locked *eed1Δ/Δ* mutant in using alternative carbon sources (such as fatty acids, carboxylic and amino acids) at lower concentrations or the absence of glucose enhances its colonization ability and pathogenicity.

Hence, metabolic adaptation and fitness of *C. albicans* during infection not only supress the activity and recognition of immune cells, but also enhance the pathogenicity and mortality in systemic infection independently of hyphal formation.

4.2. Role of Metabolism during C. albicans Interactions with Epithelial Cells

While metabolic changes in immune cells have been the subject of recent studies, our current knowledge of these changes in epithelial cells (ECs) following microbial infection is limited. ECs are not merely passive barriers to prevent the invasion of microbes at the body's exterior surfaces but are also important in maintaining the balance with resident microbial communities. There are few studies showing metabolic reprogramming in ECs during microbial infection, namely increased glycolytic activity during *Staphylococcus aureus* infection of skin keratinocytes [35].

Concerning ECs interactions with fungi, oral epithelial cells (OECs) have been found to upregulate metabolic reprogramming-related genes in response to fungal infections, including HIF1-α pathway during *C. albicans* oropharyngeal candidiasis in mice [127] or *C. parapsilosis* infection of human OECs [128]. Similar to what is observed in phagocytosed cells, *C. albicans* upregulates gluconeogenesis, the glyoxylate pathway and FAO in the late phase of interaction with OECs, which might be related to the invasion process [129]. However, further analyses should be performed to unravel the mechanisms underlying these metabolic shifts.

Additionally, vaginal EC responses to varied species of *Candida* (*C. albicans*; *C. glabrata*; *C. parapsilosis*; and *C. tropicalis*) have also been studied using dual RNA sequencing in a time course infection model for vaginal ECs, analysing both fungal and host transcriptomic profiles [130]. In this study, Pekmezovic and co-workers showed a biphasic response to *Candida* spp. in vaginal ECs. The initial response is highly uniform among *Candida* species and characterised by mitochondrial-associated type 1 interferon (IFN) signalling. Of note, most mitochondrial genes were upregulated in the early phase of *Candida* infection, and the morphology of mitochondria changed in response to the infection. Moreover, mitochondrial DNA (mtDNA) and ROS are released into the vaginal ECs cytoplasm in all *Candida* species, both acting as damage-associated molecular patterns (DAMPs). In terms of fungal transcriptome, at 3 h post-infection *C. albicans* and *C. glabrata* upregulated carbohydrate catabolic processes and stress response pathways, whilst *C. parapsilosis* upregulated, among others, genes related to amino acid metabolism, iron transport, ribosome assembly and translation. In contrast, *C. tropicalis* differentially expressed genes were mainly related to RNA processing, ribosome biogenesis and ergosterol biosynthetic processes. Unlike the early responses, the late damage-associated epithelial transcriptional response is morphology-dependent, with the hyphal-associated toxin candidalysin enhancing the host responses [130].

5. Conclusions and Future Perspectives

Nutritional environment and metabolic adaptations in both host and fungal cells are key during their interactions. Further characterising and understanding host immunometabolic responses to *Candida* infections will potentially help developing novel therapeutic strategies to modulate these responses. In addition, identifying which metabolic enzymes are essential during the activation of anti-*Candida* immunity will lead to the detection of genetic variants associated with higher susceptibility in individuals suffering from recurrent or chronic fungal infections. Likewise, modulating nutrients in the infection environment could help enhance host responses and/or hamper fungal growth. Therefore, further research on these promising fields must be carried out to expand our knowledge and design new strategies to tackle fungal infections.

Author Contributions: Writing—original draft preparation, A.P., N.B., S.D.S.N., A.H.; writing—review and editing, A.P., S.S., D.L.M. All authors have read and agreed to the published version of the manuscript.

Funding: This review was supported by the Engineering and Physical Sciences Research Council (EPSRC) EP/S001301/1 and BBSRC grant -BB/S016899/1. D.L.M. and S.S. are further supported by Unilever and Sanofi.

Institutional Review Board Statement: Not applicable.

Informed Consent Statement: Not applicable.

Data Availability Statement: Not applicable.

Conflicts of Interest: The authors declare no conflict of interest.

References

1. Seed, P.C. The Human Mycobiome. *Cold Spring Harb. Perspect. Med.* **2014**, *5*, a019810. [CrossRef]
2. Tang, J.; Iliev, I.D.; Brown, J.; Underhill, D.M.; Funari, V.A. Mycobiome: Approaches to Analysis of Intestinal Fungi. *J. Immunol. Methods* **2015**, *421*, 112–121. [CrossRef]
3. Witherden, E.A.; Shoaie, S.; Hall, R.A.; Moyes, D.L. The Human Mucosal Mycobiome and Fungal Community Interactions. *J. Fungi* **2017**, *3*, 56. [CrossRef]
4. Limon, J.J.; Skalski, J.H.; Underhill, D.M. Commensal Fungi in Health and Disease. *Cell Host Microbe* **2017**, *22*, 156–165. [CrossRef]
5. Iliev, I.D.; Leonardi, I. Fungal Dysbiosis: Immunity and Interactions at Mucosal Barriers. *Nat. Rev. Immunol.* **2017**, *17*, 635–646. [CrossRef]
6. Brown, G.D.; Denning, D.W.; Gow, N.A.R.; Levitz, S.M.; Netea, M.G.; White, T.C. Hidden Killers: Human Fungal Infections. *Sci. Transl. Med.* **2012**, *4*, 165rv13. [CrossRef]
7. Arastehfar, A.; Gabaldón, T.; Garcia-Rubio, R.; Jenks, J.D.; Hoenigl, M.; Salzer, H.J.F.; Ilkit, M.; Lass-Flörl, C.; Perlin, D.S. Drug-Resistant Fungi: An Emerging Challenge Threatening Our Limited Antifungal Armamentarium. *Antibiotics* **2020**, *9*, 877. [CrossRef]
8. Traven, A.; Naderer, T. Central Metabolic Interactions of Immune Cells and Microbes: Prospects for Defeating Infections. *EMBO Rep.* **2019**, *20*, e47995. [CrossRef]
9. O'Neill, L.A.J.; Kishton, R.J.; Rathmell, J. A Guide to Immunometabolism for Immunologists. *Nat. Rev. Immunol.* **2016**, *16*, 553–565. [CrossRef]
10. Shyer, J.A.; Flavell, R.A.; Bailis, W. Metabolic Signaling in T Cells. *Cell Res.* **2020**, *30*, 649–659. [CrossRef]
11. O'Neill, L.A.J.; Pearce, E.J. Immunometabolism Governs Dendritic Cell and Macrophage Function. *J. Exp. Med.* **2016**, *213*, 15–23. [CrossRef]
12. Lachmandas, E.; Boutens, L.; Ratter, J.M.; Hijmans, A.; Hooiveld, G.J.; Joosten, L.A.B.; Rodenburg, R.J.; Fransen, J.A.M.; Houtkooper, R.H.; van Crevel, R.; et al. Microbial Stimulation of Different Toll-like Receptor Signalling Pathways Induces Diverse Metabolic Programmes in Human Monocytes. *Nat. Microbiol.* **2016**, *2*, 16246. [CrossRef]
13. Pearce, E.L.; Pearce, E.J. Metabolic Pathways in Immune Cell Activation and Quiescence. *Immunity* **2013**, *38*, 633–643. [CrossRef]
14. Leonhardt, J.; Große, S.; Marx, C.; Siwczak, F.; Stengel, S.; Bruns, T.; Bauer, R.; Kiehntopf, M.; Williams, D.L.; Wang, Z.-Q.; et al. Candida Albicans β-Glucan Differentiates Human Monocytes Into a Specific Subset of Macrophages. *Front. Immunol.* **2018**, *9*, 2818. [CrossRef]
15. Kelly, B.; O'Neill, L.A.J. Metabolic Reprogramming in Macrophages and Dendritic Cells in Innate Immunity. *Cell Res.* **2015**, *25*, 771–784. [CrossRef]
16. Jha, A.K.; Huang, S.C.-C.; Sergushichev, A.; Lampropoulou, V.; Ivanova, Y.; Loginicheva, E.; Chmielewski, K.; Stewart, K.M.; Ashall, J.; Everts, B.; et al. Network Integration of Parallel Metabolic and Transcriptional Data Reveals Metabolic Modules That Regulate Macrophage Polarization. *Immunity* **2015**, *42*, 419–430. [CrossRef]
17. Tannahill, G.M.; Curtis, A.M.; Adamik, J.; Palsson-McDermott, E.M.; McGettrick, A.F.; Goel, G.; Frezza, C.; Bernard, N.J.; Kelly, B.; Foley, N.H.; et al. Succinate Is an Inflammatory Signal That Induces IL-1β through HIF-1α. *Nature* **2013**, *496*, 238–242. [CrossRef]
18. Pollak, N.; Dölle, C.; Ziegler, M. The Power to Reduce: Pyridine Nucleotides–Small Molecules with a Multitude of Functions. *Biochem. J.* **2007**, *402*, 205–218. [CrossRef]
19. Nagy, C.; Haschemi, A. Time and Demand Are Two Critical Dimensions of Immunometabolism: The Process of Macrophage Activation and the Pentose Phosphate Pathway. *Front. Immunol.* **2015**, *6*, 164. [CrossRef]
20. Batista-Gonzalez, A.; Vidal, R.; Criollo, A.; Carreño, L.J. New Insights on the Role of Lipid Metabolism in the Metabolic Reprogramming of Macrophages. *Front. Immunol.* **2019**, *10*, 2993. [CrossRef]
21. Ricciotti, E.; FitzGerald, G.A. Prostaglandins and Inflammation. *Arterioscler. Thromb. Vasc. Biol.* **2011**, *31*, 986–1000. [CrossRef]
22. Biswas, S.K.; Mantovani, A. Orchestration of Metabolism by Macrophages. *Cell Metab.* **2012**, *15*, 432–437. [CrossRef]
23. Wang, T.; Liu, H.; Lian, G.; Zhang, S.-Y.; Wang, X.; Jiang, C. HIF1α-Induced Glycolysis Metabolism Is Essential to the Activation of Inflammatory Macrophages. *Mediat. Inflamm.* **2017**, *2017*, 9029327. [CrossRef]
24. Li, C.; Wang, Y.; Li, Y.; Yu, Q.; Jin, X.; Wang, X.; Jia, A.; Hu, Y.; Han, L.; Wang, J.; et al. HIF1α-Dependent Glycolysis Promotes Macrophage Functional Activities in Protecting against Bacterial and Fungal Infection. *Sci. Rep.* **2018**, *8*, 3603. [CrossRef]
25. Weichhart, T.; Hengstschläger, M.; Linke, M. Regulation of Innate Immune Cell Function by MTOR. *Nat. Rev. Immunol.* **2015**, *15*, 599–614. [CrossRef]

26. Divangahi, M.; Aaby, P.; Khader, S.A.; Barreiro, L.B.; Bekkering, S.; Chavakis, T.; van Crevel, R.; Curtis, N.; DiNardo, A.R.; Dominguez-Andres, J.; et al. Trained Immunity, Tolerance, Priming and Differentiation: Distinct Immunological Processes. *Nat. Immunol.* **2021**, *22*, 2–6. [CrossRef]

27. Netea, M.G.; Joosten, L.A.B.; Latz, E.; Mills, K.H.G.; Natoli, G.; Stunnenberg, H.G.; O'Neill, L.A.J.; Xavier, R.J. Trained Immunity: A Program of Innate Immune Memory in Health and Disease. *Science* **2016**, *352*, aaf1098. [CrossRef]

28. Ifrim, D.C.; Quintin, J.; Joosten, L.A.B.; Jacobs, C.; Jansen, T.; Jacobs, L.; Gow, N.A.R.; Williams, D.L.; van der Meer, J.W.M.; Netea, M.G. Trained Immunity or Tolerance: Opposing Functional Programs Induced in Human Monocytes after Engagement of Various Pattern Recognition Receptors. *Clin. Vaccine Immunol.* **2014**, *21*, 534–545. [CrossRef]

29. Netea, M.G.; Domínguez-Andrés, J.; Barreiro, L.B.; Chavakis, T.; Divangahi, M.; Fuchs, E.; Joosten, L.A.B.; van der Meer, J.W.M.; Mhlanga, M.M.; Mulder, W.J.M.; et al. Defining Trained Immunity and Its Role in Health and Disease. *Nat. Rev. Immunol.* **2020**, *20*, 375–388. [CrossRef]

30. Domínguez-Andrés, J.; Joosten, L.A.; Netea, M.G. Induction of Innate Immune Memory: The Role of Cellular Metabolism. *Curr. Opin. Immunol.* **2019**, *56*, 10–16. [CrossRef]

31. Warburg, O.; Wind, F.; Negelein, E. The Metabolism of Tumors in The Body. *J. Gen. Physiol.* **1927**, *8*, 519–530. [CrossRef]

32. Martín-Vicente, M.; González-Riaño, C.; Barbas, C.; Jiménez-Sousa, M.Á.; Brochado-Kith, O.; Resino, S.; Martínez, I. Metabolic Changes during Respiratory Syncytial Virus Infection of Epithelial Cells. *PLoS ONE* **2020**, *15*, e0230844. [CrossRef]

33. Delgado, T.; Carroll, P.A.; Punjabi, A.S.; Margineantu, D.; Hockenbery, D.M.; Lagunoff, M. Induction of the Warburg Effect by Kaposi's Sarcoma Herpesvirus Is Required for the Maintenance of Latently Infected Endothelial Cells. *Proc. Natl. Acad. Sci. USA* **2010**, *107*, 10696–10701. [CrossRef]

34. Hopkins, E.G.D.; Roumeliotis, T.I.; Mullineaux-Sanders, C.; Choudhary, J.S.; Frankel, G. Intestinal Epithelial Cells and the Microbiome Undergo Swift Reprogramming at the Inception of Colonic Citrobacter Rodentium Infection. *MBio* **2019**, *10*, e00062-19. [CrossRef]

35. Wickersham, M.; Wachtel, S.; Wong Fok Lung, T.; Soong, G.; Jacquet, R.; Richardson, A.; Parker, D.; Prince, A. Metabolic Stress Drives Keratinocyte Defenses against Staphylococcus Aureus Infection. *Cell Rep.* **2017**, *18*, 2742–2751. [CrossRef]

36. Mani, M.; Chen, C.; Amblee, V.; Liu, H.; Mathur, T.; Zwicke, G.; Zabad, S.; Patel, B.; Thakkar, J.; Jeffery, C.J. MoonProt: A Database for Proteins That Are Known to Moonlight. *Nucleic Acids Res.* **2015**, *43*, D277–D282. [CrossRef]

37. Chen, C.; Liu, H.; Zabad, S.; Rivera, N.; Rowin, E.; Hassan, M.; Gomez De Jesus, S.M.; Llinás Santos, P.S.; Kravchenko, K.; Mikhova, M.; et al. MoonProt 3.0: An Update of the Moonlighting Proteins Database. *Nucleic Acids Res.* **2021**, *49*, D368–D372. [CrossRef]

38. Godfrey, W.H.; Kornberg, M.D. The Role of Metabolic Enzymes in the Regulation of Inflammation. *Metabolites* **2020**, *10*, 426. [CrossRef]

39. Wolf, A.J.; Reyes, C.N.; Liang, W.; Becker, C.; Shimada, K.; Wheeler, M.L.; Cho, H.C.; Popescu, N.I.; Coggeshall, K.M.; Arditi, M.; et al. Hexokinase Is an Innate Immune Receptor for the Detection of Bacterial Peptidoglycan. *Cell* **2016**, *166*, 624–636. [CrossRef]

40. Chang, C.-H.; Curtis, J.D.; Maggi, L.B.; Faubert, B.; Villarino, A.V.; O'Sullivan, D.; Huang, S.C.-C.; van der Windt, G.J.W.; Blagih, J.; Qiu, J.; et al. Posttranscriptional Control of T Cell Effector Function by Aerobic Glycolysis. *Cell* **2013**, *153*, 1239–1251. [CrossRef]

41. Millet, P.; Vachharajani, V.; McPhail, L.; Yoza, B.; McCall, C.E. GAPDH Binding to TNF-α mRNA Contributes to Posttranscriptional Repression in Monocytes: A Novel Mechanism of Communication between Inflammation and Metabolism. *J. Immunol.* **2016**, *196*, 2541–2551. [CrossRef]

42. Wygrecka, M.; Marsh, L.M.; Morty, R.E.; Henneke, I.; Guenther, A.; Lohmeyer, J.; Markart, P.; Preissner, K.T. Enolase-1 Promotes Plasminogen-Mediated Recruitment of Monocytes to the Acutely Inflamed Lung. *Blood* **2009**, *113*, 5588–5598. [CrossRef]

43. Subramanian, A.; Miller, D.M. Structural Analysis of Alpha-Enolase. Mapping the Functional Domains Involved in down-Regulation of the c-Myc Protooncogene. *J. Biol. Chem.* **2000**, *275*, 5958–5965. [CrossRef]

44. De Rosa, V.; Galgani, M.; Porcellini, A.; Colamatteo, A.; Santopaolo, M.; Zuchegna, C.; Romano, A.; De Simone, S.; Procaccini, C.; La Rocca, C.; et al. Glycolysis Controls the Induction of Human Regulatory T Cells by Modulating the Expression of FOXP3 Exon 2 Splicing Variants. *Nat. Immunol.* **2015**, *16*, 1174–1184. [CrossRef]

45. Xie, M.; Yu, Y.; Kang, R.; Zhu, S.; Yang, L.; Zeng, L.; Sun, X.; Yang, M.; Billiar, T.R.; Wang, H.; et al. PKM2-Dependent Glycolysis Promotes NLRP3 and AIM2 Inflammasome Activation. *Nat. Commun.* **2016**, *7*, 13280. [CrossRef]

46. Palsson-McDermott, E.M.; Curtis, A.M.; Goel, G.; Lauterbach, M.A.R.; Sheedy, F.J.; Gleeson, L.E.; van den Bosch, M.W.M.; Quinn, S.R.; Domingo-Fernandez, R.; Johnston, D.G.W.; et al. Pyruvate Kinase M2 Regulates Hif-1α Activity and IL-1β Induction and Is a Critical Determinant of the Warburg Effect in LPS-Activated Macrophages. *Cell Metab.* **2015**, *21*, 65–80. [CrossRef]

47. Pioli, P.A.; Hamilton, B.J.; Connolly, J.E.; Brewer, G.; Rigby, W.F.C. Lactate Dehydrogenase Is an AU-Rich Element-Binding Protein That Directly Interacts with AUF1. *J. Biol. Chem.* **2002**, *277*, 35738–35745. [CrossRef]

48. Menk, A.V.; Scharping, N.E.; Moreci, R.S.; Zeng, X.; Guy, C.; Salvatore, S.; Bae, H.; Xie, J.; Young, H.A.; Wendell, S.G.; et al. Early TCR Signaling Induces Rapid Aerobic Glycolysis Enabling Distinct Acute T Cell Effector Functions. *Cell Rep.* **2018**, *22*, 1509–1521. [CrossRef]

49. Paik, W.K.; Pearson, D.; Lee, H.W.; Kim, S. Nonenzymatic Acetylation of Histones with Acetyl-CoA. *Biochim. Biophys. Acta* **1970**, *213*, 513–522. [CrossRef]

50. Choudhary, C.; Weinert, B.T.; Nishida, Y.; Verdin, E.; Mann, M. The Growing Landscape of Lysine Acetylation Links Metabolism and Cell Signalling. *Nat. Rev. Mol. Cell Biol.* **2014**, *15*, 536–550. [CrossRef]

51. Sabari, B.R.; Zhang, D.; Allis, C.D.; Zhao, Y. Metabolic Regulation of Gene Expression through Histone Acylations. *Nat. Rev. Mol. Cell Biol.* **2017**, *18*, 90–101. [CrossRef]

52. Kaelin, W.G.; McKnight, S.L. Influence of Metabolism on Epigenetics and Disease. *Cell* **2013**, *153*, 56–69. [CrossRef]

53. Zhang, D.; Tang, Z.; Huang, H.; Zhou, G.; Cui, C.; Weng, Y.; Liu, W.; Kim, S.; Lee, S.; Perez-Neut, M.; et al. Metabolic Regulation of Gene Expression by Histone Lactylation. *Nature* **2019**, *574*, 575–580. [CrossRef]

54. Zasłona, Z.; O'Neill, L.A.J. Cytokine-like Roles for Metabolites in Immunity. *Mol. Cell* **2020**, *78*, 814–823. [CrossRef]

55. Haas, R.; Cucchi, D.; Smith, J.; Pucino, V.; Macdougall, C.E.; Mauro, C. Intermediates of Metabolism: From Bystanders to Signalling Molecules. *Trends Biochem. Sci.* **2016**, *41*, 460–471. [CrossRef]

56. Weiss, H.J.; Angiari, S. Metabolite Transporters as Regulators of Immunity. *Metabolites* **2020**, *10*, 418. [CrossRef]

57. He, W.; Miao, F.J.-P.; Lin, D.C.-H.; Schwandner, R.T.; Wang, Z.; Gao, J.; Chen, J.-L.; Tian, H.; Ling, L. Citric Acid Cycle Intermediates as Ligands for Orphan G-Protein-Coupled Receptors. *Nature* **2004**, *429*, 188–193. [CrossRef]

58. Colegio, O.R.; Chu, N.-Q.; Szabo, A.L.; Chu, T.; Rhebergen, A.M.; Jairam, V.; Cyrus, N.; Brokowski, C.E.; Eisenbarth, S.C.; Phillips, G.M.; et al. Functional Polarization of Tumour-Associated Macrophages by Tumour-Derived Lactic Acid. *Nature* **2014**, *513*, 559–563. [CrossRef]

59. Ene, I.V.; Brown, A.J.P. 14 Integration of Metabolism with Virulence in Candida Albicans. In *Fungal Genomics*; Springer: Berlin/Heidelberg, Germany, 2014; Volume 13, pp. 349–370.

60. Barelle, C.J.; Priest, C.L.; Maccallum, D.M.; Gow, N.A.R.; Odds, F.C.; Brown, A.J.P. Niche-Specific Regulation of Central Metabolic Pathways in a Fungal Pathogen. *Cell. Microbiol.* **2006**, *8*, 961–971. [CrossRef]

61. Miramón, P.; Lorenz, M.C. A Feast for Candida: Metabolic Plasticity Confers an Edge for Virulence. *PLoS Pathog.* **2017**, *13*, e1006144. [CrossRef]

62. Brown, A.J.P.; Brown, G.D.; Netea, M.G.; Gow, N.A.R. Metabolism Impacts upon Candida Immunogenicity and Pathogenicity at Multiple Levels. *Trends Microbiol.* **2014**, *22*, 614–622. [CrossRef]

63. Sandai, D.; Yin, Z.; Selway, L.; Stead, D.; Walker, J.; Leach, M.D.; Bohovych, I.; Ene, I.V.; Kastora, S.; Budge, S.; et al. The Evolutionary Rewiring of Ubiquitination Targets Has Reprogrammed the Regulation of Carbon Assimilation in the Pathogenic Yeast Candida Albicans. *MBio* **2012**, *3*, e00495-12. [CrossRef]

64. Garbe, E.; Vylkova, S. Role of Amino Acid Metabolism in the Virulence of Human Pathogenic Fungi. *Curr. Clin. Microbiol. Rep.* **2019**, *6*, 108–119. [CrossRef]

65. Ene, I.V.; Lohse, M.B.; Vladu, A.V.; Morschhäuser, J.; Johnson, A.D.; Bennett, R.J. Phenotypic Profiling Reveals That Candida Albicans Opaque Cells Represent a Metabolically Specialized Cell State Compared to Default White Cells. *MBio* **2016**, *7*, e01269-16. [CrossRef]

66. Askew, C.; Sellam, A.; Epp, E.; Hogues, H.; Mullick, A.; Nantel, A.; Whiteway, M. Transcriptional Regulation of Carbohydrate Metabolism in the Human Pathogen Candida Albicans. *PLoS Pathog.* **2009**, *5*, e1000612. [CrossRef]

67. Bonhomme, J.; Chauvel, M.; Goyard, S.; Roux, P.; Rossignol, T.; D'Enfert, C. Contribution of the Glycolytic Flux and Hypoxia Adaptation to Efficient Biofilm Formation by Candida Albicans. *Mol. Microbiol.* **2011**, *80*, 995–1013. [CrossRef]

68. Pérez, J.C.; Kumamoto, C.A.; Johnson, A.D. Candida Albicans Commensalism and Pathogenicity Are Intertwined Traits Directed by a Tightly Knit Transcriptional Regulatory Circuit. *PLoS Biol.* **2013**, *11*, e1001510. [CrossRef]

69. Basso, V.; d'Enfert, C.; Znaidi, S.; Bachellier-Bassi, S. From Genes to Networks: The Regulatory Circuitry Controlling Candida Albicans Morphogenesis. *Curr. Top. Microbiol. Immunol.* **2019**, *422*, 61–99. [CrossRef]

70. Han, T.-L.; Cannon, R.D.; Villas-Bôas, S.G. The Metabolic Basis of Candida Albicans Morphogenesis and Quorum Sensing. *Fungal. Genet. Biol.* **2011**, *48*, 747–763. [CrossRef]

71. Swoboda, R.K.; Bertram, G.; Delbrück, S.; Ernst, J.F.; Gow, N.A.; Gooday, G.W.; Brown, A.J. Fluctuations in Glycolytic MRNA Levels during Morphogenesis in Candida Albicans Reflect Underlying Changes in Growth and Are Not a Response to Cellular Dimorphism. *Mol. Microbiol.* **1994**, *13*, 663–672. [CrossRef]

72. Nantel, A.; Dignard, D.; Bachewich, C.; Harcus, D.; Marcil, A.; Bouin, A.-P.; Sensen, C.W.; Hogues, H.; van het Hoog, M.; Gordon, P.; et al. Transcription Profiling of Candida Albicans Cells Undergoing the Yeast-to-Hyphal Transition. *Mol. Biol. Cell* **2002**, *13*, 3452–3465. [CrossRef]

73. Doedt, T.; Krishnamurthy, S.; Bockmühl, D.P.; Tebarth, B.; Stempel, C.; Russell, C.L.; Brown, A.J.P.; Ernst, J.F. APSES Proteins Regulate Morphogenesis and Metabolism in Candida Albicans. *Mol. Biol. Cell* **2004**, *15*, 3167–3180. [CrossRef]

74. Lan, C.-Y.; Newport, G.; Murillo, L.A.; Jones, T.; Scherer, S.; Davis, R.W.; Agabian, N. Metabolic Specialization Associated with Phenotypic Switching in Candidaalbicans. *Proc. Natl. Acad. Sci. USA* **2002**, *99*, 14907–14912. [CrossRef]

75. Piekarska, K.; Mol, E.; van den Berg, M.; Hardy, G.; van den Burg, J.; van Roermund, C.; MacCallum, D.; Odds, F.; Distel, B. Peroxisomal Fatty Acid Beta-Oxidation Is Not Essential for Virulence of Candida Albicans. *Eukaryot. Cell* **2006**, *5*, 1847–1856. [CrossRef]

76. Ramírez, M.A.; Lorenz, M.C. Mutations in Alternative Carbon Utilization Pathways in Candida Albicans Attenuate Virulence and Confer Pleiotropic Phenotypes. *Eukaryot. Cell* **2007**, *6*, 280–290. [CrossRef]

77. Tripathi, G.; Wiltshire, C.; Macaskill, S.; Tournu, H.; Budge, S.; Brown, A.J.P. Gcn4 Co-Ordinates Morphogenetic and Metabolic Responses to Amino Acid Starvation in Candida Albicans. *EMBO J.* **2002**, *21*, 5448–5456. [CrossRef]

78. García-Sánchez, S.; Aubert, S.; Iraqui, I.; Janbon, G.; Ghigo, J.-M.; D'Enfert, C. Candida Albicans Biofilms: A Developmental State Associated with Specific and Stable Gene Expression Patterns. *Eukaryot. Cell* **2004**, *3*, 536–545. [CrossRef]
79. Vylkova, S.; Carman, A.J.; Danhof, H.A.; Collette, J.R.; Zhou, H.; Lorenz, M.C. The Fungal Pathogen Candida Albicans Autoinduces Hyphal Morphogenesis by Raising Extracellular PH. *MBio* **2011**, *2*, e00055-11. [CrossRef]
80. Min, K.; Naseem, S.; Konopka, J.B. N-Acetylglucosamine Regulates Morphogenesis and Virulence Pathways in Fungi. *J. Fungi* **2019**, *6*, 8. [CrossRef]
81. Gow, N.A.R.; Hube, B. Importance of the Candida Albicans Cell Wall during Commensalism and Infection. *Curr. Opin. Microbiol.* **2012**, *15*, 406–412. [CrossRef]
82. Bowman, S.M.; Free, S.J. The Structure and Synthesis of the Fungal Cell Wall. *Bioessays* **2006**, *28*, 799–808. [CrossRef]
83. Lee, K.K.; Munro, C.A. Carbon Metabolism in Pathogenic Yeasts (Especially Candida): The Role of Cell Wall Metabolism in Virulence. In *Molecular Mechanisms in Yeast Carbon Metabolism*; Piškur, J., Compagno, C., Eds.; Springer: Berlin/Heidelberg, Germany, 2014; pp. 141–167. ISBN 978-3-642-55013-3.
84. Taff, H.T.; Nett, J.E.; Zarnowski, R.; Ross, K.M.; Sanchez, H.; Cain, M.T.; Hamaker, J.; Mitchell, A.P.; Andes, D.R. A Candida Biofilm-Induced Pathway for Matrix Glucan Delivery: Implications for Drug Resistance. *PLoS Pathog.* **2012**, *8*, e1002848. [CrossRef]
85. Ene, I.V.; Adya, A.K.; Wehmeier, S.; Brand, A.C.; MacCallum, D.M.; Gow, N.A.R.; Brown, A.J.P. Host Carbon Sources Modulate Cell Wall Architecture, Drug Resistance and Virulence in a Fungal Pathogen. *Cell. Microbiol.* **2012**, *14*, 1319–1335. [CrossRef]
86. Rodaki, A.; Bohovych, I.M.; Enjalbert, B.; Young, T.; Odds, F.C.; Gow, N.A.R.; Brown, A.J.P. Glucose Promotes Stress Resistance in the Fungal Pathogen Candida Albicans. *Mol. Biol. Cell* **2009**, *20*, 4845–4855. [CrossRef]
87. Satala, D.; Karkowska-Kuleta, J.; Zelazna, A.; Rapala-Kozik, M.; Kozik, A. Moonlighting Proteins at the Candidal Cell Surface. *Microorganisms* **2020**, *8*, 1046. [CrossRef]
88. Jeffery, C.J. Moonlighting Proteins. *Trends Biochem. Sci.* **1999**, *24*, 8–11. [CrossRef]
89. Karkowska-Kuleta, J.; Satala, D.; Bochenska, O.; Rapala-Kozik, M.; Kozik, A. Moonlighting Proteins Are Variably Exposed at the Cell Surfaces of Candida Glabrata, Candida Parapsilosis and Candida Tropicalis under Certain Growth Conditions. *BMC Microbiol.* **2019**, *19*, 149. [CrossRef]
90. Gozalbo, D.; Gil-Navarro, I.; Azorín, I.; Renau-Piqueras, J.; Martínez, J.P.; Gil, M.L. The Cell Wall-Associated Glyceraldehyde-3-Phosphate Dehydrogenase of Candida Albicans Is Also a Fibronectin and Laminin Binding Protein. *Infect. Immun.* **1998**, *66*, 2052–2059. [CrossRef]
91. Funk, J.; Schaarschmidt, B.; Slesiona, S.; Hallström, T.; Horn, U.; Brock, M. The Glycolytic Enzyme Enolase Represents a Plasminogen-Binding Protein on the Surface of a Wide Variety of Medically Important Fungal Species. *Int. J. Med. Microbiol.* **2016**, *306*, 59–68. [CrossRef]
92. Ayón-Núñez, D.A.; Fragoso, G.; Bobes, R.J.; Laclette, J.P. Plasminogen-Binding Proteins as an Evasion Mechanism of the Host's Innate Immunity in Infectious Diseases. *Biosci. Rep.* **2018**, *38*, BSR20180705. [CrossRef]
93. Sun, J.N.; Solis, N.V.; Phan, Q.T.; Bajwa, J.S.; Kashleva, H.; Thompson, A.; Liu, Y.; Dongari-Bagtzoglou, A.; Edgerton, M.; Filler, S.G. Host Cell Invasion and Virulence Mediated by Candida Albicans Ssa1. *PLoS Pathog.* **2010**, *6*, e1001181. [CrossRef]
94. Phan, Q.T.; Myers, C.L.; Fu, Y.; Sheppard, D.C.; Yeaman, M.R.; Welch, W.H.; Ibrahim, A.S.; Edwards, J.E.; Filler, S.G. Als3 Is a Candida Albicans Invasin That Binds to Cadherins and Induces Endocytosis by Host Cells. *PLoS Biol.* **2007**, *5*, e64. [CrossRef]
95. Zhu, W.; Phan, Q.T.; Boontheung, P.; Solis, N.V.; Loo, J.A.; Filler, S.G. EGFR and HER2 Receptor Kinase Signaling Mediate Epithelial Cell Invasion by Candida Albicans during Oropharyngeal Infection. *Proc. Natl. Acad. Sci. USA* **2012**, *109*, 14194–14199. [CrossRef]
96. Denning, D.W.; Bromley, M.J. How to Bolster the Antifungal Pipeline. *Science* **2015**, *347*, 1414–1416. [CrossRef]
97. Ene, I.V.; Heilmann, C.J.; Sorgo, A.G.; Walker, L.A.; de Koster, C.G.; Munro, C.A.; Klis, F.M.; Brown, A.J.P. Carbon Source-Induced Reprogramming of the Cell Wall Proteome and Secretome Modulates the Adherence and Drug Resistance of the Fungal Pathogen Candida Albicans. *Proteomics* **2012**, *12*, 3164–3179. [CrossRef]
98. Lok, B.; Adam, M.A.A.; Kamal, L.Z.M.; Chukwudi, N.A.; Sandai, R.; Sandai, D. The Assimilation of Different Carbon Sources in Candida Albicans: Fitness and Pathogenicity. *Med. Mycol.* **2021**, *59*, 115–125. [CrossRef]
99. Moosa, M.-Y.S.; Sobel, J.D.; Elhalis, H.; Du, W.; Akins, R.A. Fungicidal Activity of Fluconazole against Candida Albicans in a Synthetic Vagina-Simulative Medium. *Antimicrob. Agents Chemother.* **2004**, *48*, 161–167. [CrossRef]
100. Mota, S.; Alves, R.; Carneiro, C.; Silva, S.; Brown, A.J.; Istel, F.; Kuchler, K.; Sampaio, P.; Casal, M.; Henriques, M.; et al. Candida Glabrata Susceptibility to Antifungals and Phagocytosis Is Modulated by Acetate. *Front. Microbiol.* **2015**, *6*, 919. [CrossRef]
101. Pellon, A.; Sadeghi Nasab, S.D.; Moyes, D.L. New Insights in Candida Albicans Innate Immunity at the Mucosa: Toxins, Epithelium, Metabolism, and Beyond. *Front. Cell. Infect. Microbiol.* **2020**, *10*, 81. [CrossRef]
102. Austermeier, S.; Kasper, L.; Westman, J.; Gresnigt, M.S. I Want to Break Free—Macrophage Strategies to Recognize and Kill Candida Albicans, and Fungal Counter-Strategies to Escape. *Curr. Opin. Microbiol.* **2020**, *58*, 15–23. [CrossRef]
103. Netea, M.G.; Brown, G.D.; Kullberg, B.J.; Gow, N.A.R. An Integrated Model of the Recognition of Candida Albicans by the Innate Immune System. *Nat. Rev. Microbiol.* **2008**, *6*, 67–78. [CrossRef]
104. Swidergall, M. Candida Albicans at Host Barrier Sites: Pattern Recognition Receptors and Beyond. *Pathogens* **2019**, *8*, 40. [CrossRef]
105. Erwig, L.P.; Gow, N.A.R. Interactions of Fungal Pathogens with Phagocytes. *Nat. Rev. Microbiol.* **2016**, *14*, 163–176. [CrossRef]

106. Moyes, D.L.; Wilson, D.; Richardson, J.P.; Mogavero, S.; Tang, S.X.; Wernecke, J.; Höfs, S.; Gratacap, R.L.; Robbins, J.; Runglall, M.; et al. Candidalysin Is a Fungal Peptide Toxin Critical for Mucosal Infection. *Nature* **2016**, *532*, 64–68. [CrossRef]
107. Naglik, J.R.; Gaffen, S.L.; Hube, B. Candidalysin: Discovery and Function in Candida Albicans Infections. *Curr. Opin. Microbiol.* **2019**, *52*, 100–109. [CrossRef]
108. Domínguez-Andrés, J.; Arts, R.J.W.; Ter Horst, R.; Gresnigt, M.S.; Smeekens, S.P.; Ratter, J.M.; Lachmandas, E.; Boutens, L.; van de Veerdonk, F.L.; Joosten, L.A.B.; et al. Rewiring Monocyte Glucose Metabolism via C-Type Lectin Signaling Protects against Disseminated Candidiasis. *PLoS Pathog.* **2017**, *13*, e1006632. [CrossRef]
109. Gonçalves, S.M.; Duarte-Oliveira, C.; Campos, C.F.; Aimanianda, V.; Ter Horst, R.; Leite, L.; Mercier, T.; Pereira, P.; Fernández-García, M.; Antunes, D.; et al. Phagosomal Removal of Fungal Melanin Reprograms Macrophage Metabolism to Promote Antifungal Immunity. *Nat. Commun.* **2020**, *11*, 2282. [CrossRef]
110. Stappers, M.H.T.; Clark, A.E.; Aimanianda, V.; Bidula, S.; Reid, D.M.; Asamaphan, P.; Hardison, S.E.; Dambuza, I.M.; Valsecchi, I.; Kerscher, B.; et al. Recognition of DHN-Melanin by a C-Type Lectin Receptor Is Required for Immunity to Aspergillus. *Nature* **2018**, *555*, 382–386. [CrossRef]
111. Weerasinghe, H.; Traven, A. Immunometabolism in Fungal Infections: The Need to Eat to Compete. *Curr. Opin. Microbiol.* **2020**, *58*, 32–40. [CrossRef]
112. Romani, L. Immunity to Fungal Infections. *Nat. Rev. Immunol.* **2011**, *11*, 275–288. [CrossRef]
113. Ballou, E.R.; Avelar, G.M.; Childers, D.S.; Mackie, J.; Bain, J.M.; Wagener, J.; Kastora, S.L.; Panea, M.D.; Hardison, S.E.; Walker, L.A.; et al. Lactate Signalling Regulates Fungal β-Glucan Masking and Immune Evasion. *Nat. Microbiol.* **2016**, *2*, 16238. [CrossRef]
114. Ene, I.V.; Cheng, S.-C.; Netea, M.G.; Brown, A.J.P. Growth of Candida Albicans Cells on the Physiologically Relevant Carbon Source Lactate Affects Their Recognition and Phagocytosis by Immune Cells. *Infect. Immun.* **2013**, *81*, 238–248. [CrossRef]
115. Danhof, H.A.; Vylkova, S.; Vesely, E.M.; Ford, A.E.; Gonzalez-Garay, M.; Lorenz, M.C. Robust Extracellular PH Modulation by Candida Albicans during Growth in Carboxylic Acids. *MBio* **2016**, *7*, e01646-16. [CrossRef]
116. Lorenz, M.C.; Bender, J.A.; Fink, G.R. Transcriptional Response of Candida Albicans upon Internalization by Macrophages. *Eukaryot. Cell* **2004**, *3*, 1076–1087. [CrossRef]
117. Rubin-Bejerano, I.; Fraser, I.; Grisafi, P.; Fink, G.R. Phagocytosis by Neutrophils Induces an Amino Acid Deprivation Response in Saccharomyces Cerevisiae and Candida Albicans. *Proc. Natl. Acad. Sci. USA* **2003**, *100*, 11007–11012. [CrossRef]
118. Jiménez-López, C.; Collette, J.R.; Brothers, K.M.; Shepardson, K.M.; Cramer, R.A.; Wheeler, R.T.; Lorenz, M.C. Candida Albicans Induces Arginine Biosynthetic Genes in Response to Host-Derived Reactive Oxygen Species. *Eukaryot. Cell* **2013**, *12*, 91–100. [CrossRef]
119. Chew, S.Y.; Chee, W.J.Y.; Than, L.T.L. The Glyoxylate Cycle and Alternative Carbon Metabolism as Metabolic Adaptation Strategies of Candida Glabrata: Perspectives from Candida Albicans and Saccharomyces Cerevisiae. *J. Biomed. Sci.* **2019**, *26*, 52. [CrossRef]
120. Lorenz, M.C.; Fink, G.R. The Glyoxylate Cycle Is Required for Fungal Virulence. *Nature* **2001**, *412*, 83–86. [CrossRef]
121. Rai, M.N.; Balusu, S.; Gorityala, N.; Dandu, L.; Kaur, R. Functional Genomic Analysis of Candida Glabrata-Macrophage Interaction: Role of Chromatin Remodeling in Virulence. *PLoS Pathog.* **2012**, *8*, e1002863. [CrossRef]
122. Fradin, C.; Kretschmar, M.; Nichterlein, T.; Gaillardin, C.; D'Enfert, C.; Hube, B. Stage-Specific Gene Expression of Candida Albicans in Human Blood. *Mol. Microbiol.* **2003**, *47*, 1523–1543. [CrossRef]
123. Tucey, T.M.; Verma, J.; Harrison, P.F.; Snelgrove, S.L.; Lo, T.L.; Scherer, A.K.; Barugahare, A.A.; Powell, D.R.; Wheeler, R.T.; Hickey, M.J.; et al. Glucose Homeostasis Is Important for Immune Cell Viability during Candida Challenge and Host Survival of Systemic Fungal Infection. *Cell Metab.* **2018**, *27*, 988–1006.e7. [CrossRef]
124. Tucey, T.M.; Verma, J.; Olivier, F.A.B.; Lo, T.L.; Robertson, A.A.B.; Naderer, T.; Traven, A. Metabolic Competition between Host and Pathogen Dictates Inflammasome Responses to Fungal Infection. *PLoS Pathog.* **2020**, *16*, e1008695. [CrossRef]
125. Camilli, G.; Griffiths, J.S.; Ho, J.; Richardson, J.P.; Naglik, J.R. Some like it Hot: Candida Activation of Inflammasomes. *PLoS Pathog.* **2020**, *16*, e1008975. [CrossRef]
126. Dunker, C.; Polke, M.; Schulze-Richter, B.; Schubert, K.; Rudolphi, S.; Gressler, A.E.; Pawlik, T.; Prada Salcedo, J.P.; Niemiec, M.J.; Slesiona-Künzel, S.; et al. Rapid Proliferation Due to Better Metabolic Adaptation Results in Full Virulence of a Filament-Deficient Candida Albicans Strain. *Nat. Commun.* **2021**, *12*, 3899. [CrossRef]
127. Kirchner, F.R.; Littringer, K.; Altmeier, S.; Tran, V.D.T.; Schönherr, F.; Lemberg, C.; Pagni, M.; Sanglard, D.; Joller, N.; LeibundGut-Landmann, S. Persistence of Candida Albicans in the Oral Mucosa Induces a Curbed Inflammatory Host Response That Is Independent of Immunosuppression. *Front. Immunol.* **2019**, *10*. [CrossRef]
128. Horváth, M.; Nagy, G.; Zsindely, N.; Bodai, L.; Horváth, P.; Vágvölgyi, C.; Nosanchuk, J.D.; Tóth, R.; Gácser, A. Oral Epithelial Cells Distinguish between Candida Species with High or Low Pathogenic Potential through MicroRNA Regulation. *Msystems* **2021**, *6*, e00163-21. [CrossRef]
129. Zakikhany, K.; Naglik, J.R.; Schmidt-Westhausen, A.; Holland, G.; Schaller, M.; Hube, B. In Vivo Transcript Profiling of Candida Albicans Identifies a Gene Essential for Interepithelial Dissemination. *Cell. Microbiol.* **2007**, *9*, 2938–2954. [CrossRef]
130. Pekmezovic, M.; Hovhannisyan, H.; Gresnigt, M.S.; Iracane, E.; Oliveira-Pacheco, J.; Siscar-Lewin, S.; Seemann, E.; Qualmann, B.; Kalkreuter, T.; Müller, S.; et al. Candida Pathogens Induce Protective Mitochondria-Associated Type I Interferon Signalling and a Damage-Driven Response in Vaginal Epithelial Cells. *Nat. Microbiol.* **2021**, *6*, 643–657. [CrossRef]

Review

Invasive *Candida* Infections in Neonates after Major Surgery: Current Evidence and New Directions

Domenico Umberto De Rose [1], Alessandra Santisi [1], Maria Paola Ronchetti [1], Ludovica Martini [1], Lisa Serafini [2], Pasqua Betta [3], Marzia Maino [4], Francesco Cavigioli [5], Ilaria Cocchi [5], Lorenza Pugni [6], Elvira Bonanno [7], Chryssoula Tzialla [8], Mario Giuffrè [9], Jenny Bua [10], Benedetta Della Torre [11], Giovanna Nardella [12], Danila Mazzeo [13], Paolo Manzoni [14], Andrea Dotta [1], Pietro Bagolan [15], Cinzia Auriti [1,*] and on behalf of Study Group of Neonatal Infectious Diseases [†]

1. Neonatal Intensive Care Unit, Medical and Surgical Department of Fetus, Newborn and Infant, "Bambino Gesù" Children's Hospital IRCCS, 00165 Rome, Italy; domenico.derose@opbg.net (D.U.D.R.); alessandra.santisi@opbg.net (A.S.); mariapaola.ronchetti@opbg.net (M.P.R.); ludovica.martini@opbg.net (L.M.); andrea.dotta@opbg.net (A.D.)
2. Neonatal Intensive Care Unit, Department of Critical Care Medicine, "A. Meyer" University Children's Hospital, 50139 Florence, Italy; lisa.serafini@meyer.it
3. Neonatology Unit, Azienda Ospedaliero-Universitaria "Policlinico-Vittorio Emanuele", 95124 Catania, Italy; mlbetta@yahoo.it
4. Neonatal Intensive Care Unit, Giovanni XXIII Hospital, 24127 Bergamo, Italy; mmaino@asst-pg23.it
5. Neonatology Unit, Ospedale dei Bambini "V. Buzzi", ASST FBF-Sacco-Buzzi, 20154 Milan, Italy; francesco.cavigioli@asst-fbf-sacco.it (F.C.); ilaria.cocchi88@gmail.com (I.C.)
6. Neonatal Intensive Care Unit, Fondazione IRCCS Ca' Granda Ospedale Maggiore Policlinico, 20122 Milan, Italy; lorenza.pugni@mangiagalli.it
7. Neonatology Unit, Azienda Ospedaliera SS. Annunziata, 87100 Cosenza, Italy; elvirabonanno@libero.it
8. Neonatal Unit and Neonatal Intensive Care Unit, Fondazione IRCCS Policlinico San Matteo, 27100 Pavia, Italy; c.tzialla@smatteo.pv.it
9. Department of Health Promotion Sciences, Maternal and Infant Care, Internal Medicine and Medical Specialties "G. D'Alessandro", University of Palermo, 90133 Palermo, Italy; mario.giuffre@unipa.it
10. Neonatal Intensive Care Unit, Institute for Maternal and Child Health IRCCS "Burlo Garofolo", 34137 Trieste, Italy; jenny.bua@burlo.trieste.it
11. Neonatal Intensive Care Unit, Santa Maria della Misericordia Hospital, 06123 Perugia, Italy; benedetta.dellatorre@ospedale.perugia.it
12. Division of Neonatology, Azienda Ospedaliero-Universitaria "Ospedali Riuniti", 71122 Foggia, Italy; giovannanardella@yahoo.it
13. Neonatology Unit, Policlinico Gaetano Martino, 98124 Messina, Italy; danilamazzeo@outlook.it
14. Division of Pediatrics and Neonatology, Department of Maternal, Neonatal, and Infant Health, Ospedale degli Infermi, ASL Biella, 13875 Ponderano, Biella, Italy; paolomanzoni@hotmail.com
15. Neonatal Surgery Unit, Medical and Surgical Department of Fetus, Newborn and Infant, "Bambino Gesù" Children's Hospital IRCCS, 00165 Rome, Italy; pietro.bagolan@opbg.net
*   Correspondence: cinzia.auriti@opbg.net; Tel.: +39-06-68592427; Fax: +39-06-68593916
†   On behalf of Study Group of Neonatal Infectious Diseases of the Italian Society of Neonatology (SIN).

Citation: De Rose, D.U.; Santisi, A.; Ronchetti, M.P.; Martini, L.; Serafini, L.; Betta, P.; Maino, M.; Cavigioli, F.; Cocchi, I.; Pugni, L.; et al. Invasive *Candida* Infections in Neonates after Major Surgery: Current Evidence and New Directions. *Pathogens* **2021**, *10*, 319. https://doi.org/10.3390/pathogens10030319

Academic Editor: Jonathan Richardson

Received: 30 January 2021
Accepted: 4 March 2021
Published: 9 March 2021

Publisher's Note: MDPI stays neutral with regard to jurisdictional claims in published maps and institutional affiliations.

Abstract: Infections represent a serious health problem in neonates. Invasive *Candida* infections (ICIs) are still a leading cause of mortality and morbidity in neonatal intensive care units (NICUs). Infants hospitalized in NICUs are at high risk of ICIs, because of several risk factors: broad spectrum antibiotic treatments, central catheters and other invasive devices, fungal colonization, and impaired immune responses. In this review we summarize 19 published studies which provide the prevalence of previous surgery in neonates with invasive *Candida* infections. We also provide an overview of risk factors for ICIs after major surgery, fungal colonization, and innate defense mechanisms against fungi, as well as the roles of different *Candida* spp., the epidemiology and costs of ICIs, diagnosis of ICIs, and antifungal prophylaxis and treatment.

Keywords: invasive *Candida* infections; invasive fungal infections; antifungal prophylaxis; newborns; surgery; neonatal surgery

1. Introduction

Yeasts are commensal microorganisms that usually colonize mucosal surfaces and skin. In particular clinical conditions, such as immune suppression, prolonged use of broad-spectrum antibiotics and/or steroids, the balance of the colonizing flora of the skin is altered and fungi express numerous factors that contribute to pathogenicity. Adherence is one of the most important factors that facilitate the colonization and dissemination of fungi, by the expression of adhesins, which facilitate binding to host substrates, including beta–integrins, on the endothelium and white blood cells. The yeast-to-hypha transition of *C. albicans* facilitates biofilm formation, tissue invasion, and dissemination of the infection [1,2]. Other virulence factors are membrane and cell wall barriers, dimorphism, biofilm formation, signal transduction pathway, proteins related to stress tolerance, hydrolytic enzymes (e.g., proteases, lipases, hemolysins), and toxin production) [3].

Therefore, fungi can lead to severe infections in the host. Invasive fungal infections (IFIs) in neonatal intensive care units (NICUs) are a substantial health problem, as they are the second most common cause of infection-related death among critically ill neonates. IFIs lead also to significant neurodevelopmental disability among survivors, representing a substantial health problem, especially among the neonates with lowest gestational age and lowest birthweight [4–6]. Critically ill patients in NICUs (and in particular preterm neonates) are at high risk of IFIs, especially if they need broad-spectrum antibiotic treatments, surgery that disrupts natural defense barriers, intravascular catheters for prolonged periods, or implantation of invasive devices to survive. Their immunological impairments are predisposed to fungal colonization and to subsequent systemic infection. Bloodstream infections due to the *Candida* species (C. spp.) are considered the most common IFIs in critically ill patients in NICUs.

In specific subgroups (e.g., abdominal surgical patients), IFIs are also frequent, but there are no epidemiological studies on the incidence of IFIs in neonates with major surgical diseases. Clinical and epidemiological studies are needed to identify preventive strategies in preterm and term infants, who undergo major surgery or specific subgroups of this category of patients.

This review aims to summarize scientific evidence about invasive *Candida* infections (ICIs) in neonates undergoing surgery.

2. Methods

This paper provides a review of the literature on ICIs in neonates after major surgery. An extensive literature search in the MEDLINE database (via PubMed) has been performed from 2000 up to 9 January 2021. The following keywords "*Candida*" OR "fungal infection" AND "surgery" AND "neonates" OR "infants" were searched as entry terms. All retrieved articles were screened, and then full texts of records deemed eligible for inclusion were assessed. References in the relevant papers were also reviewed.

Papers written in languages other than English, or not providing data about the main focus of this research (the number of neonates with ICIs after undergoing major surgery, separate data for neonates and children, and case reports and reviews) were excluded.

Major surgery is considered to be any invasive operative procedure in which a mesenchymal barrier is opened (pleural cavity, peritoneum, meninges).

An infant is considered colonized by *Candida* when a surveillance culture (such as pharyngeal or tracheal swab, urine, feces, skin swabs) develops colonies of *Candida* spp., without signs or symptoms suggestive of infection [7].

Infants with ICIs have specific or nonspecific signs or symptoms, and isolation of a *Candida* spp. is obtained from a sterile cultural site (blood, cerebrospinal fluid, peritoneal fluid) [8].

We also provided an overview of risk factors for ICIs after major surgery, the innate defense mechanisms against fungi, the role of different *Candida* spp., the epidemiology and costs of ICIs, and the antifungal prophylaxis.

3. Results

A total of 253 records were identified through literature search (via PubMed) from 2000 onwards. Among them 155 were excluded based on the titles, the abstracts, and the type (case reports and review). The remaining 71 full-text articles were assessed for eligibility. We found no studies focused on the incidence of IFIs in neonates who previously underwent major surgery.

Conversely, we found 19 studies that provided how many neonates with IFIs underwent major surgery before the onset of the infection [9–27]. The selection process is shown in Figure 1.

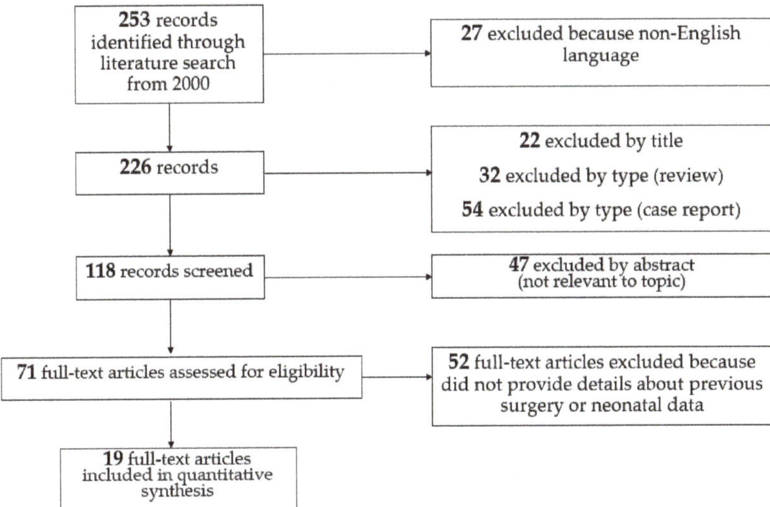

Figure 1. Literature selection of recent studies reporting incidence of previous surgery in infants with invasive fungal infections.

Of the 19 studies included in the quantitative synthesis, 8 collected patients' data retrospectively, while 11 collected it prospectively (Table 1). A total of 1637 neonates with IFIs were reported. Of these, 550 (33.6%) underwent major surgery before the onset of the infection. Abdominal surgery was not always reported by the studies, with percentages ranging from 13 up to 80. Fungal-infection-related mortality is difficult to demonstrate, therefore in-hospital overall mortality is more often reported.

Table 1. Recent studies reporting prevalence of previous surgery in infants with invasive fungal infections.

Country and Year [Ref.]	Study Type	Study Period	Inclusion Criteria	Neonates (n)	Preterm (n, %)	Central Line (n, %)	Previous Surger (n, %)	Abdominal Surgery (n, %)	Non Abdominal Surgery (n, %)	In-Hospital Mortality (n, %)	Fungal Mortality (n, %)
US, 2000 [9]	R, SC	1981–1999	ICI	96	83 VLBW (86)	63 (66)	3 (12)	NA	NA	31 (32)	11 (11)
Kuwait, 2000 [10]	R, SC	1994–1998	Positive BC	25	9 LBW (36)	NA	25 (100)	20 (80)	5 (20)	8 (25)	NA
US, 2000 [11]	P, MC	1993–1995	Positive BC	35	29 VLBW (83)	NA	13 (37)	NA	NA	8 (23)	NA
Greece, 2004 [12]	P, SC	1994–2000	ICI	59	NA	NA	23 (39)	NA	NA	17 (29)	*C. albicans*: 15/38 (39); *C. parapsilosis*: 1/9 (11); others: NA
Costa Rica, 2005 [13]	R, SC	1994–1998	Positive BC	110	46 (62)	98 (89)	79 (72)	40 (36)	39 (35)	37 (34)	29 (26)
US, 2005 [14]	P, MC	1998–2000	Positive BC	35	30 VLBW (86)	23 (66)	13 (37)	8 (23)	5 (14)	7 (20)	NA
Jordan, 2008 [15]	R, SC	1995–2006	Positive BC	24	13 (54)	19 (79)	10 (42)	10 (42)	0	13 (54)	4 (17)
Australia, 2009 [16]	P, MC	2001–2004	Positive BC	33	33 (94)	24 (89)	5 (19)	NA	NA	7 (22)	NA
China, 2013 [17]	R, SC	2004–2010	IFI	45	29 VLBW (64)	32 (71)	NA	9 (20)	NA	4 (9)	NA
Portugal, 2014 [18]	P, MC	2005–2010	IFI	44	37 (84)	44 (100)	14 (32)	NA	NA	5 (11)	5 (11)
England, 2014 [19]	R and P, MC	R: 2004–2009; P: 2010	IFI	84	79 (94)	71 (87)	NA	11 (13)	NA	26 (31)	18 (21)
China, 2015 [20]	R, SC	2006–2010	Positive BC	19	NA	16 (84)	NA	9 (47)	NA	3 (16)	NA
Italy, 2016 [21]	P, MC	2005–2015	ICI	14	12 LBW (85)	NA	2 (14)	2 (14)	0	NA for neonates with ICI	NA for neonates with ICI

Table 1. *Cont.*

Country and Year [Ref.]	Study Type	Study Period	Inclusion Criteria	Neonates (n)	Preterm (n, %)	Central Line (n, %)	Previous Surger (n, %)	Abdominal Surgery (n, %)	Non Abdominal Surgery (n, %)	In-Hospital Mortality (n, %)	Fungal Mortality (n, %)
US, 2018 [22]	P, MC	2008–2015	Positive BC	90	46 (78)	70 (78)	8 (9)	2 (2)	7 (7)	14 (16)	NA
Iran, 2018 [23]	P, SC	2014–2016	Positive BC	35	17 (49)	33 (94)	14 (40)	10 (29)	4 (11)	15 (43)	NA for neonates with ICI
Germany, 2018 [24]	P, MC	2009–2015	Need for antifungal treatment	724	724 (100)	652 (90)	NA	272 (38)	NA	71 (10)	NA
Taiwan, 2018 [25]	R, SC	2004–2015	ICI	113	NA	108 (96)	31 (27)	NA	NA	48 (43)	32 (28)
Turkey, 2019 [26]	R, SC	2007–2012	ICI	22	20	3 (6)	5 (17)	NA	NA	10 (46)	NA
France, 2019 [27]	P, MC	2010–2012	IFI treated with micafungin	31	29 (97)	NA	NA	4 (4)	NA	0	0

BC: blood culture. ICI: invasive Candida infection. IFI: invasive fungal infection. LBW: low birth weight. MC: multi-center study. NA: not available. P: prospective. R: retrospective. SC: single-center study. VLBW: very low birth weight.

4. Risk Factors for Invasive *Candida* Infections after Major Surgery

Infants in NICUs for surgical diseases are at high risk for IFIs, as a result of prematurity, the need for invasive procedures, the disruption of natural barriers due to surgery and other many risk factors (Figure 2) [28,29]. Bloodstream infection due to *Candida* spp. is considered the most common IFI in critically ill patients [30–33].

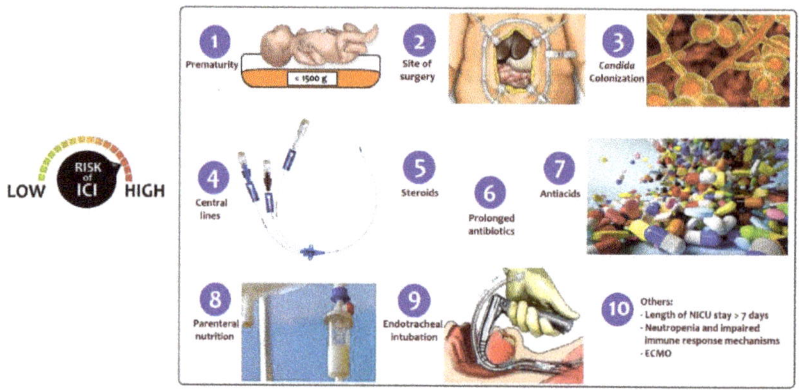

Figure 2. Risk factors for invasive *Candida* infections after major surgery.

Well-recognized risk factors associated with ICIs are:

1. **Prematurity**: Prematurity was recognized as the most common underlying condition (78%) among newborns with candidemia, with a median gestational age at birth of 25 weeks (IQR: 24–26) according to the United States' Centers for Disease Control and Prevention's (CDC's) active population-based surveillance [22]. Most preterm neonates had a very low birth weight (VLBW, 1000–1500 g) or extremely low birth weight (ELBW, <1000 g) [22]. Mortality is high in ELBW infants with ICIs: Benjamin et al. reported an overall mortality of 34% for ELBW infants with ICIs compared with 14% for ELBW infants without ICIs [34].

2. **Site of surgery:** Surgery in the 90 days before diagnosis was the most common (38%) underlying condition among infants with ICIs. The abdomen was the most common site of surgery, according to data from four US CDCs [22]. Gastrointestinal diseases, including congenital anomalies (i.e., gastroschisis, omphalocele, duodenal or ileocolic atresia/stenosis, necrotizing enterocolitis with intestinal perforation, stoma carriers in any location) predispose patients to candidemia, as a result of a compromised intestinal barrier that promotes translocation of *Candida* colonizing the gastrointestinal tract [35].

3. *Candida* **colonization**: *Candida* colonization is the most important risk factor for ICIs and is further discussed below; it can involve from 10% to 60% of preterm babies during their hospital stay in NICU [36].

4. **Use of central lines**: Despite numerous efforts in recent decades to reduce the incidence of central line associated sepsis (CLABSI) and central lines related sepsis (CRBSI) in NICUs, such infections still represent a major complication of health care assistance in those critically ill infants. Central-line-associated blood-stream infections (CLABSIs) arise from at least 48 h after CVC insertion to 48 h after CVC removal. Catheter-related blood-stream infections (CRBSIs) are bacteremias with positivity of CVC blood cultures developing at least 2 h earlier compared to peripheral blood cultures, or when the same organism is recovered from percutaneous blood culture and catheter lumen blood culture, with 3-fold greater colony count in the latter [37]. In particular, newborns undergoing major surgery in most cases have a central vascular catheter and are most susceptible to these infections. Among the germs involved in the genesis of CLABSIs, *Candida* spp. represented the third most common pathogens

(13%), after *Coagulase-Negative Staphylococci* (28%), and *Staphylococcus aureus* (19%) in a study in 304 NICUs [38]. The length of stay of indwelling catheters is a strong risk factor for CLABSI and CRBSI, while no differences have been reported between the CLABSI incidence in femoral vein catheters, peripherally inserted catheters, and umbilical venous catheters [38]. Catheter removal is recommended if a CRBSI caused by *coagulase-negative Staphylococci*, gram-negative bacilli (*Pseudomonas aeruginosa* and *Klebsiella pneumoniae*), and fungi occurs, due to the particular ability of these germs to form an intraluminal biofilm, resistant to antibiotics and/or antifungals. Biofilms on indwelling catheters may be composed of gram-positive or gram-negative bacteria or yeasts. It consists of microbial cells surrounded by a self-secreting polymer matrix, that is released into the extracellular space [39]. The matrix is composed of water, polysaccharides, proteins, lipids, and extracellular DNA. This matrix provides a protective barrier from the surrounding environment and is able to hinder the penetration of antimicrobial drugs, while also providing protection against the host's immune defense mechanisms. From this biofilm, germs are progressively released, causing the infection to persist and favoring the dissemination of microbes to additional sites in the body. The biofilm is very difficult to eradicate from the catheter, due to the difficult penetration of antimicrobial drugs into the matrix. Therefore, CVC removal is the gold standard approach in cases of CRBSI that do not respond to systemic treatment [37,40]. The best timing of central venous catheter removal in the presence of an associated and/or catheter-related *Candida* infection has been studied by many authors [40,41], which demonstrated that early catheter removal in candidemia is associated with better outcomes in terms of shorter duration of infection, reduced mortality, and reduced long-term neurologic disabilities. When catheter removal is not recommended for the patient's condition, the lock therapy with antimicrobials may be an option. This rescue therapy has shown promise as a strategy for the treatment of CRBSI due to several *Candida* species. The most promising strategies of antifungal lock therapy include the use of amphotericin B, ethanol, or echinocandins [42,43].

5. **Use of corticosteroids**: Treatment with corticosteroids is a risk factors for invasive fungal infections in the neonatal period. However, data are controversial. The addition of steroids to the antibiotic therapy in animal models increases the intestinal colonization, with an increase in the incidence of invasive infections [44]. Yu et al. reported no significant differences between neonates with ICI and their control peers reviewing medical charts of 5135 NICU admissions [17]. Length and dosage of steroid treatment may play a role in altering the risk in these infants.

6. **Use of prolonged broad-spectrum antibiotics**: Longer duration of antibiotic treatment, in particular third-generation cephalosporins, vancomycin, or carbapenems, increases the risk of ICIs [17]. One of the hypotheses for the role of cephalosporins is that their concentration within the biliary system would cause intestinal dysmicrobism, favoring the proliferation of opportunistic germs, in particular fungi. Considering the antibiotic and other drugs exposure and the risk of infection by species of *Candida*, the third generation of cephalosporins seems to be a risk factor for *Candida albicans* infection, while parenteral nutrition, lipidic emulsion, and H2 antagonists are risk factors for *Candida parapsilosis* infections [11].

7. **Use of antacids**: Inhibitors of gastric acidity such as proton-pump inhibitors (PPIs, e.g., omeprazole) are widely used to prevent and manage feeding intolerance and gastroesophageal reflux, although few data on safety and efficacy are available. However, PPIs potentially increase the risk of systemic infections and necrotizing enterocolitis (NEC), especially in preterm infants [45]. In a multicenter cohort of 743 infants, the main pathogens causing infections in infants exposed to inhibitors of gastric acidity were gram-negative-bacilli and *Candida* spp. [46].

8. **Use of parenteral nutrition**: Parenteral nutrition (PN) is often considered an ideal microbial growth medium, and lipid administration in particular poses a specific risk for microbial growth [47]. PN given without the use of appropriate filters could

contribute to potentially important extrinsic mechanism of infection in NICU patients [11]. Patients with species other than *C. albicans* were more likely to have PN than those with *C. albicans* (96.3% versus 71.4%, $p = 0.039$) [48].

9. **Endotracheal intubation and invasive devices**: Surgical and mechanical devices such as endotracheal tubes, drains, or urinary catheters may be also responsible for the nosocomial spread of pathogens. According to an epidemiologic surveillance study, two devices increased the relative risk for nosocomial infections by 2.6 times and three devices by 3.6 times [49].

10. **Others**: A length of stay in NICU >7 days was reported as one of the main potential risk factors by a multicenter IFI surveillance project (the AURORA project) [50]. Lack of, or inadequate, hand hygiene of healthcare workers has been also reported as one of the main reasons for horizontal transmission of virulent *Candida* spp. responsible for the invasive infections in critical patients, such as neonates [51]. Neutropenia, defined as neutrophil count <1500/mm^3, was found as an independent predictor of candidemia in NICUs [52,53]. Extracorporeal membrane oxygenation (ECMO) procedures and locations may contribute to acquired infection risk and the most common organisms identified were *coagulase-negative Staphylococci*, followed by *Candida*, and *Pseudomonas* species at eight children's hospitals [54].

5. *Candida* Colonization

The percentage of colonized preterm infants who develop invasive infections ranges from 5% to 30% [55], due to immunological immaturity, the immaturity of the skin and of mucous membranes, and to the need for invasive therapeutic supports. In general, all patients admitted to intensive care are exposed to fungal infections, some of them are effectively colonized, and only a minority develop systemic infections that originate from peripheral colonization. In intensive care units other than neonatal intensive care units (e.g., surgical intensive care units) the risk of colonization is greater in the presence of central venous catheters, bladder catheters, mechanical ventilation, and lack of enteral nutrition. Skin and the gastrointestinal tract are the most common sites for *Candida* colonization in preterm and term infants [35,56]. Colonization can take place either vertically, from the maternal genital tract, or horizontally, by transmission of the germ through the hands of health caregivers. The use of broad-spectrum antibiotic therapy favors *Candida* spp. colonization, even if it does not seem to condition the transition from colonization to systemic infection. Some researchers have shown that the addition of steroids to antibiotic therapy in animal models increases the intestinal colonization, with an increase in the incidence of invasive infections [44].

The frequency of colonization is inversely proportional to the neonate's gestational age and birth weight. Severely preterm infants are the most affected, experiencing invasive infections following colonization. Furthermore, the risk of invasive infections is proportional to the number of colonized body sites and their localization. More noncontiguous colonization sites are associated with a greater probability of progression to invasive infection. Therefore, preterm and full-term infants colonized in more than one body site are more likely to develop invasive *Candida* spp. infection the less contiguous the colonization sites are [57]. Colonization at two or more sites occurs similarly with *Candida albicans* and *Candida parapsilosis*, while *Candida albicans* is most frequently responsible when more than two sites are involved [58].

Isolation of *Candida* spp. from the urine of neonates can be indicative of contamination or of urinary tract infection (UTI), although any positive culture from normally sterile body fluids such as urine, peritoneal fluid, or cerebrospinal fluid is often considered as an invasive candidiasis that needs to be treated as well as candidemia [5]. To date, it is still not clear how often *Candida* UTI (defined as growth of *Candida* from urine at >10^6 CFU/L from a suprapubic aspirate or >10^7 CFU/L from a bladder catheter specimen) is a precursor of candidemia or of *Candida* infection at other sites [59]. Among 30 infants with candiduria, an active surveillance (PICNIC study) detected 4 infants who developed extra-renal dissemination of *Candida* infection. In this study, the extra-renal site was blood in 3/4 cases and the central nervous system in 1/4; involved species were *Candida albicans* (75%) and *Candida parapsilosis* (25%) [59]. Three of these infants had a congenital heart disease and were treated between candiduria and positive culture at the extra-renal site with amphotericin B, fluconazole, or both; one was a 26-weeks preterm infant. Days between positive urine culture and positive culture at the extra-renal site ranged from 2 to 41 [59].

6. Innate Defense Mechanisms against *Candida* and Surgery

Innate immunity is critical for the survival of neonates, who encounter for the first time a lot of new micro-organisms, such as *C. albicans*, which is the most common fungal pathogen found in NICUs. A wide range of genetic and epigenetic factors may influence neonatal innate immunity [60]. Dysregulation of neonatal innate immune responses increase their susceptibility to severe infections [61].

Polysaccharide structures of *C. albicans* cell wall, such as β-glucans and mannans, constitute the main pathogen-associated molecular patterns (PAMPs) involved in *Candida*–host immune system interaction [62]. In the absence of a specific antibody-mediated opsonization, that cannot be mounted by neonates, PAMPs are identified by pattern-recognition receptors (PRRs) expressed on immune cells' surfaces, as macrophage mannose receptors (MMR) and toll-like receptors (TLRs) [63]. Neonatal macrophages are capable to phagocytize *Candida* spp. using MMR, but cannot be entirely stimulated by interferon-γ (IFN-γ), considering the lack of a normal regulation of IFN-γ receptor in neonates [64].

An intact epithelium and endothelium represent important mechanical barriers against fungal invasion [62]. The formation of fungal hyphae contributes to epithelial damage and immune activation through Candidalysin, a recently discovered peptide toxin, encoded by the *ECE1* gene [65].

The intestinal mucosal barrier plays a key role in the protection against an invasion of fungal pathogens. In fact, gut cells behave not only as a physical barrier but have also an active role producing mucus and anti-microbial peptides such as β-defensins [66,67]. However, in the case of impaired barriers, *Candida*, which is usually found in the gut, may invade the intestinal epithelial barrier and translocate into the bloodstream, especially in case of abdominal surgery [67].

Furthermore, mucosal colonization by *Candida* spp. (and *C. albicans* in particular) is a major risk factor for potential life-threatening candidemia [68]. The presence of *C. albicans* stimulates the mitogen-activated protein kinase (MAPK) pathway and c-Fos activation, likely with a threshold level to activate immune response. The threshold could be pivotal in triggering an inflammatory response from a simple colonization [67].

7. Population Microdiversity and Role of Different Species of *Candida*

Candida spp. are distributed differentially according to age: *C. albicans* and *C. parapsilosis* are prevalent in neonates [69], whereas adults are mainly affected with *C. albicans* and *C. glabrata* [70]. In the largest study to date (EUROCANDY), involving 23 pediatric centres, *C. albicans* (52.5%) and *C. parapsilosis* (28%) were the predominant species, followed by *C. tropicalis*, *C. glabrata*, *C. krusei* and other rare species (including *C. dubliniensis*, *C. pulcherrima*, *C. blankii*, *C. famata*, *C. guilliermondii*, *C. lusitaniae*, *C. magnolia*, *C. orthopsilosis*, *C. zeylanoides*). *C. albicans* was prevalent among neonates (60.2%), while highest infection rates of *C. parapsilosis* were observed among infants (42%), with significantly lower prevalence in neonates (26%) [71]. Similar data were reported by a multicenter pediatric and neonatal study (involving 23 centers in the United States and 19 in 15 other countries), with 48% *C. albicans* isolates and 28% *C. parapsilosis* isolates in newborns [72]. Focusing on patients of surgical intensive care units of the EUROCANDY cohort, 72.2% episodes were due to *C. albicans* while the remaining cases were ascribed to *C. parapsilosis*. However, the number of neonates, infants, and children who underwent major surgery was not specified [71]. High-risk neonates become colonized with *Candida* spp. not only vertically during vaginal birth from their mothers, who may be receiving an azole for vaginal candidiasis, but also horizontally from colonized hospital-workers during their stay in NICU.

Although *C. albicans* remains the most common isolate in NICU, a shift to infections caused by *C. parapsilosis* and *C. tropicalis* has occurred during the last decades, and it has been associated with decreased mortality [12].

Among all ICIs, *C. albicans*, *C. parapsilosis*, *C. tropicalis*, *C. glabrata*, and *C. krusei* account for nearly 90% of isolates from blood or other sterile site cultures. Candidemia caused by other uncommon species, such as *C. guilliermondii*, and *C. lusitaniae*, is less well-known. It seems, though, to have a poorer response to antifungal treatment (frequently due to antifungal minimal inhibitory concentration -MIC- above the epidemiologic cut-off value) and a longer duration of candidemia [25].

Whereas specific inflammatory and tissue-destructive histopathologic features were found in most neonatal *C. albicans* cases, the mechanisms underlying cases of species other than *C. albicans* are still poorly understood. According to autopsy-based data, species other than *C. albicans* could involve both the gastrointestinal tract and pulmonary airways and their incidence is often underrated [73].

8. Epidemiology of Fungal Infections in NICUs

Although there is an inter-site variability in the incidence of candidemia [22,71], prevention of ICIs should be an achievable evidence-based goal for every NICU [74]. NICU and PICU admissions were considered as significant predictors for mortality, with an odds ratio of 4.67 and 8.325, respectively, in the EUROCANDY cohort [71]. However, most data involve extremely preterm infants.

In specific subgroups of patients (e.g., abdominal surgical patients), ICIs are also frequent [30–33,75], but to date there are no large epidemiological studies on the incidence of ICIs in neonates who have undergone major surgery. *Candida* spp., within four weeks from admission in intensive care units, colonize the skin and mucous membranes of about 64% of critically ill neonates and can progress to invasive infection [76].

ICIs are a major cause of morbidity and mortality among critically ill patients [31,77,78] and impose an important economic burden mainly due to prolonged ICU stay, cost of antifungal drugs, and overall use of hospital resources [79,80].

In case of nosocomial ICI outbreaks, a cluster of infections could be defined when at least two cases of severe neonatal infection (i.e., bloodstream infection) occur within a defined time interval in one center with the same pathogen species in different patients: *Candida albicans* is one of the most frequently occurring microorganisms, according to a recent German surveillance system [81].

Therefore, a nosocomial ICI outbreak within a NICU could have important clinical and economic repercussions. A contaminated environment has been identified as a possible source of the outbreak: the colonized locations included wiping cloths, faucets, sinks, an operating table, puddles in the bathroom, a ventilator, and an ultrasonic probe in a recent outbreak of *Candida parapsilosis* fungemia in a Chinese hospital [82]. An emergency plan should be promptly scheduled with environmental surveillance and comprehensive interventions, such as hand hygiene and disinfection techniques. Improving both disinfection and isolation, as well as interrupting the pathway of transmission, resulted to be the key to controlling the spread of infection [83].

New methods (such as fingerprinting analysis of *Candida* isolates) can help to identify the identical strains, to investigate suspected outbreaks and to help therapeutic decision-making [84].

9. Prophylaxis of Fungal Infections

The high-risk population of critically ill neonates benefits greatly from prompt, effective treatment and prophylactic measures. A prompt antifungal treatment is one of the most important determinants for mortality reduction. In addition, antifungal prophylaxis given to critically ill patients at high risk for ICIs may have a positive impact on patients' outcomes, given ICIs' high morbidity and mortality rates [85,86].

Fluconazole prophylaxis has been proven to be safe and effective in neonates, reducing ICIs by more than 80% and *Candida*-related mortality by 90%, especially in high-risk preterm infants, without significant side effects or emergence of resistant fungal species [75]. Considering its long half-life plasma elimination, which allows an intermittent administration schedule, fluconazole should be administered at 3 mg/kg once a day, two times a week in the first two weeks of life whereas, from the third week of life, prophylaxis should be administered every other day [4]. The benefits of prophylaxis are less clear when incidence of ICIs is lower than 2%, and the administration should be discussed case by case, in relation to the presence of risk factors for ICIs.

There is currently clear evidence on the efficacy of fluconazole prophylaxis in the prevention of ICIs in preterm infants [87–91], but not in surgical newborns. In these neonates, fluconazole prophylaxis is not clearly suggested, although they are considered at risk of ICI as explained above. A major concern regarding a larger prophylactic use of antifungal agents, even in term infants with risk factors, is the emergence of resistant species. However, resistance to fluconazole or echinocandins in newborns is reported as rare: fluconazole-resistant *C. albicans* was seen among 1.6% of the isolates, while no echinocandins-resistant *C. albicans* was observed [23].

10. Diagnosis of Invasive *Candida* Infections

Diagnosis of ICIs is very difficult in newborns, as clinical signs and symptoms of ICIs can be nonspecific and often subtle. For this clinical, radiological, and mycological evaluations should be carried out simultaneously. In addition to microbiological cultures (blood, urine, cerebrospinal fluid, peritoneal fluid, tracheal aspirate), laboratory techniques for diagnosing ICIs also include the direct microscopic examination, the histological examination of the involved tissues, the evaluation of fungal antibodies and fungal antigens (galactomannan, 1,3-beta-D-glucan) by enzyme-linked immunosorbent assay (ELISA) or by immunofluorescence, and the detection of fungal DNA by polymerase chain reaction (PCR) in blood and/or other biologic fluids. While fungi grow readily in culture media, their identification requires large volumes of blood, which are difficult to collect, especially in preterm infants. Therefore, blood cultures can be negative in a large number of patients with fungal sepsis. In addition, up to 50% of infants with positive cerebrospinal fluid (CSF) for *Candida albicans* or *Candida parapsilosis* may have negative blood cultures within seven days. This explains the complexity of diagnosing ICIs in the newborn and the need for a prompt empirical therapy at the time of diagnostic suspicion [92].

In particular, two new diagnostic molecular tools seem to be particularly promising to early diagnose ICIs, especially in the cases where a previous antifungal prophylaxis or empirical therapy could have reduced the possibility of a positive blood culture:

(a) the T2 Magnetic Resonance *Candida* Panel (T2 *Candida*, T2 Biosystems, Lexington, MA, USA) can detect five major *Candida* species (*C. albicans/C. tropicalis*, *C. parapsilosis*, and *C. krusei/C. glabrata*) directly in blood and it does not require viable organisms, with a lower time to positivity (lower than 3 h) [93,94]. T2 *Candida* can be used to efficiently diagnose or rule out candidemia even using low-volume blood specimens from pediatric patients: this could result in improved time to appropriate antifungal therapy or reduction in unnecessary empirical antifungal therapy [95].

(b) the indirect immunofluorescence assay (IFA) for *C. albicans* germ tube antibody (CAGTA) IgG is a method that enables the detection of specific IgG antibodies against antigens located on the cell wall surface of the mycelium of *Candida* spp. in human serum/plasma. Vircell Kit (Granada, Spain) and VirClia IgG Monotest (Granada, Spain) are the routine detection ways with widespread use in Europe. According to a systematic review, the diagnostic accuracy of the CAGTA assays is moderate for ICIs, and CAGTA findings should be interpreted in parallel with other biomarkers [96].

However, to best of our knowledge, there are still no studies in literature that evaluated the performance of these tests in the neonatal age only.

11. Treatment of Invasive *Candida* Infections

Mortality associated with *Candida* sepsis involves about half of infants, while survivors could develop severe long-term neurosensory impairment, including ocular, hearing, and cognitive impairment, cerebral palsy, and periventricular leukomalacia. Initial therapy is therefore often empirical, and the combination of prematurity, thrombocytopenia, and prolonged use of broad-spectrum antibiotics generally guides the initiation of empiric antifungal therapy [5].

Current Infectious Diseases Society of America guidelines recommend amphotericin B deoxycholate and fluconazole first-line therapies in infants with IC [44], while European guidelines recommend formulations of amphotericin B, fluconazole, and micafungin. Amphotericin B exists in various formulations, amphotericin B deoxycholate (D-AMPH-B), and liposomal amphotericin B (L-AMPH-B) [7]. The recommended dose for D-AMPH-B starts from 0.5–0.7 mg/kg/day up to 1.5 mg/kg/day. The recommended dose for L-AMPH-B is 3–5 mg/kg/day [97,98].

In neonates, the dose of fluconazole administered as therapy is 12 mg/kg/day regardless of birth weight or gestational age. The measurement of the blood levels reached (therapeutic drug monitoring) could help in establishing the drug concentrations actually reached during therapy. In fact, for many antifungal drugs, changes in clearance associated with changes in birth weight and gestational age of newborns, especially preterm, have been observed [99]. In fact, in full-term infants, the plasma half-life of fluconazole is approximately 70 h (30 h in adults) while in premature infants it is 73 h at birth, 53 h at 6 days of age, and 46 h at 12 days of age. These pharmacokinetic characteristics make fluconazole a more attractive candidate for the prevention of ICI, mainly in premature infants, allowing for infrequent administration [100].

Although L-AMPH-B and D-AMPH-B are the most commonly used antifungal drugs in newborns, there are no prospective randomized neonatal studies that provide reliable information on the pharmacokinetic properties of these drugs and their safety.

All of these antifungals have unsatisfactory levels of evidence to support their use in neonates and, when used in this special patient population, they have limits ranging from renal and bone marrow toxicity to uncertain optimal dosage regimens, increased resistance of some *Candida* spp. and, finally, suboptimal spread to the kidneys or brain tissue. There is a need for alternative antifungal drugs with greater specificity and reduced toxicity in neonatal populations than those commonly used in the treatment of invasive neonatal infections.

Echinocandins could have a prominent role in contexts where there is a wide use of prophylaxis with fluconazole and resistance of *Candida* strains to azoles could emerge. Pharmacokinetic studies demonstrated excellent tolerability, safety, and efficacy of echinocandins in neonates. Furthermore, with their ability to target 1,3-beta-D-glucan synthesis as a means of inhibiting excess production of extracellular matrix, echinocandins represent an attractive therapy against *Candida* biofilms formation [101]. The in vitro efficacy of echinocandins in treating catheter biofilms was confirmed by Cateau et al., who found that lock solutions of 2 and 5 mg/L, respectively, of caspofungin and micafungin used to treat biofilms forming on a silicone catheter led to a significant and persistent reduction of yeast metabolic activity of intermediate and mature biofilms [102]. Additionally, biofilm impairment mediated by echinocandins would trigger a larger proinflammatory response from phagocytes, due to an increase in 1,3-beta-D-glucan exposure [103].

Some problems could emerge in the therapy against *Candida parapsilosis*. Echinocandins have in fact a high minimum inhibitory concentration against *Candida parapsilosis*. Despite this awareness, no clinical failures have been reported to date. Consequently, the resistance of *Candida parapsilosis* to echinocandins remains unexplored. Micafungin is the echinocandin with the more reliable evidence in neonatal population. It is the only echinocandin approved for neonates and young infants. The therapy at 8 mg/kg/day achieved a high response in a phase 2 study on 35 neonates with medical and surgical underlying diseases and confirmed or suspected ICIs [8].

12. Future Research Considerations

Prospective studies are needed to determine the clinical implications of new diagnostic molecular tools (T2MR and CAGTA) in neonatal age and their potential use in antimicrobial stewardship.

Empiric antifungal therapy needs further evidence sustaining the efficacy in reducing mortality and long-term neurodevelopmental disabilities in preterm infants and other categories of patients. Neonatal pharmacokinetics and pharmacodynamics data of the most-used antifungal drugs are still inconclusive, due to the complexity of carrying out this type of studies in the neonatal age. Furthermore, the clearing time of fungal infection in neonates and the microbiological criteria used to define clearance are currently ambiguous.

Concerning neonates with major surgical needs, admitted in NICUs, there is lack of a precise assessment of the incidence of fungal colonization and invasive infections and lack of evidence that may, or may not, support the benefits of antifungal prophylaxis. As of 9 January 2021, no trials on ICIs are enrolling infants after major surgery, according to clinical trial registries such as: https://clinicaltrials.gov (accessed on 9 March 2021) and https://www.umin.ac.jp/ctr (accessed on 9 March 2021).

Therefore, we are currently recruiting study subjects in a multicenter prospective observational study to assess the real incidence of ICIs in surgical neonates and infants up to three months of life in NICUs. The study involves 13 of the major Italian NICUs and it is coordinated by our group at Bambino Gesù Children's Hospital (Rome, Italy). The primary outcome of the study is to assess the real incidence and risk factors of ICIs in neonates and infants up to three months of life requiring major surgery. We hope to provide the results of this research as soon as possible.

13. Conclusions

Infants requiring surgery carry many risk factors for candidemia and are likely to benefit from antifungal prophylaxis. To date, guidelines for the prevention of ICIs recommend intravenous or oral fluconazole prophylaxis in ELBW infants, while no specific recommendation is available for infants requiring major surgery. This finding should not be extrapolated from previous studies, and further epidemiologic data are needed to identify possible preventive strategies against candidemia in preterm and term infants who undergo major surgery.

Author Contributions: Conceptualization, D.U.D.R. and C.A.; methodology, D.U.D.R. and C.A.; formal analysis, D.U.D.R., A.S., M.P.R., L.M. and C.A.; investigation and interpretation, D.U.D.R., A.S., M.P.R., L.M. and C.A.; data curation, D.U.D.R., A.S., M.P.R., L.M. and C.A.; writing—original draft preparation, D.U.D.R., A.S., M.P.R. and L.M.; writing—review and editing, L.S., P.B. (Pasqua Betta), M.M., F.C., I.C., L.P., E.B., C.T., M.G., J.B., B.D.T., G.N., D.M., P.M., A.D., P.B. (Pietro Bagolan) and C.A.; supervision, C.A.; project administration, C.A. All authors have read and agreed to the published version of the manuscript.

Funding: This research received no external funding.

Conflicts of Interest: The authors declare no conflict of interest.

References

1. Hoyer, L.L.; Cota, E. Candida albicans Agglutinin-Like Sequence (Als) Family Vignettes: A Review of Als Protein Structure and Function. *Front. Microbiol.* **2016**, *7*, 280. [CrossRef]
2. Filler, S.G.; Sheppard, D.C. Fungal invasion of normally non-phagocytic host cells. *PLoS Pathog.* **2006**, *2*, e129. [CrossRef]
3. Staniszewska, M. Virulence Factors in Candida species. *Curr. Protein Pept. Sci.* **2020**, *21*, 313–323. [CrossRef] [PubMed]
4. Hsieh, E.; Smith, P.B.; Jacqz-Aigrain, E.; Kaguelidou, F.; Cohen-Wolkowiez, M.; Manzoni, P.; Benjamin, D.K., Jr. Neonatal fungal infections: When to treat? *Early Hum. Dev.* **2012**, *88*, S6–S10. [CrossRef]

5. Benjamin, D.K.; Stoll, B.J.; Gantz, M.G.; Walsh, M.C.; Sánchez, P.J.; Das, A.; Shankaran, S.; Higgins, R.D.; Auten, K.J.; Miller, N.A.; et al. Eunice Kennedy Shriver National Institute of Child Health and Human Development Neonatal Research Network. Neonatal candidiasis: Epidemiology, risk factors, and clinical judgment. *Pediatrics* **2010**, *126*, 1–18. [CrossRef]

6. Hope, W.W.; Castagnola, E.; Groll, A.H.; Roilides, E.; Akova, M.; Arendrup, M.C.; Arikan-Akdagli, S.; Bassetti, M.; Bille, J.; Cornely, O.A.; et al. ESCMID* guideline for the diagnosis and management of Candida diseases 2012: Prevention and management of invasive infections in neonates and children caused by *Candida* spp. *Clin. Microbiol. Infect.* **2012**, *18*, 38–52. [CrossRef]

7. Autmizguine, J.; Smith, P.B.; Prather, K.; Bendel, C.; Natarajan, G.; Bidegain, M.; Kaufman, D.A.; Burchfield, D.J.; Ross, A.S.; Pandit, P.; et al. Fluconazole Prophylaxis Study Team. Effect of fluconazole prophylaxis on Candida fluconazole susceptibility in premature infants. *J. Antimicrob. Chemother.* **2018**, *73*, 3482–3487. [CrossRef] [PubMed]

8. Auriti, C.; Goffredo, B.M.; Ronchetti, M.P.; Piersigilli, F.; Cairoli, S.; Bersani, I.; Dotta, A.; Bagolan, P.; Pai, M.P. High-dose micafungin in neonates and young infants with invasive candidiasis: Results of a phase 2 study. *Antimicrob. Agents Chemother.* **2021**, AAC.02494-20. [CrossRef]

9. Chapman, R.L.; Faix, R.G. Persistently positive cultures and outcome in invasive neonatal candidiasis. *Pediatric Infect. Dis. J.* **2000**, *19*, 822–827. [CrossRef]

10. Mokaddas, E.; Ramadan, S.; Abo el Maaty, S.; Sanyal, S. Candidemia in Pediatric Surgery Patients. *J. Chemother.* **2000**, *12*, 332–338. [CrossRef]

11. Saiman, L.; Ludington, E.; Pfaller, M.; Rangel-Frausto, S.; Wiblin, R.T.; Dawson, J.; Blumberg, H.M.; Patterson, J.E.; Rinaldi, M.; Edwards, J.E.; et al. Risk factors for candidemia in Neonatal Intensive Care Unit patients. *Pediatric Infect. Dis. J.* **2000**, *19*, 319–324. [CrossRef]

12. Roilides, E.; Farmaki, E.; Evdoridou, J.; Dotis, J.; Hatziioannidis, E.; Tsivitanidou, M.; Bibashi, E.; Filioti, I.; Sofianou, D.; Gil-Lamaignere, C.; et al. Neonatal candidiasis: Analysis of epidemiology, drug susceptibility, and molecular typing of causative isolates. *Eur. J. Clin. Microbiol. Infect. Dis.* **2004**, *23*, 745–750. [CrossRef]

13. Avila-Aguero, M.L.; Canas-Coto, A.; Ulloa-Gutierrez, R.; Caro, M.A.; Alfaro, B.; Paris, M.M. Risk factors for Candida infections in a neonatal intensive care unit in Costa Rica. *Int. J. Infect. Dis.* **2005**, *9*, 90–95. [CrossRef]

14. Shetty, S.S.; Harrison, L.H.; Hajjeh, R.A.; Taylor, T.; Mirza, S.A.; Schmidt, A.B.; Sanza, L.T.; Shutt, K.A.; Fridkin, S.K. Determining Risk Factors for Candidemia Among Newborn Infants from Population-Based Surveillance: Baltimore, Maryland, 1998–2000. *Pediatric Infect. Dis. J.* **2005**, *24*, 601–604. [CrossRef] [PubMed]

15. Badran, E.F.; Al Baramki, J.H.; Al Shamyleh, A.; Shehabi, A.; Khuri-Bulos, N. Epidemiology and clinical outcome of candidaemia among Jordanian newborns over a 10-year period. *Scand. J. Infect. Dis.* **2008**, *40*, 139–144. [CrossRef] [PubMed]

16. Blyth, C.C.; Chen, S.C.; Slavin, M.A.; Serena, C.; Nguyen, Q.; Marriott, D.; Ellis, D.; Meyer, W.; Sorrell, T.C. Australian Candidemia Study. Not Just Little Adults: Candidemia Epidemiology, Molecular Characterization, and Antifungal Susceptibility in Neonatal and Pediatric Patients. *Pediatrics* **2009**, *123*, 1360–1368. [CrossRef] [PubMed]

17. Yu, Y.; Du, L.; Yuan, T.; Zheng, J.; Chen, A.; Chen, L.; Shi, L. Risk factors and clinical analysis for invasive fungal infection in neonatal intensive care unit patients. *Am. J. Perinatol.* **2013**, *30*, 589–594. [CrossRef] [PubMed]

18. Baptista, M.I.; Nona, J.; Ferreira, M.; Sampaio, I.; Abrantes, M.; Tomé, M.T.; Neto, M.T.; Barroso, R.; Serelha, M.; Virella, D. Invasive fungal infection in neonatal intensive care units: A multicenter survey. *J. Chemother.* **2016**, *28*, 37–43. [CrossRef]

19. Oeser, C.; Vergnano, S.; Naidoo, R.; Anthony, M.; Chang, J.; Chow, P.; Clarke, P.; Embleton, N.; Kennea, N.; Pattnayak, S.; et al. Neonatal Infection Surveillance Network (neonIN). Neonatal invasive fungal infection in England 2004–2010. *Clin. Microbiol. Infect.* **2014**, *20*, 936–941. [CrossRef] [PubMed]

20. Liu, M.; Huang, S.; Guo, L.; Li, H.; Wang, F.; Zhang, Q.I.; Song, G. Clinical features and risk factors for blood stream infections of Candida in neonates. *Exp. Ther. Med.* **2015**, *10*, 1139–1144. [CrossRef] [PubMed]

21. Mesini, A.; Bandettini, R.; Caviglia, I.; Fioredda, F.; Amoroso, L.; Faraci, M.; Mattioli, G.; Piaggio, G.; Risso, F.M.; Moscatelli, A.; et al. Candida infections in paediatrics: Results from a prospective centre study in a tertiary care children' s hospital. *Mycoses* **2016**, 1–6. [CrossRef]

22. Benedict, K.; Roy, M.; Kabbani, S.; Anderson, E.J.; Farley, M.M.; Harb, S.; Harrison, L.H.; Bonner, L.; Wadu, V.L.; Marceaux, K.; et al. Neonatal and Pediatric Candidemia: Results from Population-Based Active Laboratory Surveillance in Four US Locations, 2009–2015. *J. Pediatric Infect. Dis. Soc.* **2018**, *7*, e78–e85. [CrossRef]

23. Charsizadeh, A.; Mirhendi, H.; Nikmanesh, B.; Eshaghi, H.; Makimura, K. Microbial epidemiology of candidaemia in neonatal and paediatric intensive care units at the Children's Medical Center, Tehran. *Mycoses* **2018**, *61*, 22–29. [CrossRef]

24. Fortmann, I.; Hartz, A.; Paul, P.; Pulzer, F.; Müller, A.; Böttger, R.; Proquitté, H.; Dawczynski, K.; Simon, A.; Rupp, J.; et al. German Neonatal Network. Antifungal Treatment and Outcome in Very Low Birth Weight Infants: A Population-based Observational Study of the German Neonatal Network. *Pediatric Infect. Dis. J.* **2018**, *37*, 1165–1171. [CrossRef]

25. Tsai, M.H.; Hsu, J.F.; Yang, L.Y.; Pan, Y.B.; Lai, M.Y.; Chu, S.M.; Huang, H.R.; Chiang, M.C.; Fu, R.H.; Lu, J.J. Candidemia due to uncommon Candida species in children: New threat and impacts on outcomes. *Sci. Rep.* **2018**, *8*, 1–10. [CrossRef] [PubMed]

26. Öncü, B.; Belet, N.; Emecen, A.N.; Birinci, A. Health care-associated invasive Candida infections in children. *Med. Mycol.* **2019**, *57*, 929–936. [CrossRef] [PubMed]

27. Leverger, G.; Timsit, J.F.; Milpied, N.; Gachot, B. Use of Micafungin for the Prevention and Treatment of Invasive Fungal Infections in Everyday Pediatric Care in France: Results of the MYRIADE Study. *Pediatric Infect. Dis. J.* **2019**, *38*, 716–721. [CrossRef]

28. Ostrosky-Zeichner, L.; Sable, C.; Sobel, J.; Alexander, B.D.; Donowitz, G.; Kan, V.; Kauffman, C.A.; Kett, D.; Larsen, R.A.; Morrison, V.; et al. Multicenter retrospective development and validation of a clinical prediction rule for nosocomial invasive candidiasis in the intensive care setting. *Eur. J. Clin. Microbiol. Infect. Dis.* **2007**, *26*, 271–276. [CrossRef]

29. Hermsen, E.D.; Zapapas, M.K.; Maiefski, M.; Rupp, M.E.; Freifeld, A.G.; Kalil, A.C. Validation and comparison of clinical prediction rules for invasive candidiasis in intensive care unit patients: A matched case-control study. *Crit. Care* **2011**, *15*, R198. [CrossRef]

30. Bassetti, M.; Righi, E.; Ansaldi, F.; Merelli, M.; Scarparo, C.; Antonelli, M.; Garnacho-Montero, J.; Diaz-Martin, A.; Palacios-Garcia, I.; Luzzati, R.; et al. A multicenter multinational study of abdominal candidiasis: Epidemiology, outcomes and predictors of mortality. *Intensive Care Med.* **2015**, *4*, 1601–1610. [CrossRef]

31. Kett, D.H.; Azoulay, E.; Echeverria, P.M.; Vincent, J.L. The EPIC II Group of Investigators. Candida bloodstream infections in intensive care units: Analysis of the extended prevalence of infection in intesive care unit study. *Crit. Care Med.* **2011**, *39*, 665–670. [CrossRef] [PubMed]

32. Kullberg, B.J.; Arendrup, M.C. Invasive Candidiasis. *N. Engl. J. Med.* **2015**, *373*, 1445–1456. [CrossRef] [PubMed]

33. Vincent, J.L.; Rello, J.; Marshall, J.; Silva, E.; Anzueto, A.; Martin, C.D.; Moreno, R.; Lipman, J.; Gomersall, C.; Sakr, Y.; et al. EPIC II Group of Investigators. International Study of the Prevalence and Outcomes of Infection in Intensive Care Units. *JAMA* **2009**, *302*, 2323–2329. [CrossRef] [PubMed]

34. Autmizguine, J.; Hornik, C.P.; Benjamin, D.K., Jr.; Brouwer, K.L.; Hupp, S.R.; Cohen-Wolkowiez, M.; Watt, K.M. Pharmacokinetics and Safety of Micafungin in Infants Supported with Extracorporeal Membrane Oxygenation. *Pediatric Infect. Dis. J.* **2016**, *35*, 1204–1210. [CrossRef]

35. Yan, L.; Yang, C.; Tang, J. Disruption of the intestinal mucosal barrier in Candida albicans infections. *Microbiol. Res.* **2013**, *168*, 389–395. [CrossRef]

36. Manzoni, P.; Mostert, M.; Jacqz-Aigrain, E.; Stronati, M.; Farina, D. Candida colonization in the nursery. *J. Pediatric (Rio J.).* **2012**, *88*, 187–190. [CrossRef]

37. Mermel, L.A.; Allon, M.; Bouza, E.; Craven, D.E.; Flynn, P.; O'Grady, N.P.; Raad, I.I.; Rijnders, B.J.; Sherertz, R.J.; Warren, D.K. Clinical Practice Guidelines for the Diagnosis and Management of Intravascular Catheter-Related Infection: 2009 Update by the Infectious Diseases Society of America. *Clin. Infect. Dis.* **2009**, *49*, 1–45. [CrossRef]

38. Dubbink-Verheij, G.H.; Bekker, V.; Pelsma, I.C.M.; van Zwet, E.W.; Smits-Wintjens, V.E.H.J.; Steggerda, S.J.; Te Pas, A.J.; Lopriore, E. Bloodstream infection incidence of Different central Venous Catheters in Neonates: A Descriptive Cohort Study. *Front. Pediatrics* **2017**, *5*, 1–7. [CrossRef]

39. Gominet, M.; Compain, F.; Beloin, C.; Lebeaux, D. Central venous catheters and biofilms: Where do we stand in 2017? *APMIS* **2017**, *125*, 365–375. [CrossRef]

40. Karlowicz, M.G.; Hashimoto, L.N.; Kelly, R.E.; Buescher, E.S. Should Central Venous Catheters Be Removed as Soon as Candidemia Is Detected in Neonates? *Pediatrics* **2000**, *106*, E63. [CrossRef]

41. Benjamin, D.K.; Stoll, B.J.; Fanaroff, A.A.; McDonald, S.A.; Oh, W.; Higgins, R.D.; Duara, S.; Poole, K.; Laptook, A.; Goldberg, R.; et al. Neonatal Candidiasis Among Extremely Low Birth Weight Infants: Risk Factors, Mortality Rates, and Neurodevelopmental Outcomes at 18 to 22 Months. *Pediatrics* **2006**, *117*, 84–92. [CrossRef] [PubMed]

42. Taylor, J.E.; Tan, K.; Lai, N.M.; McDonald, S.J. Antibiotic lock for the prevention of catheter-related infection in neonates. *Cochrane Database Syst. Rev.* **2015**, *6*. [CrossRef]

43. Pappas, P.G.; Kauffman, C.A.; Andes, D.R.; Clancy, C.J.; Marr, K.A.; Ostrosky-Zeichner, L.; Reboli, A.C.; Schuster, M.G.; Vazquez, J.A.; Walsh, T.J.; et al. Clinical Practice Guideline for the Management of Candidiasis: 2016 Update by the Infectious Diseases Society of America. *Clin. Infect. Dis.* **2016**, *62*, e1–e50. [CrossRef]

44. Bendel, C.M.; Wiesner, S.M.; Garni, R.M.; Cebelinski, E.; Wells, C.L. Cecal Colonization and Systemic Spread of Candida albicans in Mice Treated with Antibiotics and Dexamethasone. *Pediatric Res.* **2002**, *51*, 290–295. [CrossRef]

45. Terrin, G.; Passariello, A.; De Curtis, M.; Manguso, F.; Salvia, G.; Lega, L.; Messina, F.; Paludetto, R.; Berni Canani, R. Ranitidine is associated with infections, necrotizing enterocolitis, and fatal outcome in newborns. *Pediatrics* **2012**, *129*, e40–e45. [CrossRef]

46. Manzoni, P.; García Sánchez, R.; Meyer, M.; Stolfi, I.; Pugni, L.; Messner, H.; Cattani, S.; Betta, P.M.; Memo, L.; Decembrino, L.; et al. Exposure to Gastric Acid Inhibitors Increases the Risk of Infection in Preterm Very Low Birth Weight Infants but Concomitant Administration of Lactoferrin Counteracts This Effect. *J. Pediatrics* **2018**, *193*, 62–67.e1. [CrossRef]

47. Austin, P.D.; Hand, K.S.; Elia, M. Systematic review and meta-analyses of the effect of lipid emulsion on microbial growth in parenteral nutrition. *J. Hosp. Infect.* **2016**, *94*, 307–319. [CrossRef]

48. Caggiano, G.; Lovero, G.; De Giglio, O.; Barbuti, G.; Montagna, O.; Laforgia, N.; Montagna, M.T. Candidemia in the Neonatal Intensive Care Unit: A Retrospective, Observational Survey and Analysis of Literature Data. *Biomed. Res. Int.* **2017**, *2017*, 7901763. [CrossRef]

49. Becerra, M.R.; Tantaleán, J.A.; Suárez, V.J.; Alvarado, M.C.; Candela, J.L.; Urcia, F.C. Epidemiologic surveillance of nosocomial infections in a Pediatric Intensive Care Unit of a developing country. *BMC Pediatrics* **2010**, *10*, 66. [CrossRef]

50. Montagna, M.T.; Lovero, G.; De Giglio, O.; Iatta, R.; Caggiano, G.; Montagna, O.; Laforgia, N. AURORA Project Group. Invasive fungal infections in Neonatal Intensive Care Units of Southern Italy: A multicentre regional active surveillance (Aurora Project). *J. Prev. Med. Hyg.* **2010**, *51*, 125–130. [CrossRef]

51. De Paula Menezes, R.; Silva, F.F.; Melo, S.; Alves, P.; Brito, M.O.; de Souza Bessa, M.A.; Amante Penatti, M.P.; Pedroso, R.S.; Abdallah, V.; Röder, D. Characterization of Candida species isolated from the hands of the healthcare workers in the neonatal intensive care unit. *Med. Mycol.* **2018**, *57*, 588–594. [CrossRef]

52. Mahieu, L.M.; Van Gasse, N.; Wildemeersch, D.; Jansens, H.; Ieven, M. Number of sites of perinatal Candida colonization and neutropenia are associated with nosocomial candidemia in the neonatal intensive care unit patient. *Pediatric Crit. Care Med.* **2010**, *11*, 240–245. [CrossRef]

53. Ramy, N.; Hashim, M.; Abou Hussein, H.; Sawires, H.; Gaafar, M.; El Maghraby, A. Role of early onset neutropenia in development of candidemia in premature infants. *J. Trop. Pediatrics* **2018**, *64*, 51–59. [CrossRef]

54. Cashen, K.; Reeder, R.; Dalton, H.J.; Berg, R.A.; Shanley, T.P.; Newth, C.; Pollack, M.M.; Wessel, D.; Carcillo, J.; Harrison, R.; et al. Acquired infection during neonatal and pediatric extracorporeal membrane oxygenation. *Perfusion* **2018**, *33*, 472–482. [CrossRef]

55. Manzoni, P.; Farina, D.; Leonessa, M.; Antonielli d'Oulx, E.; Galletto, P.; Mostert, M.; Miniero, R.; Gomirato, G. Risk Factors for Progression to Invasive Fungal Infection in Preterm Neonates With Fungal Colonization. *Pediatrics* **2006**, *118*, 2359–2364. [CrossRef]

56. Kühbacher, A.; Burger-Kentischer, A.; Rupp, S. Interaction of Candida Species with the Skin. *Microorganisms* **2017**, *5*, 32. [CrossRef]

57. Manzoni, P.; Farina, D.; Leonessa, M.; Priolo, C.; Arisio, R.; Gomirato, G. Type and number of sites colonized by fungi and risk of progression to invasive fungal infection in preterm neonates in neonatal intensive care unit. *J. Perinat. Med.* **2007**, *35*, 220–226. [CrossRef]

58. Kaufman, D.A.; Gurka, M.J.; Hazen, K.C.; Boyle, R.; Robinson, M.; Grossman, L.B. Patterns of Fungal Colonization in Preterm Infants Weighing Less Than 1000 Grams at Birth. *Pediatric Infect. Dis. J.* **2006**, *25*, 733–737. [CrossRef]

59. Robinson, J.L.; Davies, H.D.; Barton, M.; O'Brien, K.; Simpson, K.; Asztalos, E.; Synnes, A.; Rubin, E.; Le Saux, N.; Hui, C.; et al. Characteristics and outcome of infants with candiduria in neonatal intensive care—A Paediatric Investigators Collaborative Network on Infections in Canada (PICNIC) study. *BMC Infect. Dis.* **2009**, *9*, 1–9. [CrossRef]

60. Yu, J.C.; Khodadadi, H.; Malik, A.; Davidson, B.; Salles, É.; Bhatia, J.; Hale, V.L.; Baban, B. Innate Immunity of Neonates and Infants. *Front. Immunol.* **2018**, *9*, 1–12. [CrossRef] [PubMed]

61. Tsafaras, G.P.; Ntontsi, P.; Xanthou, G. Advantages and Limitations of the Neonatal Immune System. *Front. Pediatrics* **2020**, *8*, 1–10. [CrossRef]

62. Netea, M.G.; Joosten, L.A.; van der Meer, J.W.; Kullberg, B.J.; van de Veerdonk, F.L. Immune defence against Candida fungal infections. *Nat. Rev. Immunol.* **2015**, *15*, 630–642. [CrossRef] [PubMed]

63. Maródi, L.; Johnston, R.B. Invasive Candida species disease in infants and children: Occurrence, risk factors, management, and innate host defense mechanisms. *Curr. Opin. Pediatrics* **2007**, *19*, 693–697. [CrossRef]

64. Maródi, L. Innate cellular immune responses in newborns. *Clin. Immunol.* **2006**, *118*, 137–144. [CrossRef] [PubMed]

65. Naglik, J.R.; König, A.; Hube, B.; Gaffen, S.L. Candida albicans—Epithelial interactions and induction of mucosal innate immunity. *Curr. Opin. Microbiol.* **2017**, *40*, 104–112. [CrossRef]

66. Naglik, J.R.; Gaffen, S.L.; Hube, B. Candidalysin: Discovery and function in Candida albicans infections. *Curr. Opin. Microbiol.* **2019**, *52*, 100–109. [CrossRef]

67. Tong, Y.; Tang, J. Candida albicans infection and intestinal immunity. *Microbiol. Res.* **2017**, *198*, 27–35. [CrossRef]

68. Richardson, J.P.; Moyes, D.L.; Ho, J.; Naglik, J.R. Candida innate immunity at the mucosa. *Semin. Cell Dev. Biol.* **2018**, *89*, 58–70. [CrossRef]

69. Kaufman, D.; Fairchild, K.D. Clinical microbiology of bacterial and fungal sepsis in very-low-birth-weight infants. *Clin. Microbiol. Rev.* **2004**, *17*, 638–680. [CrossRef]

70. Cleveland, A.A.; Farley, M.M.; Harrison, L.H.; Stein, B.; Hollick, R.; Lockhart, S.R.; Magill, S.S.; Derado, G.; Park, B.J.; Chiller, T.M. Changes in Incidence and Antifungal Drug Resistance in Candidemia: Results from Population-Based Laboratory Surveillance in Atlanta and Baltimore, 2008–2011. *Clin. Infect. Dis.* **2012**, *55*, 1352–1361. [CrossRef]

71. Warris, A.; Pana, Z.D.; Oletto, A.; Lundin, R.; Castagnola, E.; Lehrnbecher, T.; Groll, A.H.; Roilides, E.; Andersen, C.T.; Arendrup, M.C.; et al. EUROCANDY study group. Etiology and Outcome of Candidemia in Neonates and Children in Europe: An 11-year Multinational Retrospective Study. *Pediatric Infect. Dis J.* **2020**, *39*, 114–120. [CrossRef]

72. Steinbach, W.J.; Roilides, E.; Berman, D.; Hoffman, J.A.; Groll, A.H.; Bin-Hussain, I.; Palazzi, D.L.; Castagnola, E.; Halasa, N.; Velegraki, A.; et al. International Pediatric Fungal Network. Results from a prospective, international, epidemiologic study of invasive candidiasis in children and neonates. *Pediatric Infect. Dis. J.* **2012**, *31*, 1252–1257. [CrossRef]

73. Hemedez, C.; Trail-Burns, E.; Mao, Q.; Chu, S.; Shaw, S.K.; Bliss, J.M.; De Paepe, M.E. Pathology of Neonatal Non- albicans Candidiasis: Autopsy Study and Literature Review. *Pediatric Dev. Patbol.* **2019**, *22*, 98–105. [CrossRef]

74. Kaufman, D.A. "Getting to Zero": Preventing invasive Candida infections and eliminating infection-related mortality and morbidity in extremely preterm infants. *Early Hum. Dev.* **2012**, *88* (Suppl. 2), S45–S49. [CrossRef]

75. Bassetti, M.; Marchetti, M.; Chakrabarti, A.; Colizza, S.; Garnacho-Montero, J.; Kett, D.H.; Munoz, P.; Cristini, F.; Andoniadou, A.; Viale, P.; et al. A research agenda on the management of intra-abdominal candidiasis: Results from a consensus of multinational experts. *Intensive Care Med.* **2013**, *39*, 2092–2106. [CrossRef]

76. Baley, J. Neonatal Candidiasis: The Current Challenge. *Clin. Perinatol.* **1991**, *18*, 263–280. [CrossRef]

77. Baley, J.; Kliegman, R.; Boxerbaum, B.; Fanaroff, A. Fungal Colonization in the Very Low Birth Weight Infant. *Pediatrics* **1986**, *78*, 225–232.

78. Stoll, B.J.; Gordon, T.; Korones, S.B.; Shankaran, S.; Tyson, J.E.; Bauer, C.R.; Fanaroff, A.A.; Lemons, J.A.; Donovan, E.F.; Oh, W.; et al. Late-onset sepsis in very low birth weight neonates: A report from the National Institute of Child Health and Human Development Neonatal Research Network. *J. Pediatrics* **1996**, *129*, 63–71. [CrossRef]

79. Dodds Ashley, E.; Drew, R.; Johnson, M.; Danna, R.; Dabrowski, D.; Walker, V.; Prasad, M.; Alexander, B.; Papadopoulos, G.; Perfect, J. Cost of Invasive Fungal Infections in the Era of New Diagnostics and Expanded Treatment Options. *Pharmacotherapy* **2012**, *32*, 890–901. [CrossRef]

80. Harrington, R.; Kindermann, S.L.; Hou, Q.; Taylor, R.J.; Azie, N.; Horn, D.L. Candidemia and invasive candidiasis among hospitalized neonates and pediatric patients. *Curr. Med. Res. Opin.* **2017**, *33*, 1803–1812. [CrossRef]

81. Schwab, F.; Geffers, C.; Piening, B.; Haller, S.; Eckmanns, T.; Gastmeier, P. How many outbreaks of nosocomial infections occur in German neonatal intensive care units annually? *Infection* **2014**, *42*, 73–78. [CrossRef] [PubMed]

82. Qi, L.; Fan, W.; Xia, X.; Yao, L.; Liu, L.; Zhao, H.; Kong, X.; Liu, J. Nosocomial outbreak of Candida parapsilosis sensu stricto fungaemia in a neonatal intensive care unit in China. *J. Hosp. Infect.* **2018**, *100*, e246–e252. [CrossRef] [PubMed]

83. Guo, W.; Gu, H.F.; Zhang, H.G.; Chen, S.B.; Wang, J.Q.; Geng, S.X.; Li, L.; Liu, P.; Liu, X.; Ji, Y.R.; et al. An outbreak of Candida parapsilosis fungemia among preterm infants. *Genet. Mol. Res.* **2015**, *14*, 18259–18267. [CrossRef] [PubMed]

84. Asadzadeh, M.; Ahmad, S.; Al-Sweih, N.; Hagen, F.; Meis, J.F.; Khan, Z. High-resolution fingerprinting of Candida parapsilosis isolates suggests persistence and transmission of infections among neonatal intensive care unit patients in Kuwait. *Sci. Rep.* **2019**, *9*, 1340. [CrossRef] [PubMed]

85. Kollef, M.; Micek, S.; Hampton, N.; Doherty, J.A.; Kumar, A. Septic Shock Attributed to Candida Infection: Importance of Empiric Therapy and Source Control. *Clin. Infect. Dis.* **2012**, *54*, 1739–1746. [CrossRef] [PubMed]

86. Puig-Asensio, M.; Pemán, J.; Zaragoza, R.; Garnacho-Montero, J.; Martín-Mazuelos, E.; Cuenca-Estrella, M.; Almirante, B. Prospective Population Study on Candidemia in Spain (CANDIPOP) Project, Hospital Infection Study Group (GEIH), Medical Mycology Study Group (GEMICOMED) of the Spanish Society of Infectious Diseases and Clinical Microbiology (SEIMC), & Spanish Network for Research in Infectious Diseases. Impact of therapeutic strategies on the prognosis of candidemia in the ICU. *Crit. Care Med.* **2014**, *42*, 1423–1432. [CrossRef] [PubMed]

87. Kaufman, D.; Boyle, R.; Hazen, K.C.; Patrie, J.T.; Robinson, M.; Donowitz, L.G. Fluconazole prophylaxis against fungal colonization and infection in preterm infants. *N. Engl. J. Med.* **2001**, *345*, 1660–1666. [CrossRef] [PubMed]

88. Manzoni, P.; Stolfi, I.; Pugni, L.; Decembrino, L.; Magnani, C.; Vetrano, G.; Tridapalli, E.; Corona, G.; Giovannozzi, C.; Farina, D.; et al. Italian Task Force for the Study and Prevention of Neonatal Fungal Infections, & Italian Society of Neonatology. A Multicenter, Randomized Trial of Prophylactic Fluconazole in Preterm Neonates. *N. Engl. J. Med.* **2007**, *356*, 2483–2495. [CrossRef] [PubMed]

89. Healy, C.M.; Baker, C.J.; Zaccaria, E.; Campbell, J.R. Impact of Fluconazole prophylaxis on incidence and outcome of invasive Candidiasis in a Neonatal Intensive Care Unit. *J. Pediatrics* **2005**, *147*, 166–171. [CrossRef]

90. Clerihew, L.; Austin, N.; McGuire, W. Prophylactic systemic antifungal agents to prevent mortality and morbidity in very low birth weight infants. *Cochrane Database Syst. Rev.* **2007**, *4*, CD003850. [CrossRef]

91. Aliaga, S.; Clark, R.H.; Laughon, M.; Walsh, T.J.; Hope, W.W.; Benjamin, D.K.; Kaufman, D.; Arrieta, A.; Benjamin, D.K., Jr.; Smith, P.B. Changes in the Incidence of Candidiasis in Neonatal Intensive Care Units. *Pediatrics* **2014**, *133*, 236–242. [CrossRef]

92. Adams-Chapman, I.; Bann, C.M.; Das, A.; Ronald, N.; Stoll, B.J.; Walsh, M.C.; Sánchez, P.J.; Higgins, R.D.; Shankaran, S.; Watterberg, K.L.; et al. Eunice Kennedy Shriver National Institutes of Child Health and Human Development Neonatal Research Network. Neurodevelopmental Outcome of Extremely Low Birth Weight Infants with Candida infection. *J. Pediatrics* **2013**, *163*, 961–967. [CrossRef] [PubMed]

93. Patterson, T.F.; Donnelly, J.P. New Concepts in Diagnostics for Invasive Mycoses: Non-Culture-Based Methodologies. *J. Fungi.* **2019**, *5*, 9. [CrossRef] [PubMed]

94. Mylonakis, E.; Clancy, C.J.; Ostrosky-Zeichner, L.; Garey, K.W.; Alangaden, G.J.; Vazquez, J.A.; Groeger, J.S.; Judson, M.A.; Vinagre, Y.M.; Heard, S.O.; et al. T2 magnetic resonance assay for the rapid diagnosis of candidemia in whole blood: A clinical trial. *Clin. Infect. Dis.* **2015**, *60*, 892–899. [CrossRef] [PubMed]

95. Hamula, C.L.; Hughes, K.; Fisher, B.T.; Zaoutis, T.E.; Singh, I.R.; Velegraki, A. T2Candida Provides Rapid and Accurate Species Identification in Pediatric Cases of Candidemia. *Am. J. Clin. Pathol.* **2016**, *145*, 858–861. [CrossRef] [PubMed]

96. Wei, S.; Wu, T.; Wu, Y.; Ming, D.; Zhu, X. Diagnostic accuracy of Candida albicans germ tube antibody for invasive candidiasis: Systematic review and meta-analysis. *Diagn. Microbiol. Infect. Dis.* **2019**, *93*, 339–345. [CrossRef]

97. Pana, Z.-D.; Kougia, V.; Roilides, E. Therapeutic strategies for invasive fungal infections in neonatal and pediatric patients: An update. *Expert Opin. Pharm.* **2015**, *16*, 693–710. [CrossRef]

98. Bradley, J.S.; Barnett, E.D.; Cantey, J.B. (Eds.) Choosing Among Antifungal Agents: Polyenes, Azoles, and Echinocandins. In *Nelson's Pediatric Antimicrobial Therapy*, 25th ed.; Chapter 2; American Academy of Pediatrics: Itasca, IL, USA, 2019.

99. Scott, B.L.; Hornik, C.D.; Zimmerman, K. Pharmacokinetic, efficacy, and safety considerations for the use of antifungal drugs in the neonatal population. *Expert Opin. Drug Metab. Toxicol.* **2020**, *16*, 605–616. [CrossRef]

100. Saxén, H.; Hoppu, K.; Pohjavuori, M. Pharmacokinetics of fluconazole in very low birth weight infants during the first two weeks of life. *Clin. Pharmacol. Ther.* **1993**, *54*, 269–277. [CrossRef]

101. Larkin, E.L.; Dharmaiah, S.; Ghannoum, M.A. Biofilms and beyond: Expanding echinocandin utility. *J. Antimicrob. Chemother.* **2018**, *73*, i73–i81. [CrossRef]

102. Cateau, E.; Rodier, M.H.; Imbert, C. In vitro efficacies of caspofungin or micafungin catheter lock solutions on Candida albicans biofilm growth. *J. Antimicrob. Chemother.* **2008**, *62*, 153–155. [CrossRef]
103. Katragkou, A.; Roilides, E.; Walsh, T.J. Role of Echinocandins in Fungal Biofilm-Related Disease: Vascular Catheter-Related Infections, Immunomodulation, and Mucosal Surfaces. *Clin. Infect. Dis.* **2015**, *61* (Suppl. 6), S622–S629. [CrossRef] [PubMed]

Article

Metamorphic Protein Folding Encodes Multiple Anti-*Candida* Mechanisms in XCL1

Acacia F. Dishman [1,2,†], Jie He [3,†], Brian F. Volkman [1,*] and Anna R. Huppler [3,*]

1   Department of Biochemistry, Medical College of Wisconsin, Milwaukee, WI 53226, USA; adishman@mcw.edu
2   Medical Scientist Training Program, Medical College of Wisconsin, Milwaukee, WI 53226, USA
3   Department of Pediatrics, Medical College of Wisconsin, Milwaukee, WI 53226, USA; jhe@mcw.edu
*   Correspondence: bvolkman@mcw.edu (B.F.V.); ahuppler@mcw.edu (A.R.H.)
†   These authors contributed equally.

Abstract: *Candida* species cause serious infections requiring prolonged and sometimes toxic therapy. Antimicrobial proteins, such as chemokines, hold great interest as potential additions to the small number of available antifungal drugs. Metamorphic proteins reversibly switch between multiple different folded structures. XCL1 is a metamorphic, antimicrobial chemokine that interconverts between the conserved chemokine fold (an α–β monomer) and an alternate fold (an all-β dimer). Previous work has shown that human XCL1 kills *C. albicans* but has not assessed whether one or both XCL1 folds perform this activity. Here, we use structurally locked engineered XCL1 variants and *Candida* killing assays, adenylate kinase release assays, and propidium iodide uptake assays to demonstrate that both XCL1 folds kill *Candida*, but they do so via different mechanisms. Our results suggest that the alternate fold kills via membrane disruption, consistent with previous work, and the chemokine fold does not. XCL1 fold-switching thus provides a mechanism to regulate the XCL1 mode of antifungal killing, which could protect surrounding tissue from damage associated with fungal membrane disruption and could allow XCL1 to overcome candidal resistance by switching folds. This work provides inspiration for the future design of switchable, multifunctional antifungal therapeutics.

Keywords: *Candida*; *C. albicans*; XCL1; metamorphic protein; fold-switching protein; antifungal peptide

Citation: Dishman, A.F.; He, J.; Volkman, B.F.; Huppler, A.R. Metamorphic Protein Folding Encodes Multiple Anti-*Candida* Mechanisms in XCL1. *Pathogens* **2021**, *10*, 762. https://doi.org/10.3390/pathogens10060762

Academic Editor: Jonathan Richardson

Received: 11 May 2021
Accepted: 12 June 2021
Published: 17 June 2021

Publisher's Note: MDPI stays neutral with regard to jurisdictional claims in published maps and institutional affiliations.

1. Introduction

Candida albicans and other fungal pathogens cause severe and costly infections in children and adults [1–3]. High incidence of drug toxicity and the emergence of resistance limit the utility of available antifungal drug classes [4,5]. Components of the innate immune system termed antimicrobial peptides, including certain chemokines, are currently under investigation as novel antifungal agents [6,7]. Detailed understanding of the relationship between protein structure and function can promote the optimal development of these peptides as therapeutics.

In the field of structural biology, the conventional wisdom has been that each amino acid sequence folds into a single structure to carry out its biological role. However, in recent decades, proteins have been discovered that defy this norm, folding into multiple different structures and reversibly interconverting between them. Recent work has shown that these proteins, called metamorphic proteins, may be more common than initially expected [8,9]. Interest in the biological relevance of metamorphic protein folding is growing, and efforts to understand and harness protein metamorphosis for therapeutic benefit are beginning to mount [10–12].

One such metamorphic protein is the antimicrobial human chemokine XCL1. Chemokines are small, secreted immune proteins that orchestrate the migration of white blood cells under homeostatic and inflammatory conditions, some of which have antimicrobial activity [13,14]. XCL1 is unique amongst chemokines because it switches between the conserved α–β chemokine fold and an all-β alternate fold that forms a dimer [15] (Figure 1). XCL1 occupies the two folds in equal proportion under near-physiological conditions and interconverts between the two folds in the absence of a trigger [15]. The chemokine fold binds and activates XCL1's cognate G-protein coupled receptor (GPCR), XCR1, on the surface of a subset of dendritic cells [15,16]. The XCL1 alternate fold binds to glycosaminoglycans (GAGs), facilitating the formation of chemotactic concentration gradients [15,17]. It has been shown that the XCL1 alternate fold directly kills *E. coli* via physical membrane disruption, but the XCL1 chemokine fold does not [18,19]. Recent work has demonstrated that wild-type (WT) human XCL1 also kills *C. albicans* [18]. The aim of this study was to determine whether this antifungal activity is encoded by one or both of the XCL1 structures, and to further elucidate the molecular mechanism by which XCL1 kills *Candida*.

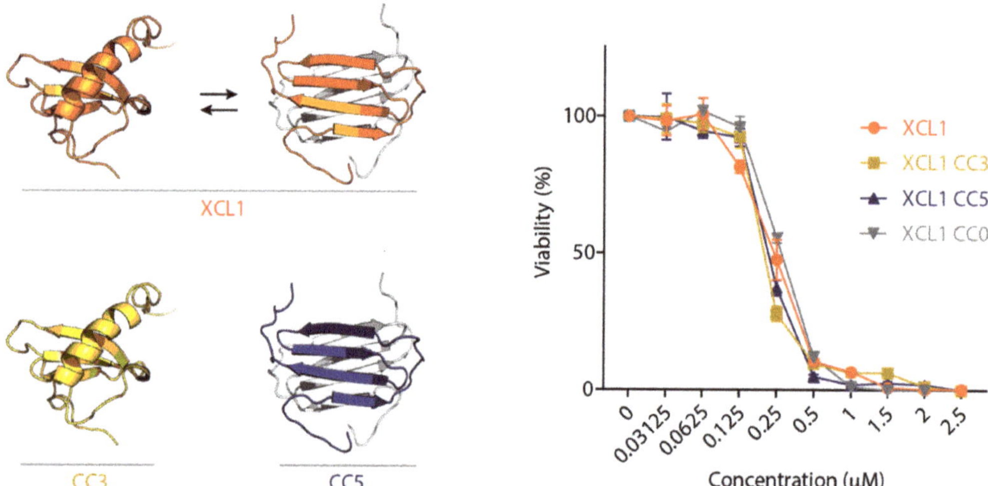

Figure 1. In vitro *C. albicans* killing activity of WT XCL1 and engineered XCL1 variants locked in the chemokine fold, alternate fold, and unfolded state. *Left:* the two XCL1 native structures (chemokine fold, left; dimeric alternate fold, right, in color (subunit A) and light grey (subunit B)). CC3, gold, is an engineered XCL1 variant that is locked in the chemokine fold. CC5, blue, is an engineered XCL1 variant that is locked in the alternate fold. WT XCL1, orange, interconverts between the two folds and occupies each fold in equal proportion under near-physiological conditions. *Right: C. albicans* killing activity of XCL1 (orange), CC3 (gold), CC5 (dark blue), and CC0 (grey). CC0 is an engineered XCL1 variant lacking the disulfide bond that has no defined folded structure.

Here, we used a panel of engineered XCL1 variants [16,17] and *Candida* killing dose-response and time course assays, adenylate kinase (AK) release assays, and propidium iodide uptake (PI) assays to show that the XCL1 chemokine fold and alternate fold both kill *C. albicans*. However, our data suggest that the two folds kill by different mechanisms. Multiple complementary assays indicate that the alternate fold kills *C. albicans* via direct membrane disruption, in agreement with previous studies [18], but the chemokine fold does not. XCL1 can provide unique therapeutic inspiration as a member of a family of antimicrobial human immune system proteins, with the added feature of switching folds to encode multiple distinct antifungal mechanisms.

2. Results

2.1. The XCL1 Chemokine Fold, Alternate Fold, and Unfolded State Kill C. albicans

Engineered XCL1 variants have been designed to lock XCL1 into the chemokine fold [16] and the alternate fold [17] by adding a new disulfide bond. The variants are named CC3 (locked chemokine fold) and CC5 (locked alternate fold) (Figure 1). Additionally, an

XCL1 variant lacking XCL1's native disulfide bond, named CC0, has no defined folded structure. We tested WT human XCL1, CC3, CC5, and CC0 for antifungal activity against *C. albicans* using a dose-response plating assay [18,20,21] and found that all of the XCL1 variants kill *C. albicans* with similar dose-response profiles (Figure 1). Previous studies suggest that the XCL1 alternate fold is capable of directly disrupting fungal membranes, but the chemokine fold is not [18]. We thus sought to determine whether the chemokine fold kills via the same mechanism as the alternate fold.

2.2. WT XCL1 Kills C. albicans Faster Than an XCL1 Variant Locked in the Chemokine Fold

If the XCL1 chemokine fold and alternate fold kill *Candida* via the same mechanism, the chemokine fold would be expected to kill with similar kinetics to WT XCL1. We performed time course killing assays for XCL1 and CC3 at a protein concentration of 1 μM against *C. albicans*, finding that CC3 kills more slowly than WT human XCL1 (Figure 2). This difference could occur if the XCL1 alternate fold kills via direct membrane disruption, which occurs quickly, while the XCL1 chemokine fold kills by a slower mechanism. To test the hypothesis that the XCL1 alternate fold kills via membrane disruption while the chemokine fold does not, we performed two complementary assays to assess the ability of the different XCL1 structures to induce membrane disruption in *C. albicans*.

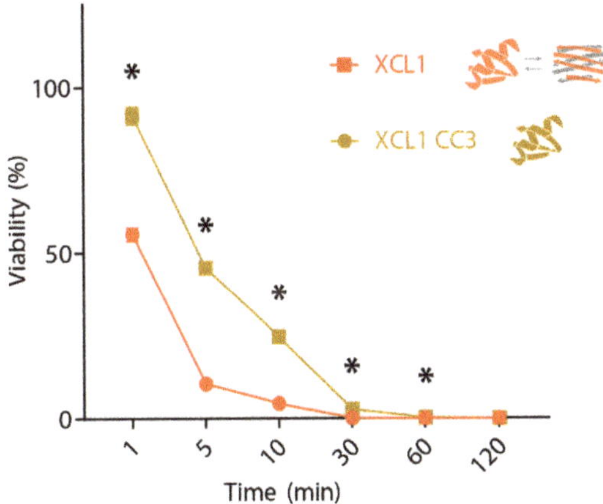

Figure 2. *Candida* killing time course assays for WT XCL1 and CC3. Each timepoint was performed in triplicate. Significant differences between XCL1 and CC3 indicated with * for $p < 0.05$.

2.3. The XCL1 Alternate Fold Induces More Intense Adenylate Kinase Release from C. albicans Than the XCL1 Chemokine Fold

To assess for membrane disruption in *C. albicans* by XCL1 and our panel of locked structural variants, we first performed an adenylate kinase release assay [22,23]. In brief, the dye-based adenylate kinase assay (AKA) measures the release of the intracellular enzyme adenylate kinase (AK) from *C. albicans* cells following protein treatment. As a positive control, we used CCL28, a human chemokine that adopts the conserved chemokine fold and is known to kill *Candida* via direct membrane disruption [24]. The protein suspension buffer, potassium phosphate buffer (PPB), was used as a negative control. We found that XCL1, CC0, and CC5 trigger the release of adenylate kinase from *C. albicans*, suggesting that the XCL1 alternate fold and unfolded state induce *C. albicans* membrane disruption (Figure 3). CC3 treatment resulted in very little adenylate kinase release, even at late time points. CC3 kills with similar potency to XCL1, CC5, and CC0 (Figure 1), but induces less AK release (Figure 3), which suggests that CC3 kills *C. albicans* by a mechanism other than membrane disruption. To confirm these findings, we performed a second complementary assay to detect candidal membrane disruption by XCL1 and CC3.

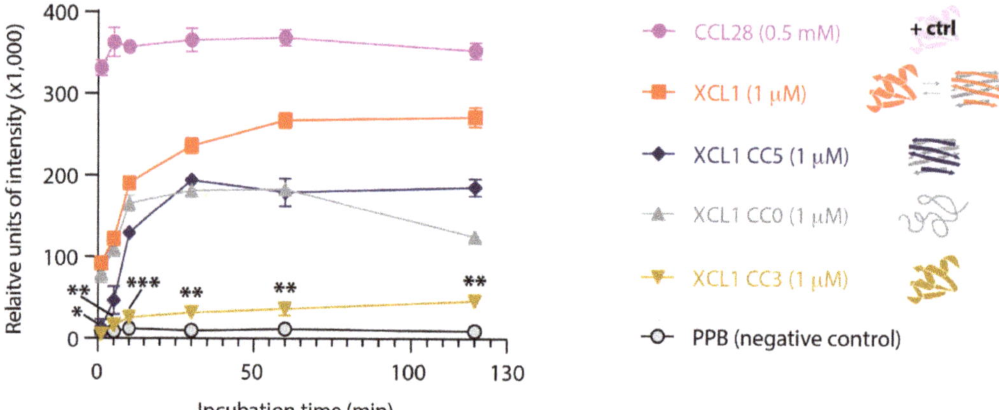

Figure 3. Adenylate kinase (AK) release from *C. albicans* as a measure of membrane disruption in response to XCL1 and structural variants. Each time point was performed in triplicate. The structural behavior of each XCL1 variant is illustrated in the legend. Significant differences between XCL1 and CC3 indicated with * for $p < 0.0005$, ** for $p < 0.00005$, and *** for $p < 0.000005$.

2.4. WT XCL1 Triggers More Propidium Iodide Uptake Than the XCL1 Locked Chemokine Fold Variant

To confirm the finding that WT XCL1 kills *Candida* via membrane disruption but CC3 does not, we performed a propidium iodide (PI) uptake assay, which detects cellular uptake of a dye that selectively labels the chromosome. Higher PI uptake indicates increased membrane disruption. As in the AKA, CCL28 was used as a positive control and PPB was used as a negative control. Paraformaldehyde kills *Candida* but does not induce membrane disruption and was included as a second negative control. We found that WT XCL1 induces significantly higher PI uptake in *C. albicans* than CC3 at 10 min, 30 min, and 60 min after the protein treatment (Figure 4). This suggests that the XCL1 alternate fold kills via membrane disruption, but the XCL1 chemokine fold does not, in concurrence with the AKA data presented in the previous section.

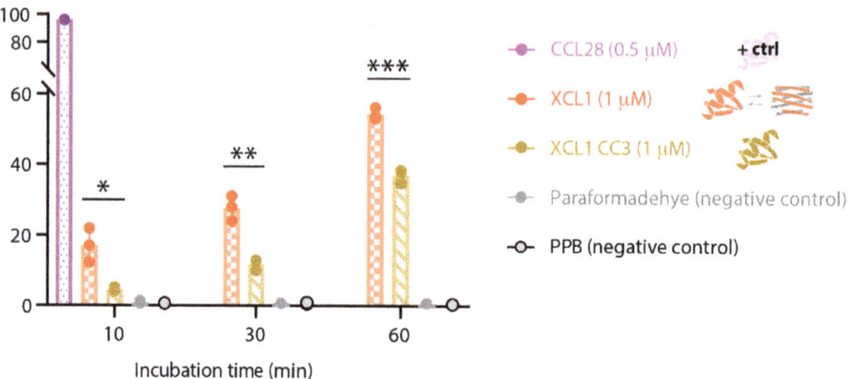

Figure 4. Propidium Iodide uptake by *C. albicans* as a measure of membrane disruption in response to XCL1, CC3, and controls. The timepoint was performed in triplicate. Each data point is represented by a circle. The structural behavior of each variant is illustrated in the legend. Significant differences between XCL1 and CC3 indicated with * for $p < 0.05$, ** for $p < 0.005$, and *** for $p < 0.0005$.

3. Discussion

Fungi such as *C. albicans* cause severe infections [1–3], but currently available treatments have limited utility due to drug toxicity and the frequent emergence of drug resistance. Thus, new, less toxic antifungal drugs are needed, and antimicrobial proteins in the human immune system, such as chemokines, can inspire the design of such novel therapies [6,7,25,26]. Many human chemokines are known to be antifungal [13,14,21], including the chemokines XCL1 and CCL28 [18]. Indeed, a recent study found that XCL1 and CCL28 are expressed highly in uterine tissue, perhaps suggesting that the antimicrobial properties of these chemokines contribute to the protection of the developing fetus from microbial pathogens [27]. XCL1 has the additional feature of being metamorphic, i.e., switching between multiple different three-dimensional structures reversibly in solution [15]. Interest in metamorphic proteins' biological and therapeutic utility has grown in recent years [9–11,28]. XCL1 can provide unique inspiration for the design of non-toxic, switchable antifungals for which the development of resistance is minimized. Such pursuits rely on understanding structure–function relationships for XCL1 antifungal activity.

The work presented here suggests that XCL1 kills *C. albicans* via two different mechanisms encoded by the two different XCL1 folds. Furthermore, the XCL1 alternate fold appears to kill *Candida* via direct membrane disruption, while the chemokine fold does not. These findings agree with previous work demonstrating that the XCL1 alternate fold induces Negative Gaussian Curvature (NGC), a topological requirement for membrane disruption, in model fungal membranes, while the chemokine fold does not [18].

Inspiration from XCL1 can inform the design of antifungal proteins that switch folds, quickly, reversibly toggling between multiple antifungal killing mechanisms, which would have distinct advantages as antifungal agents. If *Candida* evolves resistance to killing by one protein fold, interconversion to the other fold overcomes this resistance. Additionally, killing by membrane disruption may be advantageous in certain contexts but not others. Killing via a non-membrane-disruptive mechanism may modulate the resulting immune response by decreasing the release of pathogen-associated molecular patterns (PAMPs) and could protect surrounding tissue from damage. Likewise, fold-switching therapies could spare host microbial flora, for example, in the gut, that would be killed by other antifungal agents. Fold-switching proteins could also be designed that have dual functionality, with one fold encoding antifungal activity and the other encoding a distinct therapeutic function. For example, the second fold might encode anti-inflammatory or immunomodulatory activity, reducing the number of drugs required to treat fungal infections. Together, a better understanding of the molecular mechanisms of antifungal action of human metamorphic proteins can inspire the development of improved antifungal therapeutics.

4. Materials and Methods

4.1. Protein Expression and Purification

Expression of CC0, CC3, CC5, and WT XCL1 was performed as previously described [15,19,29]. In brief, proteins were expressed recombinantly with a His_6-SUMO tag (pET28a expression vector, BL21 DE3 *E. coli.*). Cultures (grown in terrific broth with 50 mg/mL Kanamycin at 37 °C) were induced with 1 mM isopropyl-b-D-thiogalactopyranoside (IPTG) once they reached an optical density of 0.5–0.7, after which they were grown for an additional 5 h at 37 °C. Cells were harvested by centrifugation ($8000 \times g$, 10 min) and stored at −80 °C. For protein purification, 50 mM sodium phosphate (pH 8.0), 300 mM sodium chloride, 10 mM imidazole, 0.1% (v/v) β-mercaptoethanol, and 1 mM phenylmethylsulfonyl fluoride (PMSF) was used to resuspend cell pellets, which were then lysed using a French press or by sonication. Centrifugation was used to collect inclusion bodies from cell lysates ($12,000 \times g$, 20 min) which were then resuspended and further purified along with the soluble fractions. Nickel column chromatography (nickel resin, Qiagen, Hilden, Germany) was used as an initial purification step. Elution fractions from the nickel columns underwent infinite dilution refolding into 20 mM Tris (pH 8.0), 200 mM sodium chloride, 10 mM cysteine, and 0.5 mM cystine. Refolding solutions were incubated at room temperature with gentle stirring overnight, then concentrated. ULP1 protease cleavage was used to remove the His$_6$-SUMO fusion tag, which was then separated from the protein of interest using either cation exchange chromatography (SP Sepharose Fast Flow resin, GE Healthcare, Chicago, IL, USA) or reverse nickel column chromatography (nickel resin, Qiagen). High-performance liquid chromatography was performed with a C18 column as a final purification step. Proteins were then frozen and lyophilized. Sample purities, identities, and homogeneities were checked with matrix-assisted laser desorption ionization time-of-flight (MALDI-TOF) spectroscopy.

4.2. Candida Killing Dose-Response Assays

Candida killing assays were performed as previously described [18,20,21]. Briefly, a single clone of *C. albicans* strain CAF2-1 was cultured in yeast peptone dextrose (YPD) medium for 16–20 h (30 °C, 250 rpm), before being washed twice and diluted to a final concentration of ~5 × 10^4 cells/mL in a low-salt, 1 mM potassium phosphate buffer, at pH 7.0 (PPB). Proteins were lyophilized and resuspended to a concentration of 400 µM in 1 mM PPB, aliquoted and stored at −70 °C until use. Proteins were serially diluted with 1 mM PPB and mixed 1:1 with the ~5 × 10^4 cells/mL *Candida* stocks to a total volume of 200 µL in a 96-well plate. Negative controls were performed using 1 µM PPB. The 96-well plates were incubated for 2 h or the indicated time with gentle shaking (80 rpm). Appropriate dilutions of the suspensions were then plated on YPD agar plates. Plates were incubated at 30 °C for 48 h, then colonies were counted. Viability is reported as the percent colony number with respect to the negative control. Assays were performed in at least duplicate, and protein conditions were tested in triplicate in each assay.

4.3. Adenylate Kinase Release Assays

Adenylate kinase leakage was measured with an adenylate kinase assay (AKA) kit (Abcam, MA, USA). *Candida* cells at 2 × 10^7 CFU/mL in PPB (100 µL) and serial dilutions of CCL28, XCL1, and indicated XCL1 variants (100 µL) were incubated for 30 min in triplicates in 96-well plates. Plates were centrifuged at 3000 rpm for 5 min, and supernatant from each well (100 µL) was transferred to the white 96-well plate and mixed with adenylate reagent (100 µL) in white 96-well flat-bottom assay plates. Luminescence was measured immediately by a SpectraMax plate reader (Molecular Devices, San Jose, CA, USA).

4.4. Propidium Iodide Uptake Assays

Candida was cultured and washed as described above. *Candida* suspensions at 5 × 10^6 CFU/mL were incubated with equal volumes of 1 mM PPB, CCL28, XCL1, or XCL1 CC3 solutions (final concentration 0.5 µM), or paraformaldehyde solution (2% paraformaldehyde, 0.5% bovine serum albumin (BSA), and 2 mM EDTA in PBS) at room temperature for 10, 30 and 60 min in triplicates. Cells were washed and resuspended with 1× FACS (PBS with 2% fetal bovine serum and 2 mM EDTA buffer) and then stained with propidium iodide (PI; Thermo Fisher Scientific, Grand Island, NY, USA) at 1 mg/mL at room temperature in the dark for 10 min. Stained cells were washed with 1× PBS and resuspended in 200 µL of the paraformaldehyde solution. Cellular PI uptake was measured by flow cytometry in the CRI flow cytometry core on a BD LSR II (BD Biosciences, San Jose, CA, USA) and analyzed with FlowJo (Tree Star, Inc., Ashland, OR, USA).

Author Contributions: Conceptualization, A.F.D., B.F.V., and A.R.H.; methodology, A.F.D. and J.H.; validation, A.F.D. and J.H.; formal analysis, J.H. and A.R.H.; resources, B.F.V. and A.R.H.; writing—original draft preparation, A.F.D.; writing—review and editing, J.H., A.R.H. and B.F.V.; visualization, A.F.D., J.H. and A.R.H.; supervision, B.F.V. and A.R.H.; funding acquisition, A.F.D., B.F.V. and A.R.H. All authors have read and agreed to the published version of the manuscript.

Funding: This work was supported in part by National Institutes of Health Grants R37 AI058072 (to B.F.V.), K08 DE026189 (to A.R.H.), and F30 CA236182 (to A.F.D). A.R.H. was also supported by the Children's Hospital of Wisconsin Research Institute. A.F.D. is a member of the NIH-supported (T32 GM080202) Medical Scientist Training Program at the Medical College of Wisconsin (MCW).

Institutional Review Board Statement: Not applicable.

Informed Consent Statement: Not applicable.

Data Availability Statement: The data presented in this study are available within the article.

Conflicts of Interest: B.F.V. has ownership interests in Protein Foundry, LLC and XLock Biosciences, LLC.

References

1. Webb, B.J.; Ferraro, J.P.; Rea, S.; Kaufusi, S.; Goodman, B.E.; Spalding, J. Epidemiology and Clinical Features of Invasive Fungal Infection in a US Health Care Network. *Open Forum Infect. Dis.* **2018**, *5*, ofy187. [CrossRef] [PubMed]
2. Warris, A.; Pana, Z.D.; Oletto, A.; Lundin, R.; Castagnola, E.; Lehrnbecher, T.; Groll, A.H.; Roilides, E.; Group, E.S. Etiology and Outcome of Candidemia in Neonates and Children in Europe: An 11-year Multinational Retrospective Study. *Pediatr. Infect. Dis. J.* **2020**, *39*, 114–120. [CrossRef] [PubMed]
3. Benedict, K.; Jackson, B.R.; Chiller, T.; Beer, K.D. Estimation of Direct Healthcare Costs of Fungal Diseases in the United States. *Clin. Infect. Dis.* **2019**, *68*, 1791–1797. [CrossRef] [PubMed]
4. Arastehfar, A.; Gabaldon, T.; Garcia-Rubio, R.; Jenks, J.D.; Hoenigl, M.; Salzer, H.J.F.; Ilkit, M.; Lass-Florl, C.; Perlin, D.S. Drug-Resistant Fungi: An Emerging Challenge Threatening Our Limited Antifungal Armamentarium. *Antibiotics* **2020**, *9*, 877. [CrossRef]
5. Brown, G.D.; Denning, D.W.; Gow, N.A.; Levitz, S.M.; Netea, M.G.; White, T.C. Hidden killers: Human fungal infections. *Sci Transl. Med.* **2012**, *4*, 165rv113. [CrossRef]
6. Basso, V.; Tran, D.Q.; Ouellette, A.J.; Selsted, M.E. Host Defense Peptides as Templates for Antifungal Drug Development. *J. Fungi* **2020**, *6*, 241. [CrossRef]
7. Fernandez de Ullivarri, M.; Arbulu, S.; Garcia-Gutierrez, E.; Cotter, P.D. Antifungal Peptides as Therapeutic Agents. *Front. Cell Infect. Microbiol.* **2020**, *10*, 105. [CrossRef]
8. Dishman, A.F.; Tyler, R.C.; Fox, J.C.; Kleist, A.B.; Prehoda, K.E.; Babu, M.M.; Peterson, F.C.; Volkman, B.F. Evolution of fold switching in a metamorphic protein. *Science* **2021**, *371*, 86–90. [CrossRef]
9. Porter, L.L.; Looger, L.L. Extant fold-switching proteins are widespread. *Proc. Natl. Acad. Sci. USA* **2018**, *115*, 5968–5973. [CrossRef]
10. Kim, A.K.; Porter, L.L. Functional and Regulatory Roles of Fold-Switching Proteins. *Structure* **2021**, *29*, 6–14. [CrossRef]
11. Dishman, A.F.; Volkman, B.F. Unfolding the Mysteries of Protein Metamorphosis. *ACS Chem. Biol.* **2018**, *13*, 1438–1446. [CrossRef] [PubMed]
12. Matsuo, K.; Kitahata, K.; Kawabata, F.; Kamei, M.; Hara, Y.; Takamura, S.; Oiso, N.; Kawada, A.; Yoshie, O.; Nakayama, T. A Highly Active Form of XCL1/Lymphotactin Functions as an Effective Adjuvant to Recruit Cross-Presenting Dendritic Cells for Induction of Effector and Memory CD8(+) T Cells. *Front. Immunol.* **2018**, *9*, 2775. [CrossRef] [PubMed]
13. Yang, D.; Chen, Q.; Hoover, D.M.; Staley, P.; Tucker, K.D.; Lubkowski, J.; Oppenheim, J.J. Many chemokines including CCL20/MIP-3alpha display antimicrobial activity. *J. Leukoc Biol.* **2003**, *74*, 448–455. [CrossRef] [PubMed]
14. Yung, S.C.; Murphy, P.M. Antimicrobial chemokines. *Front. Immunol.* **2012**, *3*, 276. [CrossRef] [PubMed]
15. Tuinstra, R.L.; Peterson, F.C.; Kutlesa, S.; Elgin, E.S.; Kron, M.A.; Volkman, B.F. Interconversion between two unrelated protein folds in the lymphotactin native state. *Proc. Natl. Acad. Sci. USA* **2008**, *105*, 5057–5062. [CrossRef]
16. Tuinstra, R.L.; Peterson, F.C.; Elgin, E.S.; Pelzek, A.J.; Volkman, B.F. An engineered second disulfide bond restricts lymphotactin/XCL1 to a chemokine-like conformation with XCR1 agonist activity. *Biochemistry* **2007**, *46*, 2564–2573. [CrossRef]
17. Fox, J.C.; Tyler, R.C.; Guzzo, C.; Tuinstra, R.L.; Peterson, F.C.; Lusso, P.; Volkman, B.F. Engineering Metamorphic Chemokine Lymphotactin/XCL1 into the GAG-Binding, HIV-Inhibitory Dimer Conformation. *ACS Chem. Biol.* **2015**, *10*, 2580–2588. [CrossRef]
18. Dishman, A.F.; Lee, M.W.; de Anda, J.; Lee, E.Y.; He, J.; Huppler, A.R.; Wong, G.C.L.; Volkman, B.F. Switchable Membrane Remodeling and Antifungal Defense by Metamorphic Chemokine XCL1. *ACS Infect. Dis.* **2020**, *6*, 1204–1213. [CrossRef]
19. Nevins, A.M.; Subramanian, A.; Tapia, J.L.; Delgado, D.P.; Tyler, R.C.; Jensen, D.R.; Ouellette, A.J.; Volkman, B.F. A Requirement for Metamorphic Interconversion in the Antimicrobial Activity of Chemokine XCL1. *Biochemistry* **2016**, *55*, 3784–3793. [CrossRef]
20. Thomas, M.A.; He, J.; Peterson, F.C.; Huppler, A.R.; Volkman, B.F. The Solution Structure of CCL28 Reveals Structural Lability that Does Not Constrain Antifungal Activity. *J. Mol. Biol.* **2018**, *430*, 3266–3282. [CrossRef] [PubMed]
21. Hieshima, K.; Ohtani, H.; Shibano, M.; Izawa, D.; Nakayama, T.; Kawasaki, Y.; Shiba, F.; Shiota, M.; Katou, F.; Saito, T.; et al. CCL28 has dual roles in mucosal immunity as a chemokine with broad-spectrum antimicrobial activity. *J. Immunol.* **2003**, *170*, 1452–1461. [CrossRef]
22. Beattie, S.R.; Krysan, D.J. Antifungal drug screening: Thinking outside the box to identify novel antifungal scaffolds. *Curr. Opin. Microbiol.* **2020**, *57*, 1–6. [CrossRef] [PubMed]
23. Krysan, D.J.; Didone, L. A high-throughput screening assay for small molecules that disrupt yeast cell integrity. *J. Biomol. Screen* **2008**, *13*, 657–664. [CrossRef] [PubMed]
24. He, J.; Thomas, M.A.; de Anda, J.; Lee, M.W.; Van Why, E.; Simpson, P.; Wong, G.C.L.; Grayson, M.H.; Volkman, B.F.; Huppler, A.R. Chemokine CCL28 Is a Potent Therapeutic Agent for Oropharyngeal Candidiasis. *Antimicrob. Agents Chemother.* **2020**, *64*. [CrossRef] [PubMed]

25. Lei, J.; Sun, L.; Huang, S.; Zhu, C.; Li, P.; He, J.; Mackey, V.; Coy, D.H.; He, Q. The antimicrobial peptides and their potential clinical applications. *Am. J. Transl. Res.* **2019**, *11*, 3919–3931.
26. Sierra, J.M.; Fuste, E.; Rabanal, F.; Vinuesa, T.; Vinas, M. An overview of antimicrobial peptides and the latest advances in their development. *Expert Opin. Biol. Ther.* **2017**, *17*, 663–676. [CrossRef]
27. Menzies, F.M.; Oldham, R.S.; Waddell, C.; Nelson, S.M.; Nibbs, R.J.B. A Comprehensive Profile of Chemokine Gene Expression in the Tissues of the Female Reproductive Tract in Mice. *Immunol. Investig.* **2020**, *49*, 264–286. [CrossRef]
28. Goodchild, S.C.; Curmi, P.M.G.; Brown, L.J. Structural gymnastics of multifunctional metamorphic proteins. *Biophys. Rev.* **2011**, *3*, 143. [CrossRef]
29. Fox, J.C.; Nakayama, T.; Tyler, R.C.; Sander, T.L.; Yoshie, O.; Volkman, B.F. Structural and agonist properties of XCL2, the other member of the C-chemokine subfamily. *Cytokine* **2015**, *71*, 302–311. [CrossRef]

Article

Fungicidal Activity of a Safe 1,3,4-Oxadiazole Derivative Against *Candida albicans*

Daniella Renata Faria [1], Raquel Cabral Melo [1], Glaucia Sayuri Arita [1], Karina Mayumi Sakita [1], Franciele Abigail Vilugron Rodrigues-Vendramini [1], Isis Regina Grenier Capoci [1], Tania Cristina Alexandrino Becker [2], Patrícia de Souza Bonfim-Mendonça [1], Maria Sueli Soares Felipe [3], Terezinha Inez Estivalet Svidzinski [1] and Erika Seki Kioshima [1,*]

[1] Laboratory of Medical Mycology, Department of Clinical Analysis and Biomedicine, State University of Maringá (UEM), Maringá, Paraná 87020-900, Brazil; renata.daniella@gmail.com (D.R.F.); raquelcmelo97@gmail.com (R.C.M.); glauciasayuria@gmail.com (G.S.A.); karina.msakita@gmail.com (K.M.S.); francieleavr@gmail.com (F.A.V.R.-V.); isiscapoci@gmail.com (I.R.G.C.); patbonfim.09@gmail.com (P.d.S.B.-M.); terezinha.svidzinski@gmail.com (T.I.E.S.)
[2] Laboratory of General Pathology, Department of Basic Health Sciences, State University of Maringá, Maringá (UEM), Maringá, Paraná 87020-900, Brazil; tcabecker@uem.br
[3] Program of Genomic Sciences and Biotechnology, Catholic University of Brasilia, Brasília 70790-160, Brazil; msueliunb@gmail.com
* Correspondence: eskioshima@gmail.com or eskcotica@uem.br; Tel.: +55-44-3011-4810

Citation: Faria, D.R.; Melo, R.C.; Arita, G.S.; Sakita, K.M.; Rodrigues-Vendramini, F.A.V.; Capoci, I.R.G.; Becker, T.C.A.; Bonfim-Mendonça, P.d.S.; Felipe, M.S.S.; Svidzinski, T.I.E.; et al. Fungicidal Activity of a Safe 1,3,4-Oxadiazole Derivative Against *Candida albicans*. *Pathogens* **2021**, *10*, 314. https://doi.org/10.3390/pathogens10030314

Academic Editors: Lawrence S. Young and Jonathan Richardson

Received: 26 January 2021
Accepted: 1 March 2021
Published: 7 March 2021

Publisher's Note: MDPI stays neutral with regard to jurisdictional claims in published maps and institutional affiliations.

Abstract: *Candida albicans* is the most common species isolated from nosocomial bloodstream infections. Due to limited therapeutic arsenal and increase of drug resistance, there is an urgent need for new antifungals. Therefore, the antifungal activity against *C. albicans* and in vivo toxicity of a 1,3,4-oxadiazole compound (LMM6) was evaluated. This compound was selected by in silico approach based on chemical similarity. LMM6 was highly effective against several clinical *C. albicans* isolates, with minimum inhibitory concentration values ranging from 8 to 32 µg/mL. This compound also showed synergic effect with amphotericin B and caspofungin. In addition, quantitative assay showed that LMM6 exhibited a fungicidal profile and a promising anti-biofilm activity, pointing to its therapeutic potential. The evaluation of acute toxicity indicated that LMM6 is safe for preclinical trials. No mortality and no alterations in the investigated parameters were observed. In addition, no substantial alteration was found in Hippocratic screening, biochemical or hematological analyzes. LMM6 (5 mg/kg twice a day) was able to reduce both spleen and kidneys fungal burden and further, promoted the suppresses of inflammatory cytokines, resulting in infection control. These preclinical findings support future application of LMM6 as potential antifungal in the treatment of invasive candidiasis.

Keywords: *Candida albicans*; candidiasis; 1,3,4-oxadiazole; drug discovery; antifungal agents; drug resistance; toxicity; biofilm

1. Introduction

Candida spp. is the most common cause of nosocomial bloodstream infections by fungal, responsible for over 90% of these cases [1,2]. Immunocompromised patients are the most critically affected, with mortality rates that can reach 60% [3,4]. *C. albicans* remains the most frequent species worldwide, isolated between 20–70% in clinical specimens [5,6]. Among the numerous factors associated with virulence in *C. albicans* which contributes to the high rates of infection and mortality, biofilm is likely to be one of the most important and clinically relevant factors [7]. This fungal organization is able to disrupt host immune response and also the action of antifungal agents on these structures [8]. Therefore, seeking for new and effective treatments against biofilm-associated yeast become necessary.

Currently fungal infections treatment is based on polyenes, azoles or echinocandins [9,10]. Although these agents demonstrate high levels of antifungal activity, their use has serious limitations, in particular due to toxicity, poor tolerability, drug interactions and a narrow activity spectrum [10–12]. Moreover, despite the rates are still low, azole and echinocandin resistance has already been reported in isolates of *C. albicans* [13,14]. This scenario has forced the search for alternatives to conventional antimicrobial therapy. The combination of compounds has potential advantages over monotherapy in terms of reducing dose-related toxicity, possibility of action on more than one target and improved antifungal activity [15,16]. Combination therapy may be a solution for antifungal drug resistance.

Another strategy is in silico techniques that have explored virtual screening of chemical libraries against specific targets for drug discovery, reducing time and costs [17,18]. One promising target has been studied is the thioredoxin reductase (Trr1), a flavoenzyme which mainly confers oxidative stress resistance [19]. Recently, the antifungal activity of two compounds of oxadiazoles class (LMM5 and LMM11) which acts on *C. albicans* Trr1 target, were described. These compounds have shown promise against important pathogenic fungi, such as *Candida* spp., *Cryptococcus neoformans* and *Paracoccidioides* spp. with low toxicity in vitro and in vivo [20–23]. In this study, we evaluated the antifungal activity against *C. albicans* as well as the toxicity in murine model of LMM6 compound, which also belongs to the oxadiazole class and it was selected by in silico approach based on similarity to the LMM11 [20].

2. Results

2.1. Fungicidal Activity of LMM6

The susceptibility profile of 30 clinical isolates and reference strains are presented in Table 1. Similar values of MIC for LMM6 (8–32 µg/mL) were reported. All *C. albicans* tested were susceptible to AMB and CAS (100%—31/31). For FLC, 96.8% (30/31) were susceptible and 3.2% (1/31) was resistant. Whereas for ITC, 76.4% (24/31) were susceptible, 19.4% (6/31) dose-dependent-susceptible (SDD) and 3.2% (1/31) was considered as resistant. To confirm the obtained LMM6 antifungal activity, a quantitative analysis was also performed (Figure 1A). The compound effect on CFU was detected at 8–256 µg/mL concentrations with the best activity observed between 64–256 µg/mL in which there was a CFU $\geq 5 \log_{10}$ reduction ($p < 0.05$). MCF results revealed a dose-dependent activity profile for LMM6. The reference strain (Figure 1B) and all clinical isolates (Supplementary Material, Figure S1) showed a notable reduction of fungal growth from 16 and 0.5 µg/mL, respectively, in relation to the positive control.

LMM6 showed better inhibitory effect than FLC (conventional antifungal) from 6 h in time-kill curve assay (Figure 1C). This activity was sustained for 48 h. Moreover, LMM6 showed fungicidal profile at three concentrations (32, 64 and 128 µg/mL). The reduction in the number of viable cells was $\geq 4 \log_{10}$ (> 99.9%), as compared to control, from 24 h. As expected, the FLC showed a fungistatic profile with CFU reduction $\leq 2 \log_{10}$, as compared to control.

Table 1. Antifungal susceptibility of 30 clinical isolates and reference strain to LMM6 and conventional antifungal drugs.

	MIC (µg/mL)			Interpretation † N (%)		
Antifungal Drugs	**Range**	**MIC$_{50}$**	**MIC$_{90}$**	**Susceptible**	**SDD**	**Resistent**
Amphotericin B	0.03–0.5	0.125	0.25	31 (100)	0 (0)	0 (0)
Caspofungin	0.03–0.25	0.125	0.25	31 (100)	0 (0)	0 (0)
Fluconazole	0.06 > 64	0.25	0.25	30 (96.8)	0 (0)	1 (3.2)
Itraconazole	0.03 > 16	0.125	0.25	24 (77.4)	6 (19.4)	1 (3.2)
LMM6	8.0–32	16	32	-	-	-

Abbreviations; MIC: minimal inhibitory concentration; SDD: susceptible-dose-dependent; † MIC interpretation were established by CLSI document M27-S4; MIC$_{50}$ and MIC$_{90}$ were defined as antifungal concentration capable of inhibiting 50% and 90% growth of the isolates, respectively.

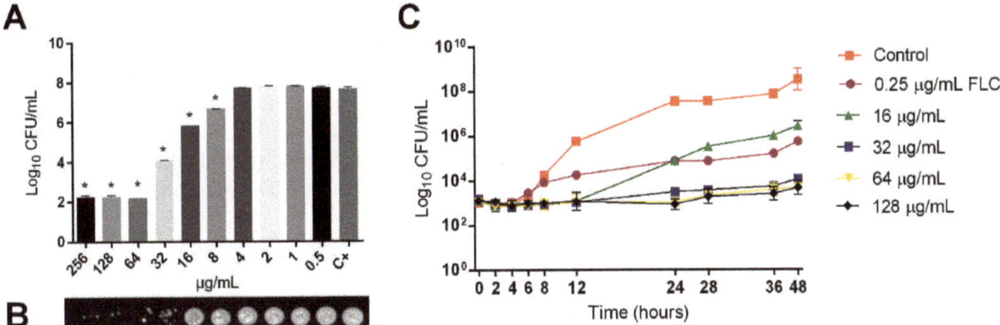

Figure 1. Fungicidal activity of LMM6 against reference strain. (**A**) Quantitative analysis by logarithm reduction of colony forming units. (**B**) Minimum fungicidal concentration (MFC). (**C**) Time-kill curves plotted from $\log_{10}$ CFU/mL versus time (0–48 h) for fluconazole conventional drug control (FLC; MIC 0.25 µg/mL) and LMM6 at concentrations 16, 32, 64 and 128 µg/mL, indicating fungicidal profile of new compound. C+ or Control (drug-free control composed medium plus inoculum). Each data point represents the mean ± standard deviation (error bars). * Values of $p < 0.05$ were considered statistically significant.

2.2. Synergistic Effect between LMM6 and Conventional Antifungals

LMM6 exhibited synergistic interaction with both fungicidal conventional drugs, resulting in a FIC index < 1 for reference strain (AMB: 0.53 and CAS: 0.56) and for one clinical isolate (AMB: 0.75 and CAS: 0.56) (Table 2). However, LMM6 when combined with fungistatic conventional drugs, no synergistic effect was exhibited, resulting in a FIC index > 1 for reference strain (FLC and ITC: 2) and clinical isolate (FLC and ITC: 1.5). The synergic effect of LMM6 with AMB or CAS were confirmed by the presence of blue areas (positive ΔE), in Bliss independence surface analysis (Figure 2A,B,E,F). For the combination of LMM6 with FLC or ITC, red areas were prevalence, which indicates negative ΔE, featuring a without effect combination for these drugs (Figure 2C,D,G,H).

Table 2. Synergism between LMM6 and conventional antifungals drugs against *C. albicans* by the checkerboard method.

Strains	Combinations	FIC^A	FIC^B	FIC Index	Interpretation
Reference strain	AMB/LMM6	0.5	0.031	0.531	Synergistic
	CAS/LMM6	0.5	0.063	0.563	Synergistic
	FLC/LMM6	1	1	2	No effect
	ITC/LMM6	1	1	2	No effect
SangHUMCa7	AMB/LMM6	0.5	0.25	0.75	Synergistic
	CAS/LMM6	0.5	0.063	0.563	Synergistic
	FLC/LMM6	1	0.5	1.5	No effect
	ITC/LMM6	0.5	1	1.5	No effect

Abbreviations; AMB: amphotericin B; CAS: caspofungin; FLC: fluconazole; ITC: itraconazole; FIC index: fractional inhibitory concentration index, calculated as the sum of FIC^A plus FIC^B; FIC^A: $\text{MIC}_{\text{drug conventional}}$ in combination/$\text{MIC}_{\text{drug conventional}}$ alone; FIC^B: MIC_{LMM6} in combination/MIC_{LMM6} alone. SangHUMCa7 is a clinical isolate from hospitalized patient blood that has been identified by classical methods and confirmed by MALDI TOF-MS.

2.3. LMM6 Anti-Biofilm Effect

Given the clinical relevance of biofilm in invasive fungal infections, the effect of LMM6 on *C. albicans* biofilm structure in formation was investigated. SEM showed that LMM6 was able to disrupt the biofilm growth and cause morphological changes in the cells (Figure 3A–C). In the two highest concentrations of LMM6, 64 µg/mL (Figure 3A) and 32 µg/mL (Figure 3B), the biofilm was characterized by disorganization of structure, presence of deformities in yeast cell, membrane and cell wall irregularities and cell extravasation, when compared to control (Figure 3D). In addition, the reduction in CFU (Figure 3E) was statistically significant ($p < 0.05$) at the three concentrations tested, 16 μg/mL (± 3 log10), 32 μg/mL (± 5 log10), and 64 μg/mL (± 6.5 log10), in relation to untreated control. To check the effect of LMM6 on total biofilm biomass, staining by crystal violet was carried out. All concentrations tested exhibited a statistically significant reduction ($p < 0.05$) of the total biomass (Figure 3F), 16 μg/mL (± 30%), 32 μg/mL (± 60%) and 64 μg/mL (± 80%) compared to the treated control. The morphological, CFU and total biomass alteration was dose dependent.

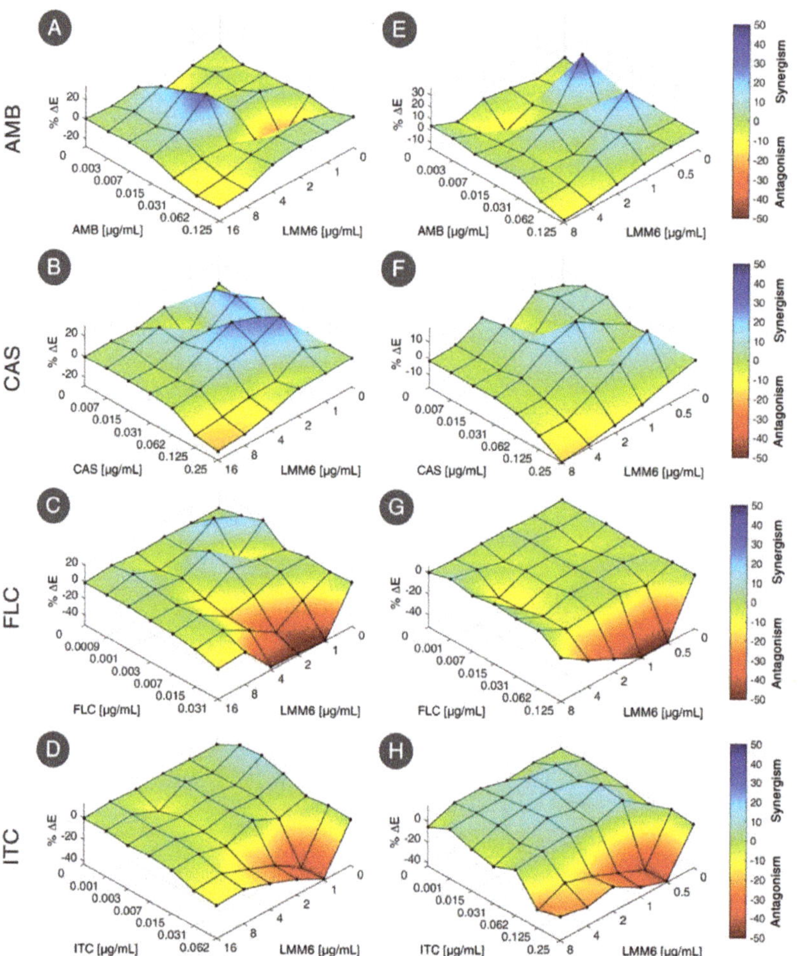

Figure 2. Bliss independence surface analysis for interaction of LMM6 with antifungal conventional drugs (AMB: amphotericin B; CAS: caspofungin; FLC: fluconazole; ITC: itraconazole). Evaluated effect against reference strain (**A–D**) and one clinical isolate from hospitalized patient blood (SangHUMCa7; (**E–H**)). The x axes represent the antifungal conventional drugs and y axes the LMM6. The magnitude of interactions is directly related to percent ΔE (%ΔE; z axes). Interactions with positive %ΔE represent synergistic effect statistically significant whereas that combinations with negative %ΔE indicate antagonism or no effect. The experimental data were analyzed independently using the Combenefit software.

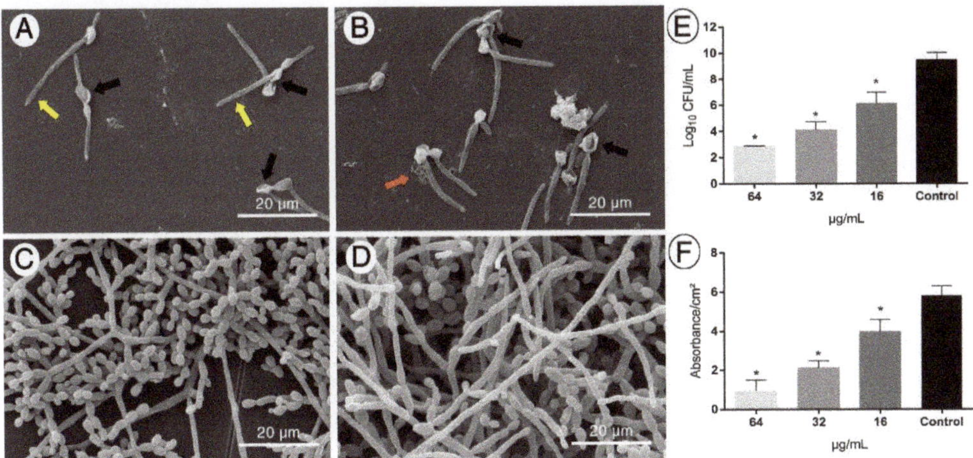

Figure 3. Anti-biofilm effect of LMM6 against reference strain. (**A–D**) Scanning electron microscopy (SEM); Representation of the analysis of at least 20 fields. (**E**) Biofilm inhibition by logarithmic reduction of colony forming units (CFU). (**F**) Reduction of total biomass evaluated by staining with crystal violet. LMM6 at concentrations of 64 µg/mL (**A**), 32 µg/mL (**B**) and 16 µg/mL (**C**) were added to the pre-adhered yeast (2h) in polystyrene plate and incubated for 24 h at 37 °C for analysis. Control (**D**): Untreated biofilm containing only RPMI 1640 medium. Black arrows indicate deformities on the cells; Red arrow shown cell extravasation and yellow arrow are membrane and cell wall irregularities. Each data represents the mean ± standard deviation (error bars). * $p < 0.05$, statistically significant reduction in relation to control. The bar in the images corresponds to 20 µm. Magnification × 5000.

2.4. LMM6 Low Toxicity in Male Balb/c Mice

The acute toxicity in mice was evaluated by intraperitoneal or intravenous LMM6 administration. Single dose of LMM6 caused no death in male mice during 14 days of observation. The reduction of locomotion marked by lethargy and piloerection was observed in all groups treated (LMM6 or vehicle). However, these mild behavioral changes returned to normal after 24 h (Supplementary Material, Table S2). Increased heart rate was also noted in all groups, including healthy mice, and returned to normal soon after manipulation.

The relative weight of organs (brain, heart, kidneys, liver, and spleen), subsequent to euthanasia, showed no significant changes after treatment with LMM6 or vehicle (Figure 4). Regardless of the administration route, the vehicle was able to increase the lung weight ($p < 0.05$), as compared to the healthy group (Figure 4E). Macroscopical observations of the organs demonstrated no changes in their color as well as texture. Body weight was not affected by the treatments. No significant difference was recorded when comparing treated mice with healthy (Supplementary Material, Figure S2).

The results of biochemical analysis in acute toxicity assay are summarized in Figure 5. Both, LMM6 and vehicle did not cause significant alteration in AST, ALT, CRE, UR and GLU levels when compared to healthy group. Exposure to LMM6 at high concentrations by two administration routes did not lead to liver or renal toxicity. The high dose effect of LMM6 on hematological parameters is presented in Table 3. According to the findings, MCHC index was statistically different ($p < 0.05$) in mice treated (LMM6 or vehicle) when compared to healthy group. The platelet counts only differed for mice treated with vehicle (IP) ($p < 0.05$). Others hematological parameters, as total RBC count, hematocrit, hemoglobin, MCV and MCH were within normal limits and showed no significant change in the analyzed groups. In differential count (Table 4), the leukocytes values increased significantly ($p < 0.05$) in mice treated with vehicle (IV). However, all other counts, neutrophils, monocytes, lymphocytes and eosinophils remained similar among all treated groups with healthy group. Fortunately, this result did not indicate any adverse trend associated with LMM6 treatment and suggests that its application does not promote any abnormalities of blood cells and components in the blood fluids.

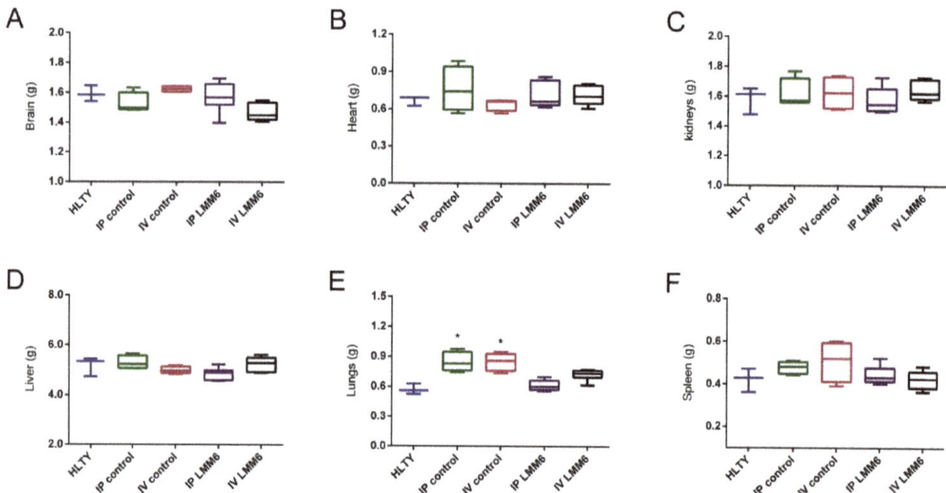

Figure 4. LMM6 effect on organs relative weight of male Balb/c mice in acute toxicity study. (**A**) Brain weight; (**B**) Heart weight; (**C**) Kidneys weight; (**D**) Liver weight; (**E**) Lungs weight and (**F**) spleen weight. HLTY: healthy animals; IP control: treated intraperitoneally with the vehicle; IV control: treated intravenously with the vehicle; IP LMM6: treated with 50 mg/kg of LMM6 intraperitoneally; IV LMM6: treated with 25 mg/kg of LMM6 intravenously. Organs weight were determined following 14 days exposure to high LMM6 concentration in single dose. Each data represents the mean $\pm$ standard deviation (error bars). * $p < 0.05$, statistically significant changes in relation to mice.

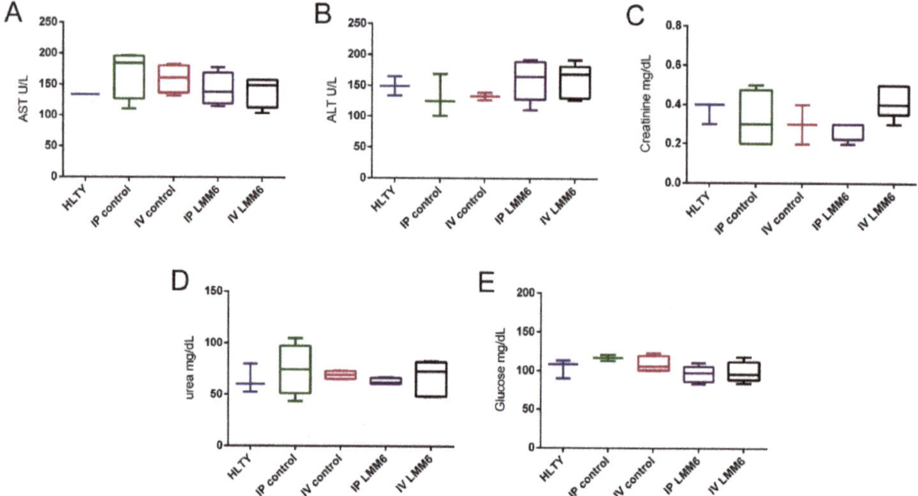

Figure 5. LMM6 effect on biochemical parameters of male Balb/c mice in acute toxicity study. (**A**) Serum levels of aspartate aminotransferase; (**B**) Serum levels of alanine aminotransferase; (**C**) Serum levels of creatinine; (**D**) Serum levels of urea and (**E**) serum levels of glucose. HLTY: healthy animals; IP control: treated intraperitoneally with the vehicle; IV control: treated intravenous with the vehicle; IP LMM6: treated with 50 mg/kg of LMM6 intraperitoneally; IV LMM6: treated with 25 mg/kg of LMM6 intravenous. Biochemical parameters were determined following 14 days exposure to high LMM6 concentration in single dose. Each data represents the mean $\pm$ standard deviation (error bars).

Table 3. Hematological parameters of male Balb/c mice exposed to high LMM6 concentration in acute toxicity assessment.

Hematological Parameters	Analyzed Groups				
	Healthy	**IP Control**	**IV Control**	**IP LMM6**	**IV LMM6**
Total RBC (10^6/mm^3)	8.39 ± 0.53	8.96 ± 0.92	9.13 ± 0.32	8.96 ± 0.58	8.31 ± 1.28
Haematocrit (%)	43.66 ± 0.58	45 ± 1	45.5 ± 1	44.43 ± 0.53	46 ± 0
Hemoglobin (g/dL)	21.48 ± 0.64	20.31 ± 0.40	20.90 ± 0.69	20.65 ± 0.58	21.01 ± 0.21
MCV (fL)	52.15 ± 3.08	50.31 ± 5.11	49.86 ± 0.98	49.77 ± 3.67	53.46 ± 3.40
MCH (pg)	25.64 ± 1.15	22.83 ± 2.06	22.89 ± 0.46	23.05 ± 1.49	24.02 ± 1.44
MCHC (%)	49.21 ± 1.83	44.88 ± 0.50 *	45.92 ± 0.73 *	46.49 ± 1.37 *	45.27 ± 1.03 *
Platelet (10^3/mm^3)	355.66 ± 71.28	570.00 ± 125.79 *	406.00 ± 42.68	402.71 ± 48.33	394.00 ± 77.24

Hematological parameters were determined following 14 days exposure to LMM6 in single dose. Abbreviations; Healthy: normal mice; IP control: treated intraperitoneally with the vehicle; IV control: treated intravenous with the vehicle; IP LMM6: treated with 50 mg/kg of LMM6 intraperitoneally; IV LMM6: treated with 25 mg/kg of LMM6 intravenous. RBC: Red blood cells; MCV: mean corpuscular volume; MCH: mean corpuscular hemoglobin; MCHC: mean corpuscular hemoglobin concentration. Values represent the mean $\pm$ SD. * $p < 0.05$, statistically significant changes compared with healthy control.

Table 4. Differential count of peripheral blood leukocytes of male Balb/c mice exposure to high LMM6 concentration in acute toxicity assessment.

	Leukogram					
Groups	**Leukocytes**	**Neutrophils**	**Monocytes**	**Lymphocytes**	**Eosinophils**	**Basophils**
	10^3/mm^3			% (10^3/mm^3)		
Healthy	4.5 ± 1.9	18 ± 4.36 (0.83 ± 0.62)	1 ± 1 (0.05 ± 0.05)	78.66 ± 5.14 (3.52 ± 1.47)	2.33 ± 0.58 (0.11 ± 0.06)	-
IP control	7.9 ± 2.0	17.25 ± 1.70 (1.36 ± 0.35)	0.5 ± 1 (0.04 ± 0.09)	80 ± 3.74 (6.37 ± 1.82)	2.25 ± 1.5 (0.16 ± 0.12)	-
IV control	10.0 ± 2.6 *	20 ± 3.91 (2.05 ± 0.77)	0.5 ± 0.58 (0.05 ± 0.06)	77.75 ± 4.19 (7.77 ± 2.02)	1.75 ± 1.5 (0.15 ± 0.12)	-
IP LMM6	5.4 ± 2.4	18.14 ± 4.45 (1.06 ± 0.62)	1 ± 0.63 (0.06 ± 0.06)	78.86 ± 4.88 (4.21 ± 1.70)	2.14 ± 1.21 (0.11 ± 0.11)	-
IV LMM6	6.0 ± 1.9	19.66 ± 2.94 (1.14 ± 0.25)	0.66 ± 0.51 (0.05 ± 0.04)	77.66 ± 2.16 (4.71 ± 1.65)	2 ± 0.82 (0.13 ± 0.08)	-

Differential count of peripheral blood leukocytes was realized following 14 days exposure to LMM6 in single dose. Abbreviations; Healthy: normal mice; IP control: treated intraperitoneally with the vehicle; IV control: treated intravenous Differential count of peripheral blood leukocytes were realized following 14 days exposure to LMM6 in single dose. Abbreviations; Healthy: normal mice; IP control: treated intraperitoneally with the vehicle; IV control: treated intravenous with the vehicle; IP LMM6: treated with 50 mg/kg of LMM6 intraperitoneally; IV LMM6: treated with 25 mg/kg of LMM6 intravenous. Values represent the mean $\pm$ SD. * $p < 0.05$, statistically significant changes compared with healthy control.

2.5. Efficacy of LMM6 on the Treatment of Systemic Candidiasis in a Murine Model

The in vivo efficacy of LMM6 was also determined. The groups treated twice a day for 5 days with LMM6 (5 mg/kg) or FLC (5 mg/kg) presented significant reduction in fungal burden in the kidney (CFU; ~1.80 $\log_{10}$ and ~3.60 $\log_{10}$, respectively) and spleen (CFU; 0.9 $\log_{10}$ and ~1.20 $\log_{10}$, respectively) when compared to control group ($p < 0.05$; Figure 6A,B). Moreover, visible white lesions covering the kidneys surface were observed in the control group, while in LMM6-treated mice or FLC, the kidneys were apparently healthy (Figure 6C). Histopathological analysis showed tissue damage with increased presence of inflammatory infiltrates and a large amount of yeasts and hyphae in the control group (Figure 6D, a-a'), which decreased considerably in the mice treated with LMM6 (Figure 6D, b-b'). For the FLC group, the fungal cells were not visible in the tissue (Figure 6D, c-c').

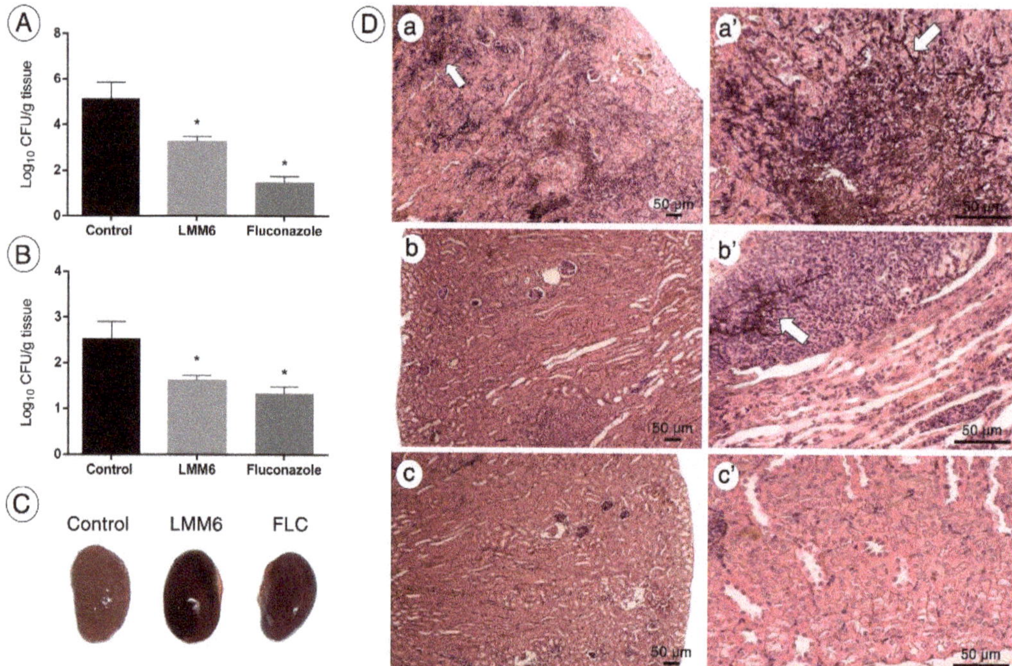

Figure 6. In vivo effect of LMM6 on treatment of mice with systemic candidiasis by *C. albicans*. The mice were infected with reference strain (5×10^5 yeast cells) and treated twice a day for 5 days via intraperitoneal injection. Control: treated with PBS and vehicle (DMSO 1% and Pluronic F-127 0.2%); LMM6: treated with LMM6 compound (5 mg/kg) and FLC: treated with fluconazole (5 mg/kg). (**A**) Fungal burden in the kidneys of the mice. (**B**) Fungal burden in the spleen of the mice. Colony Forming Units ($\log_{10}$ CFU) per gram of organ. Each data point represents the mean $\pm$ standard deviation (error bars). * Values of $p < 0.05$ were considered statistically significant. (**C**) Surfaces of the kidneys of untreated mice (control) covered with *Candida* lesions, while the kidneys of the LMM6-treated or FLC mice with healthy appearance. (**D**) Kidney histological section stained with hematoxylin eosin and Grocott-Gomori. The bar in the images corresponds to 50 μm. White arrows indicate the presence of fungus in the tissue that were easily spotted in the kidneys of control group (a and a′), whereas few were detected in the kidneys of LMM6-treated mice (b and b′). No fungus was observed on the FLC (c and c′). Representative kidney histopathological sections from 5 mice per group.

Treatment with LMM6 was able to reduce pro-inflammatory cytokine levels, similar to the action of FLC (Figure 7). Four cytokines (IL2, IL6, IFN-γ and TNF-α) were detected in both serum and kidney. In serum, cytokines IL-2 and IFN-γ showed no significant differences in all groups tested. In the kidneys, IL-2 concentration was reduced for FLC group and IFN-γ decreased for LMM6 ($p < 0.05$). In both serum and kidney, IL-6 and TNF-α presented reduction when treated with LMM6 and FLC as compared to control ($p < 0.05$).

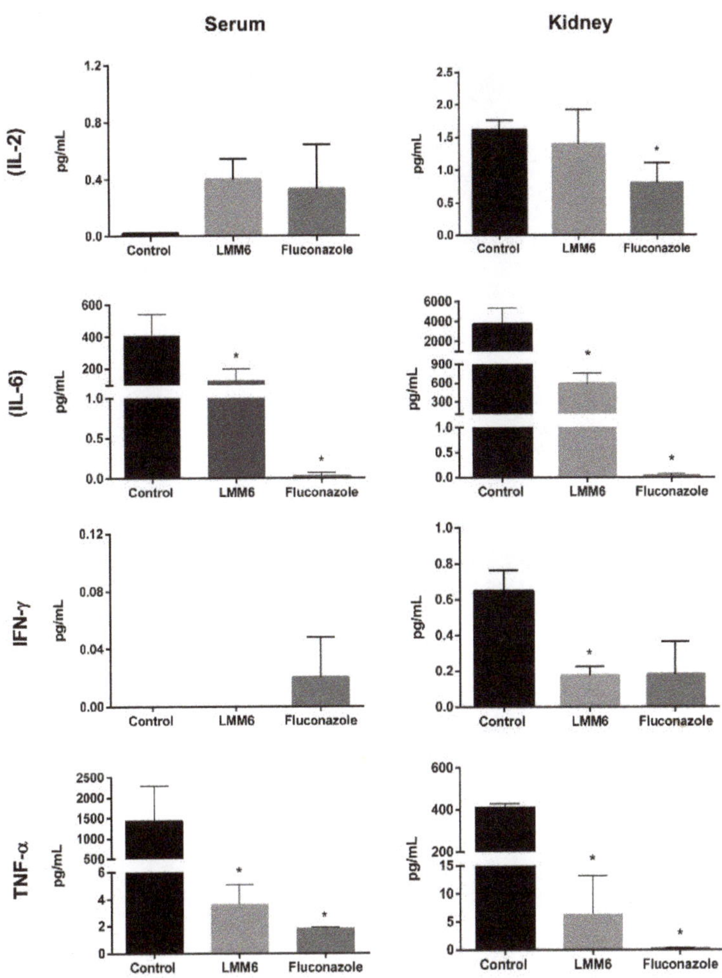

Figure 7. Cytokine detection in the kidney and serum of mice infected systemically with *C. albicans* and treated with LMM6, fluconazole or phosphate buffered saline (PBS). Control: treated with PBS and vehicle (DMSO 1% and Pluronic F-127 0.2%); LMM6: treated with LMM6 compound5 mg/kg; Fluconazole: treated with 5 mg/kg; IL-2: Interleukin-2; IL-6: Interleukin-6; IFN-γ: interferon-γ; TNF-α: tumor necrosis factor-α. The BD^{TM} Cytometric Bead Array (CBA) Mouse Inflammation Kit was used and submitted to BD FACSCalibur™ flow cytometer. Each data point represents the mean ± standard deviation (error bars). * Values of $p < 0.05$ were considered statistically significant in relation to control.

3. Discussion

Limited therapeutic arsenal and increase of drug resistance has intensified the search for new antifungals [24]. Our group had sought new therapeutic options by in silico approaches, as virtual screening based on compounds similarity. LMM6 was identified as a direct analogue of a 1,3,4-oxadiazole compound (LMM11), which targets Trr1 from *C. albicans* [17]. This flavoenzyme has been shown an important target for the new drugs development. *TRR1* gene is essential and conserved in several pathogenic fungi. Additionally, it is absent in humans, that can contribute with the development of selective drugs against pathogens [25]. 1,3,4-oxadiazole-based drugs have been widely studied by researchers [21–23,26–29]. These compounds have a high therapeutic power and broad spectrum of action, such as anticancer, antifungal, antibacterial, antitubercular, anti-inflammatory, antineuropathic, antihypertensive, antihistaminic, antiparasitic, antiobesity, antiviral activity, among others [29].

In this study, we demonstrated a more promising effect of LMM6, as compared to others 1,3,4-oxadiazoles previously tested in our laboratory [21–23]. The LMM6 antifungal activity in vitro clearly reveals reduction in the number of *C. albicans* viable cells (CFU quantitative analysis) in which it was confirmed by growth decrease by CFM. Evidence shows that the fungicidal therapy with echinocandins or amphotericin B yield higher initial cure rates and reduced microbial persistence than fungistatic therapy with triazoles in treatment of invasive *Candida* infection [30]. Our findings indicate that LMM6 presented fungicidal profile with dose and time dependent activity. The fungicidal activity of LMM6 is excellent in comparison with its analogue, which showed fungistatic activity against *C. albicans* [22]. The substitution of the furan-2-yl radical present in LMM11 by 4-fluorophenyl in LMM6 was sufficient to enhance the antifungal effect. Although the MIC90 value was unchanged, the MFC values were reduced from 128 µg/mL (LMM11) to 16ug/mL (LMM6), i.e., the presence of the aromatic ring containing a halogen atom was essential for the fungicidal effect [21,22]. Halogenated compounds are known to show more promising antifungal activity [31]. These results indicate that LMM6 could serve as a scaffold for other substitutions, enhancing the antifungal effects.

The complex biofilm structure contributes to infection persistence, high mortality rates, and is also linked to the enhanced capability of *C. albicans* to resist antifungal [7,8,32]. LMM6 was able to affect the structure of the *C. albicans* biofilm, acting both on yeast reduction and extracellular matrix. These findings were confirmed by SEM, which showed disorganization of architecture, cell extravasation, deformities and irregularities on the yeast cells. In addition, this compound can exert an influence on the membrane surface of the cells preventing adhesion, compromising biofilm formation and leading to surface detachment, as seen in another study [33].

The compounds combination is an excellent strategy due to advantages to use lower drug concentrations, wider spectrum of efficacy and reduction of adverse and toxic effects [15,16,34]. This approach has been explored against resistant *C. albicans* strains [35–37]. Strong synergistic effects of LMM6 with fungicidal conventional drugs (AMB and CAS) were observed in this investigation. These results represent an attractive prospect for the development of new strategies to manage candidiasis treatment. Differently, the combinations between LMM6 and FLC or ITC have no synergistic effect. The absence of synergistic interaction between fungistatic and fungicidal compounds has been described in the literature. In vitro and in vivo studies have shown that the combination fluconazole and amphotericin B has an indifferent or antagonistic effect on different *Candida* species. [34,38].

Acute single-dose toxicity testing using small animal models is imperative in order to predict the adverse effects of a new therapeutic compound. In this study, we demonstrated that both intraperitoneal and intravenous administration of LMM6 were well tolerated by the mice, causing no modifications in clinical status, mortality or general behavior of animals at the tested doses. Changes in the body weights of mice have been used as an indicator of toxic effects of chemicals and drugs [39,40]. Likewise, decreases or increases organs relative weight are also important and sensitive parameter for toxicology studies [41]. Both, body weight and organ relative weight such as brain, heart, kidneys, liver, lungs and spleen were normal indicating no toxic effect in treated groups with LMM6.

The liver is one of the main target organs for drugs or chemicals due to its important role in the metabolism of these compounds [42]. Elevated serum levels of enzymes such as ALT and AST have been directly linked with liver injury [43]. Kidneys are also frequently sites for drug toxicity because of its function in eliminating xenobiotics and metabolites [44]. Creatinine and urea are commonly used as indicators of renal function [45]. Although not very sensitive and specific, alterations in these parameters may indicate kidney damage [45,46]. We highlight that no significant changes were found in the biochemical analyses between the groups treated with LMM6 and healthy control.

Similarly, hematological system analyses may assist in the determination of adverse effects of developing drugs [47]. During the trial, LMM6 did not affect the various hematological parameters evaluated, except for MCHC, which number decreased significantly in all groups. Despite the difference between the groups, the MCHC index remained high in relation to reference standard considered normal for this species [48,49].

The in vivo effect of LMM6 treatment seems to be very promising for controlling *C. albicans* infection. Intraperitoneal administration of LMM6 (5 mg/kg, twice a day) was able to reduce both spleen and kidneys fungi burden. Kidneys are one of the main target organs of interest in disseminated candidiasis because besides the other organs do not show persistent colonization by *Candida*, the renal fungal load is directly related to lethality in mice [50–52]. Although the results of the fungal burden have been very encouraging, experiments to determine whether LMM6 confers greater animal survival during fungal infection must be performed. In addition, histopathological findings suggest a qualitative reduction of inflammatory infiltrates in mice treated with LMM6 as compared to control. This effect was associated with decrease of fungal load in this organ. Although neutrophils are crucial in the response against fungal pathogens, *C. albicans*-induced excessive infiltration into renal tissue during systemic candidiasis can be deleterious and result in kidney failure and/or mortality [51,53].

Pro-inflammatory cytokines are also highly associated with the response of host against *C. albicans* infection [54]. Increased levels of IL-2 suggest that LMM6 contributes to the protection against disseminated candidiasis. In parallel, LMM6 treatment caused reductions in TNF-α and IL-6 cytokines compared to control animals. Suppresses inflammatory cytokines promoted homeostatic restoration in mice similarly to FLC. Similar data were found by Basso et al., in which the treatment efficacy was associated with modulation of pathologic inflammation [55]. These findings can overcome immune response-mediated tissue damage being an advantage in the treatment of *Candida* infections.

4. Materials and Methods

4.1. Strains and Growth Conditions

A total of thirty clinical isolates of *C. albicans* from hospitalized patients (collection of fungal culture approved by the Human Research Ethic Committee no. 2.748.843) and one reference strain *C. albicans* ATCC 90028 were used. The clinical isolates were (Supplementary Material, Table S1) identified by classic methods [56] and confirmed with Matrix-assisted laser-desorption/ionization time-of-flight mass spectrometry (MALDI TOF-MS [57]). Prior to each experiment, the strains were cultivated on Sabouraud dextrose agar (SDA; Difco™, MI, USA), and the cellular density was adjusted using a Neubauer chamber. Experiments were performed with the reference strain, except for the antifungal susceptibility and checkerboard assay.

4.2. Chemical Compounds

4-[cyclohexyl(ethyl)sulfamoyl]N[5-(4-fluorophenyl)-1,3,4-oxadiazol-2-yl]benzamide (LMM6) was purchased from Life Chemicals Inc. (Burlington, ON, Canada; F2832-0106). Dimethyl sulfoxide (DMSO) was used to prepare LMM6 stock solution (50 µg/mL). Prior to each experiment, LMM6 was solubilized with non-ionic surfactant from Sigma–Aldrich (St Louis, MO, USA; Pluronic® F-127 at 0.02%). Diluents were used as control. The conventional antifungals amphotericin B (AMB) and caspofungin (CAS) were acquired commercially from Sigma–Aldrich (St Louis, MO, USA); itraconazole (ITC) and fluconazole (FLC) were obtained from Pfizer (New York, NY, USA).

4.3. Antifungal Susceptibility Testing

The minimum inhibitory concentration (MIC) of LMM6 (0.5–256 µg/mL), AMB (0.032–16 µg/mL), CAS (0.032–16 µg/mL), FLC (0.125–64 µg/mL) and ITC (0.032–16 µg/mL) was determined for 30 clinical isolates and reference strain, based on broth microdilution method, according to document M-27A3 (CLSI) with modifications [58]. The

LMM6 and drugs conventional concentrations were prepared in RPMI-1640 medium (Gibco/Invitrogen, Grand Island, NY, USA) and incubated in 96-well plates with yeast (2–3×10^3 cells/mL) for 24 h at 35 °C. Two controls were considered: negative (only medium without inoculum) and positive (medium plus inoculum). The LMM6 MIC values were determined by measuring the absorbance at 405 nm in a SpectraMax ® Plus 384 plate reader (Molecular Devices, Sunnyvale, CA, USA) and defined as the lowest concentration able to inhibit the growth $\geq$ 80% in relation to positive control. For conventional antifungal agents, MIC values were determined visually: for azoles (FLC and ITC), it was defined as the concentration which resulted in 50% reduction of fungal growth and for polyenes (AMB) and echinocandins (CAS), it was defined as the concentration that was not visible fungal growth in relation to the positive control. The conventional antifungals MICs were interpreted according to the M27-S4 document [59].

The minimum fungicidal concentration (MFC) of LMM6 was determined for all clinical isolates and reference strain by subculture in SDA. The plates were incubated at 35 °C for 24 h. The MFC was defined as the lowest LMM6 concentration at which no colony growth was visible. A quantitative analysis of LMM6 antifungal activity was also performed for each concentration LMM6 tested. After 24 h, aliquots were serially diluted in phosphate buffered saline (PBS), subcultured on SDA medium without the compound and incubated at 35 °C for 24 h for colony-forming units (CFU) counting.

4.4. Time-Kill Curve Assay

LMM6 at concentrations of 16, 32, 64 and 128 µg/mL were incubated in 24-well plates at 35 °C with yeast suspension (2–3×10^3 cells/mL) in RPMI-1640. Two controls were prepared: drug-free (only culture medium) and conventional antifungal (FLC; 0.25 µg/mL). At predetermined time points (0, 2, 4, 6, 8, 12, 24, 28, 36, and 48 h), aliquots were obtained from each condition tested, serially diluted in PBS, plated on SDA and incubated at 35 °C for 24 h for CFU determination. Mean counts ($\log_{10}$ CFU/milliliter) were plotted as a function of time for each concentration of LMM6 or FLC tested. The fungicidal activity was defined as a $\geq$ 99.9% (3 $\log_{10}$) reduction in numbers of CFU and fungistatic activity was defined as a < 99.9% reduction in growth compared to the control [60,61].

4.5. Checkerboard Assay and Bliss-Independent Interactions Analysis

Synergistic interaction between LMM6 and conventional antifungals (AMB, CAS, ITC or FLC) were performed against one clinical isolate (SangHUMCa7 clinical isolate from hospitalized patient blood that has been identified by classical methods and confirmed by MALDI TOF-MS) and reference strain by checkerboard assay [62]. The inhibition of fungal growth was determined by measuring the absorbance at 405 nm on SpectraMax ® Plus 384 plate reader (Molecular Devices, Sunnyvale, CA, USA) after 24 h. The analysis of the combination of the compounds was carried out calculating the fractional inhibitory concentration (FIC) index, which is defined as the sum of $FIC^A + FIC^B$ ($MIC_{\text{drug conventional}}$ in combination/$MIC_{\text{drug conventional}}$ alone + MIC_{LMM6} in combination/MIC_{LMM6} alone). Drug interactions were classified as FIC values < 0.5 indicate strongly synergistic effect, FIC < 1 synergistic effect, FIC = 1 additive effect, 1 < FIC < 2 no effect and FIC > 2 antagonistic effect [63]. In addition, all data were analyzed by Combenefit software to obtain Bliss-independent interactions [64]. Interactions with positive %ΔE represent synergistic effect statistically significant.

4.6. Effects of LMM6 on Biofilm Formation

The anti-biofilm activity was determined as previously described by Tobaldini-Valerio et al. [65]. A cell suspension prepared in RPMI 1640 medium at a density of 1×10^5 cells/mL was added in 96-well plates (200 µL) and incubated at 37 °C on a shaker at 120 rpm for 2 h to allow adhesion phase. After the incubation, the medium was aspirated from the wells and non-adherent cells were removed by washing with sterile PBS. Subsequently, 200 µL of LMM6 without surfactant F-127, at concentrations 16, 32 and 64 µg/mL were added to form biofilm and incubated at 37 °C/120 rpm/24 h. Untreated biofilm containing only RPMI 1640 medium was used as control. Biofilm inhibition was evaluated by total biomass (crystal violet staining; CV), viability cells (determination of CFU) and scanning electron microscopy (SEM). Before any evaluation, the total medium was removed from wells and the biofilm washed once with PBS to ensure the removal of unadhered cells and the residual of LMM6.

4.6.1. Determination of Total Biomass by Crystal Violet

To assess biofilm total biomass, the wells were fixed with methanol 100% for 15 min. After methanol removal, the 96-well plates were dried at room temperature. CV (0.1% v/v) was added to the wells (100 μL) for 5 min. The wells were washed twice with sterile distilled water and 200 μL of acetic acid (33% v/v) was added to dissolve the stain. The absorbance at 570 nm was measured in a SpectraMax $^®$ Plus 384 plate reader (Molecular Devices, Sunnyvale, CA, USA).

4.6.2. Quantification of Viable Biofilm Cells

To determine the number of viable cells, the wells were scraped with PBS using a tip until getting a final volume of 300 μL. The suspensions were vigorously vortexed for 1 min to disaggregate biofilm matrix. Serial dilutions were made in PBS, plated onto SDA and incubated for 24 h at 35 °C for quantification of $\log_{10}$ CFU/mL.

4.6.3. Effect of LMM6 on Biofilm Structure

The effect of LMM6 on biofilm structure was evaluated by SEM. The bottom of the 96-well plate was detached, and the biofilm was fixed by immersion in 2.5% glutaraldehyde (Merck KGaA, Darmstadt, HE, Germany) diluted in 2% paraformaldehyde and in 0.1 M sodium cacodylate buffer (Sigma–Aldrich, MO, USA). The samples were dehydrated in an ethanol series (70%, 80%, 90% and 100%) and coated with gold (Baltec SDC 050 sputter coater) for observation using a scanning electron microscope FEI Quanta 250 (Hillsboro, OR, USA) at 5000× magnification [66]. The images are representative of at least 20 fields.

4.7. Ethics Statement

All animal procedures were performed according to Brazil's National Council for the Control of Animal Experimentation (CONCEA) and approved by the Ethics Committee for Animal Use of the State University of Maringá, PR, Brazil (protocol number CEUA 3855010719). The animals were kept with free access to water and food, in a controlled animal facility having a constant temperature of 22–24 °C, relative humidity of 50–60% and a 12 h light/dark cycle.

4.8. Evaluation of the Acute Toxicity

The LMM6 acute toxicity in a single dose was evaluated according to the guide for conducting non-clinical toxicology and pharmacological safety studies necessary for drug development (Brazilian Health Surveillance Agency; ANVISA [67]). Inbred male Balb/c mice (n = 23), 6–7 weeks old, weighed 20–30 g, were randomly divided into five groups: HLTY group (healthy animals that just received PBS; n = 3); IP control group (treated intraperitoneally with the vehicle; PBS, DMSO 1%, and Pluronic F-127 0.2%; n = 4); IV control group (treated through the lateral tail vein/intravenous with the vehicle (PBS, DMSO 1%, and Pluronic F-127 0.2%; n = 4); IP LMM6 group (treated with 50 mg/kg of LMM6 intraperitoneally; n = 6) and IV LMM6 group (treated with 25 mg/kg of LMM6 through the lateral tail vein/intravenous; n = 6). The dose administered was calculated according to the body weight for each mice.

The animals were monitored by Hippocratic screening at times 0, 15, 30, 60, 120, 240 min and daily in which were observed clinical and behavioral parameters that included general appearance, motor coordination (touch and tail response, righting reflex and ataxy), muscle tone (paws, body, grip strength), reflexes, lethargy, central nervous systems activity

(CNS; tremors, convulsions, muscle contractions, sedation, hypnosis, and anesthesia) and autonomic nervous system (ANS; urination, defecation, piloerection, rate of respiration, heart rate), over a period of 14 days post treatment. Body weight of mice was recorded daily during the whole experiment. After the 14th day, mice were humanely anesthetized for blood collection and then euthanized with isoflurane vaporizer. The organs, brain, heart, kidneys, liver, lungs and spleen were removed, weighed and the relationship between organ weight and body weight (RW) of each mice was established for all groups. Relative weight (RW) = organ weight (g)/body weight on sacrifice day × 100 [68].

Biochemical and Hematological Analyzes

Blood collection was done from retro-orbital sinus in microtubes with and without Ethylenediaminetetraacetic acid (EDTA 10%) for hematological and biochemical analysis, respectively. The blood without the EDTA was left at 37° C to coagulate, centrifuged at 5000 rpm, 20 °C for 5 min to obtain serum and stored at −80 °C. Aspartate aminotransferase (AST), alanine aminotransferase (ALT), creatinine (CRE), urea (UR) and glucose (GLU) were determined according to the manufacturer's instructions using standard diagnostic kits (Gold Analisa Diagnóstica, Brazil) and a semi-automatic biochemistry analyzer BIOPLUS-2000 (Barueri, SP, BR).

Hematology parameters were performed using standard hematological manual methods [69,70]. The analyzes started immediately after blood collection and included: Red blood cells count (RBC; erythrocytes) with hayem liquid, white blood cell count (WBC; leukocytes) with türk solution; platelet count (PLT) by the method of Brecher and Cronkite [71]; hemoglobin concentration (Hb) based on the cyanmethemoglobin method [72] and hematocrit (Ht) by the microhematocrit method. Blood smears were prepared with 20 µL whole blood and May-Grunwald-Giemsa-stained for differential leukocyte count. Using the RBC, Ht and Hb measurement, the hematimetric indexes were calculated as follows: mean corpuscular volume (MCV= Ht × 10/RBC); mean corpuscular hemoglobin (MCH = Hb × 10/RBC) and mean corpuscular hemoglobin concentration (MCHC = Hb/Ht × 100).

4.9. LMM6 Antifungal Activity in a Murine Model of Systemic Candidiasis

Systemic candidiasis model was established in female inbred Balb/c mice (n = 18), 6–7-week-old with ± 20 g of weight, according to previously described protocols [22,73]. After injection of 100 µL (5×10^5 yeast cells) of reference strain by the lateral tail vein, the mice were randomly divided into three experimental groups with 6 animals each: LMM6 (treated with LMM6 compound 5 mg/kg), FLC (treated with fluconazole 5 mg/kg) and Control (treated with diluent: PBS buffer, DMSO and Pluronic® F-127). The antifungal treatment started after 3 h of infection and it was conducted twice a day for 5 days via intraperitoneal. The animals were anesthetized for blood collection and then humanely euthanized by isoflurane vaporizer. The left kidney and spleen were aseptically removed for determination of fungal burden. The organs were weighed and then macerated with 1 mL of lysis buffer (200 mM NaCl, 5 mM EDTA, 10 mM Tris, 10% glycerol v/v, pH 8.30). The homogenates were serially diluted, plated on SDA and incubated for 24 h at 35 °C for CFU counts. Mean CFU counts were normalized by the weight of tissue sample (g).

4.9.1. Cytokines Detection by Flow Cytometry

The kidney homogenates were centrifuged at 14,000 rpm, 4 °C for 15 min and the supernatant transferred to microtubes containing protease inhibitor (GE Healthcare; Chicago, IL, USA). Blood collected from mice retro-orbital sinus was left at 37 °C to coagulate and centrifuged at 5000 rpm, 20 °C for 5 min to obtain serum. The samples were stored at −80 °C prior to analysis. Analysis of the kidney supernatant and blood for detection systemic and local cytokines, respectively, were conducted by BD™ Cytometric Bead Array (CBA) Mouse Inflammation Kit (BD Bioscience, San Jose, CA, USA) as per the manufacturer's instructions and was analyzed on BD FACSCalibur™ (BD Bioscience, San Jose, CA, USA) flow cytometer. The following cytokines were measured: Interleukin-2 (IL-2),

Interleukin-4 (IL-4), Interleukin-6 (IL-6), Interleukin-10 (IL-10), Interleukin-17a (IL-17a), interferon-γ (IFN-γ) and tumor necrosis factor-α (TNF-α). The results for the standard curves of each cytokines and samples were generated using FCAP Array software v3.0 (BD Biosciences, San Jose, CA, USA).

4.9.2. Histopathological Analysis

Immediately after euthanasia, the right kidney of mice was collected and immersed in paraformaldehyde 4% for fixation during 24 h. The organs were preserved in 100% ethanol until processing. Posteriorly, the samples were embedded in paraffin, sectioned longitudinally at 5 μm and stained with hematoxylin eosin (HE) and Grocott-Gomori (GG) for detection of inflammatory areas and fungi in situ. Histopathology images from tissues stained were obtained using a binocular light microscope (Motic BA310- camera Moticam 5), at × 200 and × 600 magnification. In qualitative analysis of fungal cells and inflammatory infiltrates, at least 20 fields of three histological sections were analyzed for each group.

4.10. Statistical Analysis

The data were evaluated as the mean ± standard deviation (SD) using Prism 6.0 software (GraphPad, San Diego, CA, USA). Reduction of kidney fungal burden on in vivo treatment was analyzed using unpaired Student's *t*-test and the other assays with one-way analysis of variance (ANOVA) using the Bonferroni. Values of $p < 0.05$ were considered statistically significant.

5. Conclusions

In conclusion, this is the first time the antifungal activity of LMM6 both in vitro and in vivo against *C. albicans* is described. These findings suggest important therapeutic potential of LMM6 due to its fungicidal ability, inhibition of biofilm formation, synergistic effect with conventional drugs and capacity to reduce renal fungal burden in a murine model of disseminated candidiasis. Immunomodulatory activity also increases the protective effects of LMM6 against *C. albicans* infection and maintains homeostasis. LMM6 compound has no hepatotoxic or nephrotoxic effects and does not interfere with the biochemical and hematological parameters in the mice, being safe for future applications as antifungal agent or in association with conventional drugs for the treatment of candidiasis.

6. Patents

Kioshima ES, Svidzinski TIE, Bonfim-Mendonça PS, et al. Composição farmacêutica baseada em compostos 1,3,4-oxadiazólicos e seu uso na preparação de medicamentos para tratamento de infecções sistêmicas. Brazil patent BR 10 2018 009020 8. 03 May 2018.

Supplementary Materials: The following are available online at https://www.mdpi.com/2076-0817/10/3/314/s1. Table S1: Specimen/source of 30 clinical isolates of *C. albicans* and antifungal susceptibility profile to LMM6 and conventional antifungal drugs; Table S2: General appearance and behavioral observations of male Balb/c mice exposed to high LMM6 concentration in the acute toxicity study; Figure S1: Minimum fungicidal concentration of LMM6 for 30 clinical isolates of *C. albicans*. After exposure of yeast to increased LMM6 concentrations (0.25–128 μg/mL) for 24 h, aliquots of 3 μL of the solution were transferred to SDA plates and incubated at 35 °C. Representative photograph of three independent experiments; Figure S2: Comparative body weight changes in male Balb/c mice during acute toxicity experimentation. The body weight of mice was weighed daily for 14 days following single dose administration of LMM6. Healthy: normal mice; IP control: treated intraperitoneally with the vehicle; IV control: treated intravenous with the vehicle; IP LMM6: treated with 50 mg/kg of LMM6 intraperitoneally; IV LMM6: treated with 25 mg/kg of LMM6 intravenous.

Author Contributions: Conceptualization, D.R.F., R.C.M. and E.S.K.; formal analysis, D.R.F., R.C.M. and I.R.G.C.; funding acquisition, P.d.S.B.-M., M.S.S.F., T.I.E.S. and E.S.K.; investigation, D.R.F., R.C.M., G.S.A., K.M.S., F.A.V.R.-V., I.R.G.C. and T.C.A.B.; methodology, D.R.F., R.C.M. and K.M.S.; project administration, D.R.F. and E.S.K.; resources, T.C.A.B., P.d.S.B.-M., M.S.S.F., T.I.E.S. and E.S.K.;

supervision, P.d.S.B.-M., T.I.E.S. and E.S.K.; validation, D.R.F., R.C.M., G.S.A., K.M.S., F.A.V.R.-V. and I.R.G.C.; visualization, D.R.F. and E.S.K.; writing—original draft, D.R.F. writing—review and editing, D.R.F., G.S.A. and E.S.K. All authors have read and agreed to the published version of the manuscript.

Funding: This work was supported by the National Council for Scientific and Technological Development (CNPq; http://www.cnpq.br, accessed on 5 March 2021) [552276/2011-1 to ESK and TIES]; and the Coordination for the Improvement of Higher Education Personnel (CAPES) to DRF (Finance Code 001). The funders had no role in study design, data collection and analysis, decision to publish, or preparation of the manuscript.

Institutional Review Board Statement: The study was conducted according to the Brazil's National Council guidelines for the Control of Animal Experimentation (CONCEA) and approved by the Ethics Committee for Animal Use of the State University of Maringá, PR, Brazil (protocol number CEUA 3855010719 in 5 July 2019). The collection of fungal culture also was approved by the Human Research Ethic Committee no. 2.748.843 in 2 July 2018.

Informed Consent Statement: Not applicable.

Data Availability Statement: The data presented in this study are available within the article or supplementary material.

Conflicts of Interest: The authors declare no conflict of interest.

References

1. Camargo, L.F.; Marra, A.R.; Pignatari, A.C.; Sukiennik, T.; Behar, P.P.; Medeiros, E.A.; Ribeiro, J.; Girão, E.; Correa, L.; Guerra, C.; et al. Nosocomial bloodstream infections in a nationwide study: Comparison between solid organ transplant patients and the general population. *Transpl. Infect. Dis.* **2015**, *17*, 308–313. [CrossRef]
2. Braga, I.A.; Campos, P.A.; Gontijo-Filho, P.P.; Ribas, R.M. Multi-hospital point prevalence study of healthcare-associated infections in 28 adult intensive care units in Brazil. *J. Hosp. Infect.* **2018**, *99*, 318–324. [CrossRef] [PubMed]
3. Ortega-Loubon, C.; Cano-Hernández, B.; Poves-Alvarez, R.; Muñoz-Moreno, M.F.; Román-García, P.; Balbás-Alvarez, S.; de la Varga-Martínez, O.; Gómez-Sánchez, E.; Gómez-Pesquera, E.; Lorenzo-López, M.; et al. The Overlooked Immune State in Candidemia: A Risk Factor for Mortality. *J. Clin. Med.* **2019**, *8*, 1512. [CrossRef]
4. Ghrenassia, E.; Mokart, D.; Mayaux, J.; Demoule, A.; Rezine, I.; Kerhuel, L.; Calvet, L.; De Jong, A.; Azoulay, E.; Darmon, M. Candidemia in critically ill immunocompromised patients: Report of a retrospective multicenter cohort study. *Ann. Intensive Care* **2019**, *9*, 62. [CrossRef] [PubMed]
5. Guinea, J. Global trends in the distribution of *Candida* species causing candidemia. *Clin. Microbiol. Infect.* **2014**, *20*, 5–10. [CrossRef] [PubMed]
6. Cortés, J.A.; Corrales, I.F. Invasive candidiasis: Epidemiology and risk factors. In *Fungal Infection*; InTechOpen: London, UK, 2018. [CrossRef]
7. Tascini, C.; Sozio, E.; Corte, L.; Sbrana, F.; Scarparo, C.; Ripoli, A.; Bertolino, G.; Merelli, M.; Tagliaferri, E.; Corcione, A.; et al. The role of biofilm forming on mortality in patients with candidemia: A study derived from real world data. *Infect. Dis.* **2018**, *50*, 214–219. [CrossRef] [PubMed]
8. Sandai, D.; Tabana, Y.M.; Ouweini, A.E.; Ayodeji, I.O. Resistance of *Candida albicans* biofilms to drugs and the host immune system. *Jundishapur J. Microbiol.* **2016**, *9*, e37385. [CrossRef]
9. Pappas, P.G.; Kauffman, C.A.; Andes, D.R.; Clancy, C.J.; Marr, K.A.; Ostrosky-Zeichner, L.; Reboli, A.C.; Schuster, M.G.; Vazquez, J.A.; Walsh, T.J.; et al. Clinical practice guideline for the management of candidiasis: 2016 update by the Infectious Diseases Society of America. *Clin. Infect. Dis.* **2016**, *62*, e1–e50. [CrossRef]
10. Von Lilienfeld-Toal, M.; Wagener, J.; Einsele, H.; Cornely, O.A.; Kurzai, O. Invasive fungal infection. *Dtsch. Arztebl. Int.* **2019**, *116*, 271–278. [CrossRef]
11. Laniado-Laborín, R.; Cabrales-Vargas, M.N. Amphotericin B: Side effects and toxicity. *Rev. Iberoam. Micol.* **2009**, *26*, 223–227. [CrossRef]
12. Brüggemann, R.J.; Alffenaar, J.W.; Blijlevens, N.M.; Billaud, E.M.; Kosterink, J.G.; Verweij, P.E.; Burger, D.M. Clinical relevance of the pharmacokinetic interactions of azole antifungal drugs with other coadministered agents. *Clin. Infect. Dis.* **2009**, *48*, 1441–1458. [CrossRef] [PubMed]
13. Whaley, S.G.; Berkow, E.L.; Rybak, J.M.; Nishimoto, A.T.; Barker, K.S.; Rogers, P.D. Azole antifungal resistance in *Candida albicans* and emerging non-*albicans Candida* species. *Front. Microbiol.* **2017**, *7*, 2173. [CrossRef]
14. Castanheira, M.; Deshpande, L.M.; Davis, A.P.; Rhomberg, P.R.; Pfaller, M.A. Monitoring antifungal resistance in a global collection of invasive yeasts and molds: Application of CLSI epidemiological cutoff values and whole-genome sequencing analysis for detection of azole resistance in *Candida albicans*. *Antimicrob. Agents Chemother.* **2017**, *61*, e00906–e00917. [CrossRef]

15. Campitelli, M.; Zeineddine, N.; Samaha, G.; Maslak, S. Combination antifungal therapy: A review of current data. *J. Clin. Med. Res.* **2017**, *9*, 451–456. [CrossRef]

16. Scorzoni, L.; de Paula, E.; Silva, A.C.; Marcos, C.M.; Assato, P.A.; de Melo, W.C.; de Oliveira, H.C.; Costa-Orlandi, C.B.; Mendes-Giannini, M.J.; Fusco-Almeida, A.M. Antifungal therapy: New advances in the understanding and treatment of mycosis. *Front. Microbiol.* **2017**, *8*, 36. [CrossRef] [PubMed]

17. Batool, M.; Ahmad, B.; Choi, S. A structure-based drug discovery paradigm. *Int. J. Mol. Sci.* **2019**, *20*, 2783. [CrossRef]

18. Mariappan, G.; Kumari, A. Virtual screening and its applications in drug discovery process. In *Computer Applications in Drug Discovery and Development*; IGI Global: Hershey, PA, USA, 2019; pp. 101–126. [CrossRef]

19. Arnér, E.S.; Holmgren, A. Physiological functions of thioredoxin and thioredoxin reductase. *Eur. J. Biochem.* **2000**, *267*, 6102–6109. [CrossRef]

20. Kioshima, E.S.; Svidzinski, T.I.; Bonfim-Mendonça, P.S.; Capoci, I.R.; Faria, D.R.; Sakita, K.M.; Morelli, F.; Rodrigues, F.A.; Felipe, M.S.; Chaucanés, C.P.; et al. Composição Farmacêutica Baseada em Compostos 1,3,4-Oxadiazólicos e Seu Uso na Preparação de Medicamentos para Tratamento de Infecções Sistêmicas. Brazil Patent BR 10 2018 009020 8, 3 May 2018.

21. Rodrigues-Vendramini, F.A.; Faria, D.R.; Arita, G.S.; Capoci, I.R.; Sakita, K.M.; Caparroz-Assef, S.M.; Becker, T.; Bonfim-Mendonça, P.S.; Felipe, M.S.; Svidzinski, T.I.; et al. Antifungal activity of two oxadiazole compounds for the paracoccidioidomycosis treatment. *PLoS Negl. Trop. Dis.* **2019**, *13*, e0007441. [CrossRef]

22. Capoci, I.R.; Sakita, K.M.; Faria, D.R.; Rodrigues-Vendramini, F.A.; Arita, G.S.; de Oliveira, A.G.; Felipe, M.S.; Maigret, B.; Bonfim-Mendonça, P.S.; Kioshima, E.S.; et al. Two new 1,3,4-oxadiazoles with effective antifungal activity against *Candida albicans*. *Front. Microbiol.* **2019**, *10*, 2130. [CrossRef] [PubMed]

23. Faria, D.R.; Sakita, K.M.; Capoci, I.R.; Arita, G.S.; Rodrigues-Vendramini, F.A.; de Oliveira Junior, A.G.; Felipe, M.S.; Bonfim-Mendonça, P.S.; Svidzinski, T.I.; Kioshima, E.S. Promising antifungal activity of new oxadiazole against *Candida krusei*. *PLoS ONE* **2020**, *15*, e0227876. [CrossRef] [PubMed]

24. Campoy, S.; Adrio, J.L. Antifungals. *Biochem. Pharmacol.* **2017**, *133*, 86–96. [CrossRef] [PubMed]

25. Abadio, A.K.; Kioshima, E.S.; Leroux, V.; Martins, N.F.; Maigret, B.; Felipe, M.S. Identification of new antifungal compounds targeting thioredoxin reductase of *Paracoccidioides* genus. *PLoS ONE* **2015**, *10*, e0142926. [CrossRef] [PubMed]

26. Dhara, D.; Sunil, D.; Kamath, P.R.; Ananda, K.; Shrilakshmi, S.; Balaji, S. New oxadiazole derivatives: Synthesis and appraisal of their potential as antimicrobial agents. *Lett. Drug. Des. Discov.* **2018**, *15*, 21–30. [CrossRef]

27. Çavuşoğlu, B.K.; Yurttaş, L.; Cantürk, Z. The synthesis, antifungal and apoptotic effects of triazole-oxadiazoles against *Candida* species. *Eur. J. Med. Chem.* **2018**, *144*, 255–261. [CrossRef] [PubMed]

28. Karaburun, A.Ç.; Çavuşoğlu, B.K.; Çevik, U.A.; Osmaniye, D.; Sağlık, B.N.; Levent, S.; Özkay, Y.; Atlı, Ö.; Koparal, A.S.; Kaplancıklı, Z.A. Synthesis and antifungal potential of some novel benzimidazole-1,3,4-oxadiazole compounds. *Molecules* **2019**, *24*, 191. [CrossRef]

29. Verma, G.; Khan, M.F.; Akhtar, W.; Alam, M.M.; Akhter, M.; Shaquiquzzaman, M. A review exploring therapeutic worth of 1,3,4-oxadiazole tailored compounds. *Mini Rev. Med. Chem.* **2019**, *19*, 477–509. [CrossRef] [PubMed]

30. Kumar, A.; Zarychanski, R.; Pisipati, A.; Kumar, A.; Kethireddy, S.; Bow, E.J. Fungicidal versus fungistatic therapy of invasive *Candida* infection in non-neutropenic adults: A meta-analysis. *Mycology* **2018**, *9*, 116–128. [CrossRef]

31. Sánchez-Calvo, J.M.; Barbero, G.R.; Guerrero-Vásquez, G.; Durán, A.G.; Macías, M.; Rodríguez-Iglesias, M.A.; Molinillo, J.M.; Macías, F.A. Synthesis, antibacterial and antifungal activities of naphthoquinone derivatives: A structure–activity relationship study. *Med. Chem. Res.* **2016**, *25*, 1274–1285. [CrossRef]

32. Li, W.S.; Chen, Y.C.; Kuo, S.F.; Chen, F.J.; Lee, C.H. The impact of biofilm formation on the persistence of candidemia. *Front. Microbiol.* **2018**, *9*, 1196. [CrossRef]

33. Monteiro, D.R.; Feresin, L.P.; Arias, L.S.; Barão, V.A.; Barbosa, D.B.; Delbem, A.C. Effect of tyrosol on adhesion of *Candida albicans* and *Candida glabrata* to acrylic surfaces. *Med. Mycol.* **2015**, *53*, 656–665. [CrossRef]

34. Ostrosky-Zeichner, L. Combination antifungal therapy: A critical review of the evidence. *Clin. Microbiol. Infect.* **2008**, *14*, 65–70. [CrossRef] [PubMed]

35. Hossain, M.A.; Reyes, G.H.; Long, L.A.; Mukherjee, P.K.; Ghannoum, M.A. Efficacy of caspofungin combined with amphotericin B against azole-resistant *Candida albicans*. *J. Antimicrob. Chemother.* **2003**, *51*, 1427–1429. [CrossRef]

36. Sun, L.; Liao, K.; Hang, C. Caffeic acid phenethyl ester synergistically enhances the antifungal activity of fluconazole against resistant *Candida albicans*. *Phytomedicine* **2018**, *40*, 55–58. [CrossRef]

37. Li, L.; Zhang, T.; Xu, J.; Wu, J.; Wang, Y.; Qiu, X.; Zhang, Y.; Hou, W.; Yan, L.; An, M.; et al. The synergism of the small molecule ENOblock and fluconazole against fluconazole-resistant *Candida albicans*. *Front. Microbiol.* **2019**, *10*, 2071. [CrossRef] [PubMed]

38. Khalifa, H.O.; Majima, H.; Watanabe, A.; Kamei, K. In Vitro Characterization of Twenty-One Antifungal Combinations against Echinocandin-Resistant and -Susceptible *Candida glabrata*. *J. Fungi* **2021**, *7*, 108. [CrossRef] [PubMed]

39. Tonholo, D.R.; Maltarollo, V.G.; Kronenberger, T.; Silva, I.R.; Azevedo, P.O.; Oliveira, R.B.; Souza, L.C.; Tagliati, C.A. Preclinical toxicity of innovative molecules: In vitro, in vivo and metabolism prediction. *Chem. Biol. Interact.* **2020**, *315*, 108896. [CrossRef] [PubMed]

40. Dongmo, O.L.; Epoh, N.J.; Tadjoua, H.T.; Yousuf, S.; Telefo, P.B.; Tapondjou, L.A.; Choudhary, M.I. Acute and sub-acute toxicity of the aqueous extract from the stem bark of *Tetrapleura tetrapteura* Taub. (Fabaceae) in mice and rats. *J. Ethnopharmacol.* **2019**, *236*, 42–49. [CrossRef] [PubMed]

41. Sellers, R.S.; Morton, D.; Michael, B.; Roome, N.; Johnson, J.K.; Yano, B.L.; Perry, R.; Schafer, K. Society of Toxicologic Pathology position paper: Organ weight recommendations for toxicology studies. *Toxicol. Pathol.* **2007**, *35*, 751–755. [CrossRef]

42. Issa, N.T.; Wathieu, H.; Ojo, A.; Byers, S.W.; Dakshanamurthy, S. Drug metabolism in preclinical drug development: A survey of the discovery process, toxicology, and computational tools. *Curr. Drug Metab.* **2017**, *18*, 556–565. [CrossRef]

43. Ramaiah, S.K. Preclinical safety assessment: Current gaps, challenges, and approaches in identifying translatable biomarkers of drug-induced liver injury. *Clin. Lab. Med.* **2011**, *31*, 161–172. [CrossRef]

44. Pazhayattil, G.S.; Shirali, A.C. Drug-induced impairment of renal function. *Int. J. Nephrol. Renovasc. Dis.* **2014**, *7*, 457–468. [CrossRef] [PubMed]

45. Salazar, J.H. Overview of urea and creatinine. *Lab. Med.* **2014**, *45*, e19–e20. [CrossRef]

46. Ronco, C.; Bellomo, R.; Kellum, J.A. Acute kidney injury. *Lancet* **2019**, *394*, 1949–1964. [CrossRef]

47. Arika, W.N.; Nyamai, D.W.; Musila, M.N.; Ngugi, M.P.; Njagi, E.N. Hematological markers of in vivo toxicity. *J. Hematol. Thrombo. Dis.* **2016**, *4*, 236. [CrossRef]

48. Santos, E.W.; Oliveira, D.C.; Hastreiter, A.; Silva, G.B.; Beltran, J.S.; Tsujita, M.; Crisma, A.R.; Neves, S.M.; Fock, R.A.; Borelli, P. Hematological and biochemical reference values for C57BL/6, Swiss Webster and BALB/c mice. *Braz. J. Vet. Res. Anim. Sci.* **2016**, *53*, 138–145. [CrossRef]

49. Barbosa, B.S.; Praxedes, É.A.; Lima, M.A.; Pimentel, M.M.; Santos, F.A.; Brito, P.D.; Lelis, I.C.; Macedo, M.F.; Bezerra, M.B. Haematological and biochemical profile of Balb-c mice. *Acta Sci. Vet.* **2017**, *45*, 1477. [CrossRef]

50. Spellberg, B.; Ibrahim, A.S.; Edwards, J.E., Jr.; Filler, S.G. Mice with disseminated candidiasis die of progressive sepsis. *J. Infect. Dis.* **2005**, *192*, 336–343. [CrossRef] [PubMed]

51. Lionakis, M.S.; Lim, J.K.; Lee, C.C.; Murphy, P.M. Organ-specific innate immune responses in a mouse model of invasive candidiasis. *J. Innate Immun.* **2011**, *3*, 180–199. [CrossRef]

52. Jae-Chen, S.; Young-Joo, J.; Seon-Min, P.; Seok, S.K.; Jung-Hyun, S.; Jung-Il, C. Mechanism underlying renal failure caused by pathogenic *Candida albicans* infection. *Biomed. Rep.* **2015**, *3*, 179–182. [CrossRef] [PubMed]

53. Pappas, P.G.; Lionakis, M.S.; Arendrup, M.C.; Ostrosky-Zeichner, L.; Kullberg, B.J. Invasive candidiasis. *Nat. Rev. Dis. Primers* **2018**, *4*, 18026. [CrossRef] [PubMed]

54. Antachopoulos, C.; Roilides, E. Cytokines and fungal infections. *Br. J. Haematol.* **2005**, *129*, 583–596. [CrossRef] [PubMed]

55. Basso, V.; Tran, D.Q.; Schaal, J.B.; Tran, P.; Eriguchi, Y.; Ngole, D.; Cabebe, A.E.; Park, A.Y.; Beringer, P.M.; Ouellette, A.J.; et al. Rhesus theta defensin 1 promotes long term survival in systemic candidiasis by host directed mechanisms. *Sci. Rep.* **2019**, *9*, 16905. [CrossRef]

56. Yarrow, D. Methods for the isolation, maintenance and identification of yeasts. In *The Yeasts, A Taxonomic Study*; Elsevier: Amsterdam, The Netherlands, 1998; pp. 77–100. [CrossRef]

57. Cassagne, C.; Normand, A.C.; L'Ollivier, C.; Ranque, S.; Piarroux, R. Performance of MALDI-TOF MS platforms for fungal identification. *Mycoses* **2016**, *59*, 678–690. [CrossRef]

58. CLSI. *Reference Method for Broth Dilution Antifungal Susceptibility Testing of Yeasts*, 3rd ed.; CLSI Document M27-A3; Clinical and Laboratory Standards Institute: Wayne, PA, USA, 2008.

59. CLSI. *Reference Method for Broth Dilution Antifungal Susceptibility Testing of Yeasts*; 4th Informational Supplement; CLSI Document M27-S4; Clinical and Laboratory Standards Institute: Wayne, PA, USA, 2012.

60. Klepser, M.E.; Wolfe, E.J.; Jones, R.N.; Nightingale, C.H.; Pfaller, M.A. Antifungal pharmacodynamic characteristics of fluconazole and amphotericin B tested against *Candida albicans*. *Antimicrob. Agents Chemother.* **1997**, *41*, 1392–1395. [CrossRef]

61. Scorneaux, B.; Angulo, D.; Borroto-Esoda, K.; Ghannoum, M.; Peel, M.; Wring, S. SCY-078 is fungicidal against *Candida* species in time-kill studies. *Antimicrob. Agents Chemother.* **2017**, *61*, e01961-16. [CrossRef]

62. Khan, M.S.; Malik, A.; Ahmad, I. Anti-candidal activity of essential oils alone and in combination with amphotericin B or fluconazole against multi-drug resistant isolates of *Candida albicans*. *Med. Mycol.* **2012**, *50*, 33–42. [CrossRef] [PubMed]

63. Mor, V.; Rella, A.; Farnoud, A.M.; Singh, A.; Munshi, M.; Bryan, A.; Naseem, S.; Konopka, J.B.; Ojima, I.; Bullesbach, E.; et al. Identification of a new class of antifungals targeting the synthesis of fungal sphingolipids. *mBio* **2015**, *6*, e00647-15. [CrossRef] [PubMed]

64. Di Veroli, G.Y.; Fornari, C.; Wang, D.; Mollard, S.; Bramhall, J.L.; Richards, F.M.; Jodrell, D.I. Combenefit: An interactive platform for the analysis and visualization of drug combinations. *Bioinformatics* **2016**, *32*, 2866–2868. [CrossRef] [PubMed]

65. Tobaldini-Valerio, F.K.; Bonfim-Mendonça, P.S.; Rosseto, H.C.; Bruschi, M.L.; Henriques, M.; Negri, M.; Silva, S.; Svidzinski, T.I. Propolis: A potential natural product to fight *Candida* species infections. *Future Microbiol.* **2016**, *11*, 1035–1046. [CrossRef] [PubMed]

66. De Oliveira, A.G.; Spago, F.R.; Simionato, A.S.; Navarro, M.O.; da Silva, C.S.; Barazetti, A.R.; Cely, M.V.; Tischer, C.A.; San Martin, J.A.; Andrade, C.G.; et al. Bioactive organocopper compound from *Pseudomonas aeruginosa* inhibits the growth of *Xanthomonas citri* subsp. *citri*. *Front. Microbiol.* **2016**, *7*, 113. [CrossRef]

67. Agência Nacional de Vigilância Sanitária (ANVISA). Guia Para a Condução de Estudos não Clínicos de Toxicologia e Segurança Farmacológica Necessários ao Desenvolvimento de Medicamentos. 2013. Available online: https://bit.ly/2OA6uWr (accessed on 5 February 2020).

68. Kifayatullah, M.; Mustafa, M.S.; Sengupta, P.; Sarker, M.M.; Das, A.; Das, S.K. Evaluation of the acute and sub-acute toxicity of the ethanolic extract of *Pericampylus glaucus* (Lam.) Merr. in BALB/c mice. *J. Acute. Dis.* **2015**, *4*, 309–315. [CrossRef]

69. Briggs, C.; Bain, B.J. Basic haematological techniques. In *Dacie and Lewis Practical Haematology*; Elsevier: London, UK, 2017; pp. 18–49.
70. Araujo, A.V. Estudos Pré-Clínicos de Toxicidade Aguda e de Doses Repetidas da Fosfoetanolamina Sintética. Master's Thesis, Universidade de São Paulo, São Paulo, Brazil, 2017. Available online: https://www.teses.usp.br/teses/disponiveis/5/5160/tde-15032018-092543/publico/AlineVieiraPinheiroDeAraujo.pdf (accessed on 1 June 2020).
71. Brecher, G.; Cronkite, E.P. Morphology and enumeration of human blood platelets. *J. Appl. Physiol.* **1950**, *3*, 365–377. [CrossRef] [PubMed]
72. Van Kampen, E.J.; Zijlstra, W.G. Determination of hemoglobin and its derivatives. *Adv. Clin. Chem.* **1965**, *8*, 141–187. [CrossRef] [PubMed]
73. Wong, S.S.; Kao, R.Y.; Yuen, K.Y.; Wang, Y.; Yang, D.; Samaranayake, L.P.; Seneviratne, C.J. In vitro and in vivo activity of a novel antifungal small molecule against *Candida* infections. *PLoS ONE* **2014**, *9*, e85836. [CrossRef] [PubMed]

pathogens

Article

Antifungal Activity of Capridine β as a Consequence of Its Biotransformation into Metabolite Affecting Yeast Topoisomerase II Activity

Iwona Gabriel *, Kamila Rząd, Ewa Paluszkiewicz and Katarzyna Kozłowska-Tylingo

Department of Pharmaceutical Technology and Biochemistry, Gdańsk University of Technology, 80-233 Gdańsk, Poland; kamila.rzad@pg.edu.pl (K.R.); ewa.paluszkiewicz@pg.edu.pl (E.P.); katarzyna.kozlowska-tylingo@pg.edu.pl (K.K.-T.)
* Correspondence: iwogabri@pg.edu.pl; Tel.: +48-583486078; Fax: +48-583471144

Abstract: In the last few years, increasing importance is attached to problems caused by fungal pathogens. Current methods of preventing fungal infections remain unsatisfactory. There are several antifungal compounds which are highly effective in some cases, however, they have limitations in usage: Nephrotoxicity and other adverse effects. In addition, the frequent use of available fungistatic drugs promotes drug resistance. Therefore, there is an urgent need for the development of a novel antifungal drug with a different mechanism of action, blocking of the fungal DNA topoisomerases activity appear to be a promising idea. According to previous studies on the m-AMSA moderate inhibitory effect on fungal topoisomerase II, we have decided to study Capridine β (also acridine derivative) antifungal activity, as well as its inhibitory potential on yeast topoisomerase II (yTOPOII). Results indicated that Capridine β antifungal activity depends on the kind of strains analyzed (MICs range 0.5–64 µg mL^{-1}) and is related to its biotransformation in the cells. An investigation of metabolite formation, identified as Capridine β reduction product (IE1) by the fungus *Candida albicans* was performed. IE1 exhibited no activity against fungal cells due to an inability to enter the cells. Although no antifungal activity was observed, in contrast to Capridine β, biotransformation metabolite totally inhibited the yTOPOII-mediated relaxation at concentrations lower than detected for m-AMSA. The closely related Capridine β only slightly diminished the catalytic activity of yTOPOII.

Keywords: *Candida albicans*; acridine; antifungal; topoisomerase; inhibitor

Citation: Gabriel, I.; Rząd, K.; Paluszkiewicz, E.; Kozłowska-Tylingo, K. Antifungal Activity of Capridine β as a Consequence of Its Biotransformation into Metabolite Affecting Yeast Topoisomerase II Activity. *Pathogens* **2021**, *10*, 189. https://doi.org/10.3390/pathogens10020189

Academic Editor: Jonathan Richardson

Received: 22 December 2020
Accepted: 4 February 2021
Published: 9 February 2021

Publisher's Note: MDPI stays neutral with regard to jurisdictional claims in published maps and institutional affiliations.

1. Introduction

Acridine derivatives are a class of compounds with a broad spectrum of biological activity and are very interesting for scientists. Many compounds containing the acridine chromophore were synthesized and tested, and the aminoacridines found wide use, both as antibacterial agents and as antimalarials [1]. The planar acridine scaffold, an important pharmacophore, is also a source of new compounds with anti-tumor activity [2]. Surprisingly, although acridine and acridone derivatives are widely analyzed as antibacterial or anticancer agents, only a few reports have demonstrated their antifungal activity [3]. The acridine antimicrobial mode of action, apart from being DNA-targeting drugs, may also be associated with an inhibitory effect on bacterial gyrase or topoisomerase activity. DNA topoisomerases are essential enzymes that catalyze topological changes in DNA. Those enzymes play key roles in replication, transcription, recombination, and chromosome condensation [4]. Camptothecin, its derivatives and novel noncamptothecins target eukaryotic type IB topoisomerases. Human type IIA topoisomerases are the targets of the commonly used anticancer agents etoposide, anthracyclines, and mitoxantrone. Bacterial type II topoisomerases (gyrase and Topo IV) are targeted by quinolones and aminocoumarin antibiotics [5,6]. Functional topoisomerase II is also essential for the growth of yeast [7,8].

The differences between fungal and human topoisomerases have also led to the suggestion that this class of enzymes may become potential targets for the development of novel antifungal agents [9,10]. Moreover, previous studies suggested that fungal and mammalian topoisomerases respond in a different manner to 4′-(9-acridinylamino)methanesulfon-m-anisidide (m-AMSA) (Figure 1), etoposide, and its derivatives A-80198 and A-75272 (a tricyclic quinolone) [11]. Thus probably, there are sufficient biochemical differences between those enzymes to obtain selectivity for fungi over human cells. What is more, all known yeast topo I amino acid sequences contain a characteristic insert absent in the mammalian enzyme [12,13]. The function of this insertion is not known but the development of selective fungal-topo inhibitors is plausible.

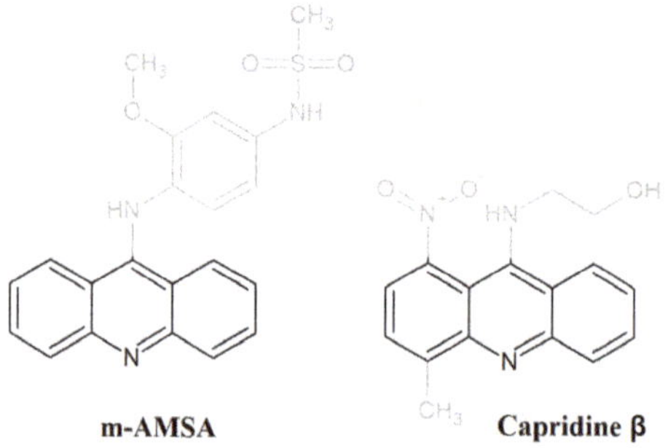

Figure 1. M-AMSA and Capridine β (C-1748) structures with a stressed acridine chromophore.

1-Nitro-9-aminoacridine derivative Capridine β, 9[2′-hydroxyethylamino]-4-methyl-1-nitroacridine (also known as C-1748) (Figure 1), similar to m-AMSA, exhibited anti-cancer and antifungal properties [14–16]. Moreover, previous studies, performed for Capridine β in cancer cell cultures and human xenograft animal models indicated a high therapeutic index and low cytotoxicity with the potential for clinical development [14,15,17–19]. Both compounds, m-AMSA, as well as Capridine β, contain an acridine chromophore (Figure 1).

Due to the fact that m-AMSA, a 9-aminoacridine derivative, is a human topoisomerase II poison and has been shown to have a moderate inhibitory effect on fungal topoisomerase II [11] along with a high therapeutic index, low cytotoxicity of Capridine β [14,15,17–19], antifungal activity [16], and acridine chromophore similarity to m-AMSA, we have decided to deeply analyze its antifungal potential as well as mechanism of action.

2. Results and Discussion

2.1. Antifungal Activity

As Capridine β antifungal activity has been previously observed [16], we have decided to analyze the m-AMSA in vitro antifungal activity against five corresponding strains. Minimal inhibitory concentrations (MICs) of the studied compounds determined by the microplate serial dilution method are shown in Table 1.

Table 1. Antifungal activity of Capridine β, IE1, and m-AMSA. * MIC$_{90}$: Minimal inhibitory concentrations at which 90% of cells were inhibited. The experiments were performed at least in five replicates.

Compound	* MIC$_{90}$ µg mL^{-1}				
	Candida albicans ATCC 10231	*Candida glabrata* ATCC 90030	*Candida krusei* ATCC 6258	*Candida parapsilosis* ATCC 22019	*Saccharomyces cerevisiae* ATCC 9763
Capridine β	1	8	8	64	0.5
IE1	>64	64	64	64	64
m-AMSA	>64	>64	>64	>64	>64
Amphotericin B	0.5	1	1	1	0.5

Results indicated that analyzed acridine derivatives antifungal activity depends on the kind of strains analyzed. Among them, the most active compound was Capridine β. A comparison of *C. albicans* ATCC 10231 growth kinetics in the absence and presence of Capridine β and Amphotericin B at concentrations corresponding to $\frac{1}{2} \times$ MIC, $1 \times$ MIC and $2 \times$ MIC indicates its high antifungal efficacy, slightly lower than that of Amphotericin B activity (Figure S1A, Supplementary Materials). As far as m-AMSA is concerned, the analysis of growth kinetics shows that much higher concentrations (32 as well as 64 µg mL^{-1}) do not result in a 90% growth inhibition (Figure S1B, Supplementary Materials).

2.2. Biotransformation of Capridine β in Fungal Cells

Due to the significant activity of Capridine β against fungal cells, we have decided to examine whether the original form of the compound or the product of biotransformation is responsible for its activity. Fungal degradation of acridine compounds under an aerobic condition is not well described. Biotransformation experiments were conducted by incubating *C. albicans* cells with Capridine β followed by a small-scale disruption using zirconium-glass beads. Results of high-performance liquid chromatography (HPLC) analyses of the cell free extracts obtained after 20 h of compound (0.2 mM) incubations with *C. albicans* ATCC 10231 are presented in Figure 2.

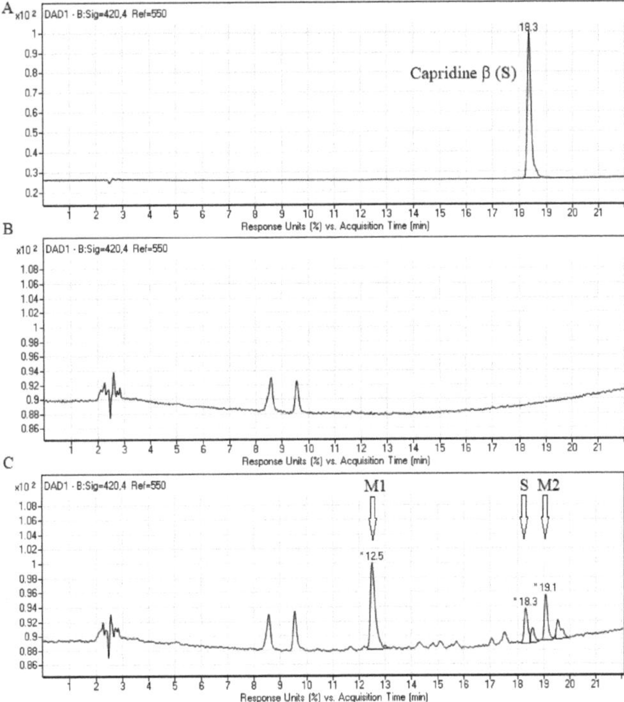

Figure 2. Metabolism of Capridine β (substrate) in *C. albicans* ATCC 10231 cells. RP-HPLC profiles of the 0.1 mM Capridine β (**A**), as well as free cell extracts obtained after incubation of the cells without (**B**) or with (**C**) the studied compound (0.2 mM Capridine β) for 20 h at 30 °C. The cell layer was washed several times and centrifuged, and the final pellet was resuspended in 60% methanol, then disrupted with zirconium-glass beads and centrifuged. The resulting solution was subjected to the RP-HPLC analysis. * Retention times of Capridine β (S) 18.3 min. and metabolites (M1) 12.5 min. and (M2) 19.1 min. are indicated. The symbols of the substrate (S) and its metabolites correspond to those in Figure 3. All the experiments were performed at least in triplicate, and the representative chromatograms are shown.

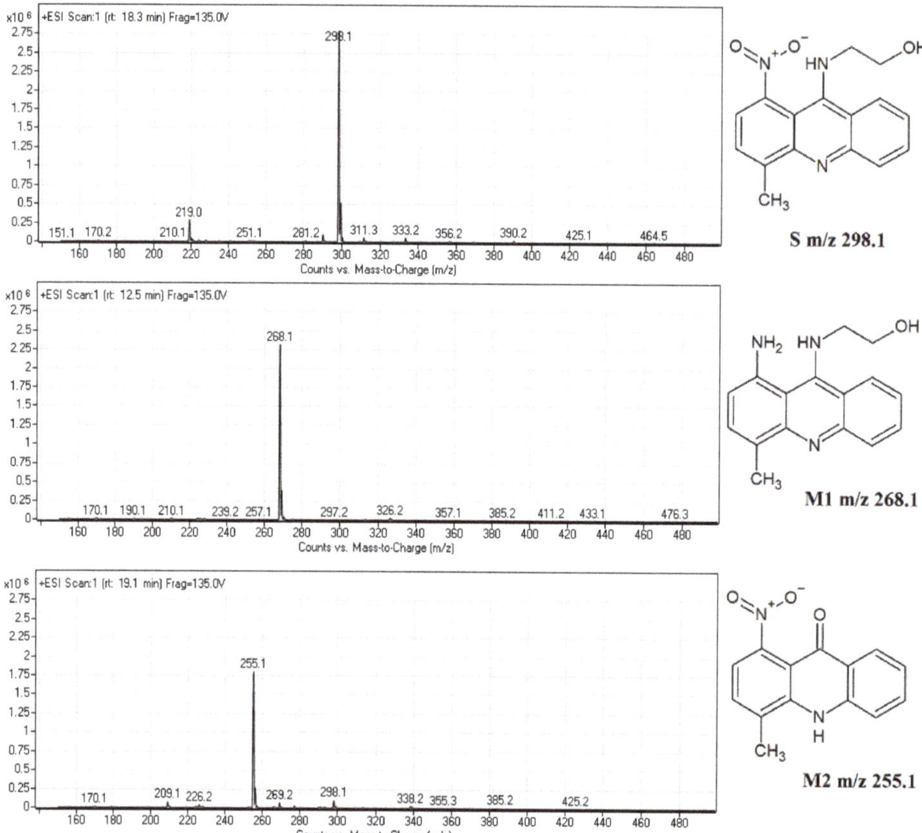

Figure 3. The liquid chromatography mass spectrometry (LC-MS) analysis of three main peaks (r.t. 18.3, 12.5, and 19.1 min) identified as Capridine β (*m*/*z* 298.1) (S) and two main metabolic products M1 (*m*/*z* 268.1) and M2 (*m*/*z* 255.1). All the experiments were performed at least in triplicate, and the representative chromatograms are shown.

The metabolic profile of the studied compound, obtained by HPLC, indicated the presence of three main peaks at retention times (r.t.) 12.5 min (M1), 18.3 and 19.1 min (M2) (Figure 2C). The retention time of 18.3 min was identical to that identified for Capridine β (substrate, Figure 2A). The result indicated that Capridine β was undoubtedly biotransformed into at least two major products M1 and M2.

Based on the liquid chromatography mass spectrometry (LC-MS) data (Figure 3), two metabolites were predicted to be 1-amino-9-hydroxyethylaminoacridine (M1) and 1-nitroacridinone (M2), mammalian metabolites were also previously reported [15,20].

In contrast to the metabolism of the studied compound in human hepatocellular carcinoma cell lines under normoxia, the metabolites profile observed in *C. albicans* seems to be similar to that obtained for mammalian cells under hypoxia or reducing conditions [15,20]. The main fungal metabolic product M1 (*m*/*z* 268.1), 1-amino-9-hydroxyethylaminoacridine was detected as the main metabolic product of human cancer cell lines only under a low level of oxygen. The second fungal metabolite M2 (*m*/*z* 255.1), 1-nitroacridinone, was previously identified only under reducing conditions with 1,4-dithiothreitol, not found with the studied human enzymatic systems and in HepG2 cells [15].

To additionally prove that the main fungal metabolic product M1 was properly identified we have decided to synthesize the reduced Capridine β form (IE1) and perform the RP-HPLC and LC-MS analysis (Figure 4).

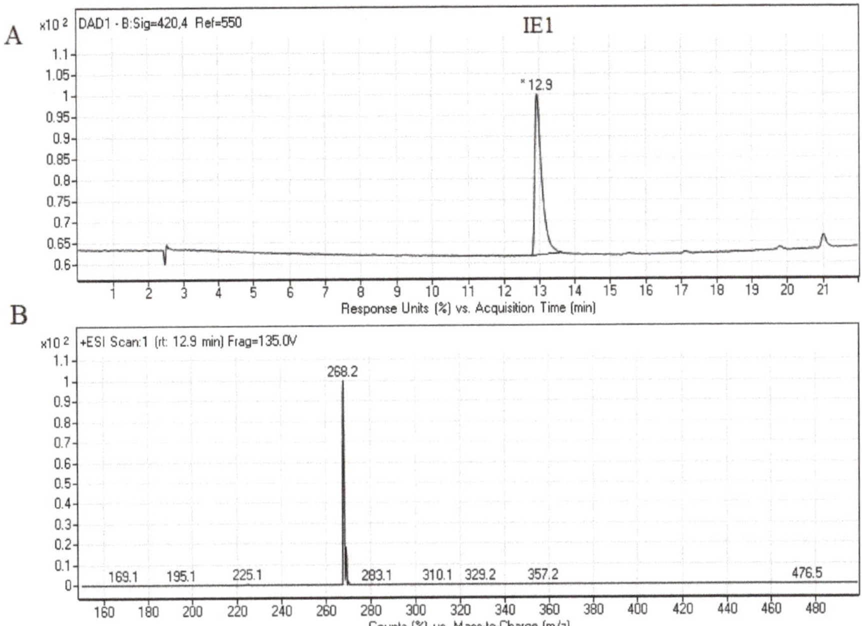

Figure 4. RP-HPLC profiles of the 0.1 mM IE1 (**A**) and its liquid chromatography mass spectrometry (LC-MS) analysis (**B**). * Retention time of IE1 12.9 min. is indicated.

The retention time and the molecular ion *m/z* 268.2 (Figure 4) found for IE1 were almost identical to those presented for product M1 (Figures 2C and 3). Summing up, we demonstrated that the metabolism of Capridine β in fungal cells differs from that observed for mammalian cells under normoxia and gave two main metabolites, one of them was undoubtedly identified as the reduced form of Capridine β.

2.3. Biological Activity of Capridine β Reduced Form (IE1), Identified as M1 Metabolite

The Capridine β reduced form (IE1), identified as M1 metabolite, was tested for its in vitro antifungal activity against five referenced ATCC strains (Table 1). High MIC values 64 µg mL^{-1} were determined for all the analyzed strains except for *C. albicans* ATCC 10231 (MIC > 64 µg mL^{-1}), which indicated small or no antifungal activity.

We next examined the accumulation of the starting compound Capridine β and its reduced form IE1 in the *C. albicans* ATCC 10231 strain using fluorescence microscopy (Figure 5).

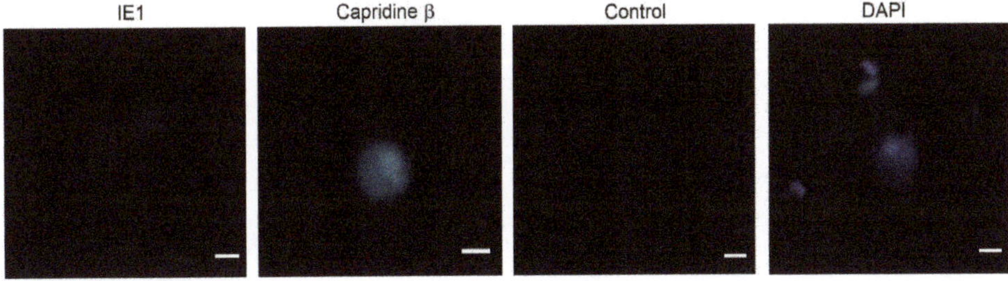

Figure 5. Capridine β and DAPI accumulation in *Candida albicans* ATCC 10231 cells. The cells were incubated in the dark without (control) or with the denoted derivatives (50 µM, 180 min, 37 °C). Compounds fluorescence was observed under a fluorescence microscope (excitation wavelength 400–420 nm, emission >450 nm, ×100). Scale indicated 5 µm.

The microscopic analysis revealed that the efficient accumulation of Capridine β in *C. albicans* cells was observed. No accumulation was detected for its reduction product (IE1). Among the analyzed derivatives, fluorescence intensity in the cell did not depend on their own spectral properties in aqueous solutions within the excitation range used for microscopic studies (Figure 6).

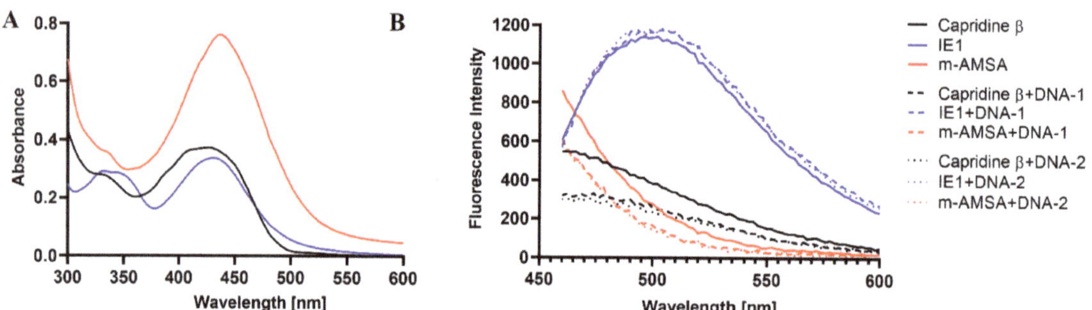

Figure 6. (**A**) UV/vis absorption spectrum of 0.1 mM Capridine β, IE1 or m-AMSA in PBS. (**B**) Background-corrected fluorescence spectra of 0.05 mM Capridine β, IE1, and m-AMSA at the excitation wavelength of 410 nm without (solid line) and with DNA-1 (dashed line) and DNA-2 (dotted line) sequences. Data shown are corrected for background fluorescence and represent averages of at least three replicate samples.

The background-corrected fluorescence spectrum of IE1, obtained in vitro, indicates the highest fluorescence intensity of this compound compared to others (Figure 6) at the excitation wavelength used for microscopic analysis (410 nm) and the same emission range (>450 nm). As it can be seen in Figure 5, the pattern of accumulation for Capridine β seems to be similar to DAPI (4,6-diamidino-2-phenylindole). DAPI staining of *C. albicans* is commonly used to localize the nucleus [21]. Thus, probably the Capridine β targeted structure is a nucleus. According to the previously published studies, an interaction with DNA was observed [14,15,18,22]. Despite the higher IE1 fluorescence intensity, than measured for Capridine β (Figure 6), no fluorescence for that compound was detected within the fungal cells, neither in the cytoplasm nor in the nucleus (Figure 5). Hence, no antifungal activity seems to be the result of an impossibility of reaching the molecular target.

Additional fluorescence studies of Capridine β also revealed its interaction with DNA. In the presence of two different DNA sequences, compound fluorescence quenching was observed (Figure 6). The same effect was observed for m-AMSA. Differently, the presence of DNA enhanced fluorescence intensity of IE1. Thus, probably for that compound a different mode of action for the drug-DNA interaction was observed. As previously reported, when the drugs are bound to DNA, a significant increase in the fluorescence emission is normally observed. In the case of groove binding agents, electrostatic, hydrogen bonding or hydrophobic interactions are involved and the molecules are close to the sugar-phosphate backbone, being possible to observe a decrease in the fluorescence intensity in the presence of the DNA [23].

Acridines and its derivatives are considered to be antimicrobial agents although their activity is obviously determined on the efficient accumulation in bacterial or fungal cells. As described previously for imidazoacridinone C-1311 and its nine derivatives, only three that entered fungal cells showed a phototoxic antifungal activity (C-1330, C-1415, and C-1558) [24]. Despite the high efficiency of C-1311 as an anticancer compound that intercalates into DNA and inhibits human topoisomerase II [25], it was unable to accumulate in *C. albicans* cells and no antifungal activity was observed [24].

On the basis of the fact that the m-AMSA (aminoacridine derivative) inhibitory effect on fungal topoisomerase II was previously observed [11], we have decided to analyze the influence of Capridine β, as well as its reduced form IE1 on the yeast topoisomerase II (yTOPOII) activity. The relaxation of supercoiled plasmid DNA by yTOPOII was studied in the presence of different concentrations of both compounds (Figure 7).

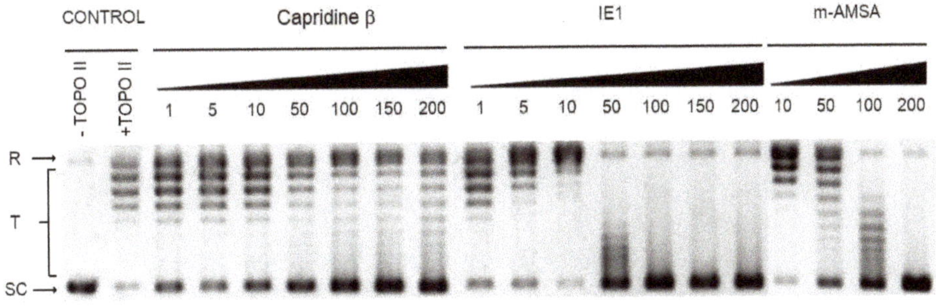

Figure 7. Inhibition of the catalytic activity of purified yeast DNA topoisomerase II by m-AMSA, Capridine β, and IE1 as measured by relaxation. Supercoiled pBR322 plasmid DNA (lane 1) was relaxed by purified yeast topoisomerase II in the absence (lane 2) or presence of m-AMSA at 10, 50, 100 or 200 µM or with analyzed compounds at 1, 5, 10, 50, 100, 150 or 200 µM. The resulting topological forms of DNA were separated by gel electrophoreses. SC: Supercoiled DNA; R: Relaxed DNA; T: DNA topoisomers. Data shown are typical of three independent experiments.

The most effective IE1 totally inhibited the yeast topoisomerase II-mediated relaxation at concentrations lower than detected for m-AMSA (Figure 7). The inhibition activity of the analyzed compounds was determined by densitometry quantification of the transition from supercoiled to relaxed forms and was expressed in relation to the control. The half maximal effective concentration (EC50) refers to the concentration of a drug, which affected the relaxation in 50%. EC50 determined for IE1 and m-AMSA was 14.1 ± 1.2 µM and >200 µM, respectively. Interestingly, the closely related Capridine β slightly diminished the catalytic activity of topoisomerase II but no complete inhibition was observed in the tested concentration range.

Acridine and acridone derivatives are widely analyzed as human topoisomerase inhibitors for cancer chemotherapy. M-AMSA in Figure 1 was the first synthetic drug approved for clinical usage that was shown to act as a topoisomerase inhibitor [26]. The molecular mechanism of antitumor triazoloacridinone C-1305 and imidazoacridinone C-1311, both acridine derivatives also indicated its intercalation with DNA as well as the formula of a topo II-stabilizing complex [27,28]. The success of anticancer and antibacterial drugs as DNA topoisomerases inhibitors highlights the potential of topoisomerases from fungal cells as targets for the development of novel antifungals. Fungal topoisomerases might be sufficiently distinct from their human counterparts to enable selective targeting but in order to be active, the antifungal topoisomerase inhibitor needs to enter into fungal cells to reach their intracellular targets. As demonstrated by Kwok et al., no antifungal activity of etoposide was observed [9], although its inhibitory effect on *C. albicans* DNA topoisomerase II was previously reported [11].

Results obtained for Capridine β indicate that not only an efficient accumulation, but also the biotransformation into metabolite that can affect fungal topoisomerase II are important with respect to its antifungal activity.

3. Materials and Methods

3.1. Chemical Synthesis of Capridine β and IE1

3.1.1. General

The products were obtained as the hydrochloride salt. Their structures were confirmed using spectral methods: Mass spectrometry ESI-MS and proton nuclear magnetic resonance (1H NMR). The purity of these compounds was ascertained using thin-layer chromatography (TLC). Melting points were determined on a Stuart SMP30 capillary apparatus. Mass spectra were recorded using an Agilent 6470A triple quadrupole LC/MS system with electrospray ionization source (ESI) in a SCAN mode. Samples were prepared as 1 µg/mL solutions in water and were supplied in 1 µL aliquots to the mass spectrometer in the mixture of acetonitrile:water:formic acid (38:57:5 $v/v/v$) at a flow rate of 500 µL/min. 1H NMR spectra were recorded on a Varian VXR-S spectrometer operating at 500 MHz. Chemical shifts are reported as δ units in ppm downfield from internal tetramethylsilane. NMR abbreviations used are as follows: m.p.—melting point, br.s—broad signal, s—singlet, d—doublet, t—triplet, m—multiple. The results of the elemental analyses for individual elements fit within ±0.4% of theoretical values.

3.1.2. The Synthesis of Capridine β

9-(2′-Hydroxyethylamino)-4-methyl-1-nitroacridine (Capridine β) was prepared according to the previously reported procedures [29,30].

9-(2′-hydroxyethylamino)-4-methyl-1-nitroacridine (Capridine β) 1H NMR (Me$_2$SO-d$_6$) δ: 8.45 (d, J = 8.3 Hz, 1H), 8.15 (d, J = 8.3 Hz, 1H), 8.08 (d, J = 7.8 Hz, 1H), 7.89 (t, J = 7.8 Hz, 1H), 7.78 (d, J = 7.8 Hz, 1H), 7.54 (t, J = 7.4 Hz, 1H), 3.61–3.75 (m, 4H); ESI-MS [M+H$^+$] C$_{16}$H$_{15}$N$_3$O$_3$—298.2.

3.1.3. The Synthesis of IE1

Capridine β (0.3 mmol) was hydrogenated in the presence of 10% Pd/C (catalytic quantities) in 5 mL methanol by passing gaseous hydrogen through them at room temperature for 24 h. After the time of hydrogenation, the catalyst was filtered off and the solvent evaporated. The product in the form of a base was dissolved in methanol (10 mL) and acidified by the HCl/diethyl ether. After diethyl ether, adding the desired product was obtained.

9-(2′-Hydroxyethylamino)-1-amino-4-methylacridine (IE1). Yield 92%; m.p. 221–223 °C; 1H NMR (500 MHz, DMSO-d$_6$+TFA) δ: 11.35 (s, 1H), 8.31 (d, J = 8.5 Hz, 1H), 8.08 (d, J = 8.5 Hz, 1H), 7.81 (t, J = 7.7 Hz, 1H), 7.40 (d, J = 8.0 Hz, 1H), 7.35 (t, J = 7.7 Hz, 1H), 6.79 (d, J = 7.7 Hz, 1H), 3.95–4.00 (m, 2H), 3.64–3.67 (m, 2H), 2.43 (s, 3H); ESI-MS [M+H$^+$] C$_{16}$H$_{17}$N$_3$O—268.1.

3.2. Microorganisms Strains and Growth Conditions

The following fungal strains were used: *C. albicans* ATCC 10231, *C. glabrata* ATCC 90030, *C. krusei* ATCC 6258, *C. parapsilosis* ATCC 22019, *S. cerevisiae* ATCC 9763, Fungal strains used in this investigation were routinely grown 18 h at 30 °C in a YPG liquid medium (1% yeast extract, 1% peptone, 2% glucose) in a shaking incubator. For growth on solid media, 1.5% agar was added to the YPG medium.

3.3. Metabolism Assays with C. albicans Cells

For studies on cellular metabolism, *Candida albicans* ATCC 10231 strains were grown in the YPG medium overnight (16–18 h) at 30 °C, washed twice with sterile water and resuspended in the fresh YPG medium at a cell density of 2×10^8 cells mL^{-1}. In addition, 20 µL of the stock solution of Capridine β in DMSO was added to 980 µL of cell cultures to obtain 0.2 mM final concentrations. The control cells were treated with the same amount of solvent (DMSO). The cells were incubated at 30 °C for 20 h. After incubation, the cells were harvested by centrifugation (4000 rpm, 10 min) and washed with water. Then, the cells were resuspended in 0.5 mL of 60% MeOH at 4 °C and disrupted with the use of zirconium-glass beads (0.5 mm in diameter) by vigorous shaking in four 5-min cycles interrupted by 2-min ice-cooling. The samples were then centrifuged (15 min, 12,000 rpm, 4 °C) and filtered (0.22 µm, PES) prior to the HPLC analysis.

3.4. Chromatographic Analysis

The LC-DAD-MS system consisted of a liquid chromatograph, a degasser, a binary pomp, an auto-sampler, and a column oven was combined with a diode array detector (DAD) and MS detector with an electrospray source (AJS ESI) and quadrupole analyzer (1260 Infinity II and 6470 Triple Quad LC/MS, Agilent Technologies, Waldbronn, Germany). The ChemStation software was used to control the LC-MS system and for data processing. The column effluent passed a DAD before arriving in the MS interface.

Chromatographic separations were performed on a Zorbax SB-C18 column (250 mm × 4.6 mm, 5 µm, Agilent Technologies, Santa Clara, CA, USA). For the separation, a gradient of mobile phase A (0.05M $HCOONH_4$ in water, pH 7.0) and mobile phase B (100% methanol) was used. The gradient profile was set as follows: 0 min—15% effluent B, 20 min—80% effluent B, 22 min—100% effluent B, 23 min—100% effluent B, 24 min—15% effluent B, 30 min—15% effluent B. The flow rate was 1 mL/min, the column temperature was 25 °C, and the injection volume was 20 µL.

The elution of sample components were monitored at 380 nm, as it has been shown previously for C-1748 and its metabolites in HepG2 cells [15,20] and at 420 nm.

The electrospray source was operated in a positive mode and the interface condition were as follows: Gas temperature 300 °C and flow 5 L/min, sheath gas temperature 250 °C and flow 11 L/min, nebulizer 45 psi, capillary voltage of 3500 V.

The data were collected in a MS scanning mode (MS2 SCAN) with the range 150–700 (m/z).

3.5. Antimicrobial Activity Assay

Antifungal in vitro activity was determined by the modified M27-A3 specified by the CLSI [31]. Wells containing serially diluted examined compounds and compound-free controls were inoculated with 12 h cultures of tested strains to the final concentration of 10^4 fungi colony-forming units (CFU)/mL. Plates were incubated for 24 h at 37 °C and growth was then quantified by measuring an optical density at 600 nm, using a microplate reader (TECAN Spark 10 M; Tecan Group Ltd., Männedorf, Switzerland). The MIC was defined as the lowest drug concentration in which at least a 90% decrease in turbidity, in comparison to the drug-free control, was observed. The antifungal activity was determined in a RPMI-1640 medium buffered to a pH value of 7.0. The final concentration of the compound solvent (DMSO) did not exceed 2.5% volume of the final suspension in each well, and did not influence the growth of the microorganism.

3.6. Acridine Derivatives Accumulation in Microbial Cells

Candida albicans ATCC 10231 strains were grown in the YPG medium overnight (16–18 h) at 30 °C, washed twice with a sterile phosphate buffered saline (PBS) and re-suspended in RPMI 1640 at a cell density of 2×10^6 cells mL^{-1}. The inoculum (500 µL) was added to 500 µL RPMI 1640 with 100 µM of the tested compounds. After 180 min, 200 µL of the cells suspension was washed three times with a sterile phosphate buffered saline (PBS), re-suspended in 25 µL 90% (v/v) glycerol/10% (v/v) 1× PBS, and transferred to a microscopic slide. The cells were examined with the Olympus BX-60 fluorescence microscope (excitation wavelength 400–410 nm, emission > 455 nm, ×100) equipped with the Olympus XC50 digital camera and cellSens Dimension imaging software.

3.7. Yeast Topoisomerase II Relaxation Assay and Inhibition

The inhibition of yeast Topoisomerase II was analyzed according to the relaxation assay kit from Inspiralis (Inspiralis Ltd., NR4 7GJ, Norwich, UK). Briefly, 500 ng of supercoiled pBR322 DNA, 1 mM ATP, 1–200 µM of the analyzed compound were mixed with a reaction buffer (1 mM Tris.HCl (pH 7.9), 10 mM KCl, 0.5 mM MgCl 2, 0.2% (v/v) glycerol). The reaction was initiated by the addition of an enzyme, allowed to proceed at 30 °C for 30 min and terminated by the addition of 40% (w/v) sucrose, 100 mM Tris-HCl pH 8, 10 mM EDTA, 0.5 mg mL^{-1} Bromophenol Blue. A two-step extraction with chloroform:isoamyl alcohol (24:1) and butanol water were made and mixtures were analyzed on the 1% agarose gel in a $1\times$ TAE buffer, 3 h, 4.5 V cm^{-1}. The gel was stained in a GelRed $3\times$ staining solution for 30 min and photographed with a Gel Doc XR+ Gel Documentation System (Bio-Rad Laboratories, Inc., 1000 Alfred Nobel Drive, Hercules, CA, USA). The relaxation inhibition effectivity (EC50) of the analyzed compounds was determined by densitometry quantification of the transition from supercoiled to relaxed forms and was expressed in relation to the control.

3.8. UV-Vis and Fluorescence Analysis

The UV-visible absorption spectra were recorded using a TECAN Spark 10M (Spark 10M; Tecan Group Ltd., Männedorf, Switzerland) at room temperature with a 1 cm path cell. Compounds at concentrations of 0.1 mM were dissolved in a phosphate-buffered saline (PBS), pH 7.4. Fluorescence spectra were recorded with the use of the TECAN Spark 10M microplate mode. In addition, 100 µL of the analyzed compounds at 0.1 mM concentrations were mixed with 100 µL of MQ water or 0.1 mM DNA sequences: DNA-1 5′→3′: CGATATCG (Tm: 24.0 °C, HPLC grade) and DNA-2 5′→3′: CCCTAGGG (Tm: 28.0 °C, HPLC grade) dissolved in MQ water. The fluorescence spectra of Capridine β, IE1, and m-AMSA with or without DNA were recorded at the excitation wavelength of 410 nm, within the excitation range used for microscopic studies.

4. Conclusions

DNA topoisomerases are enzymes that catalyze changes in the spatial structure of DNA and play an important role in replication, transcription, and recombination. Beyond their normal functions, those enzymes are significant molecular targets in antimicrobial and anticancer chemotherapy. Our results indicate that their fungal counterpart may also become a promising antifungal target. The evaluation of biological properties of Capridine β indicated that despite its high antifungal activity, the ability to enter the cell and biotransformation into a product that influences the effectiveness of fungal topoisomerase is crucial for its activity. Summing up, the search for antifungal drug candidates targeting topoisomerases among acridine derivatives is undoubtedly worth continuing.

Supplementary Materials: The following are available online at https://www.mdpi.com/2076-0817/10/2/189/s1, Figure S1: Growth kinetics of *C. albicans* ATCC 10231 cells in RPMI-1640 medium containing either m-AMSA, Capridine β or Amphotericin B.

Author Contributions: Conceptualization, biotransformation studies, supervision of synthetic as well as biological studies and manuscript preparation, I.G.; methodology, topoisomerase inhibition studies, and MICs determination, K.R.; methodology and chemical synthesis, E.P.; HPLC and ESI-MS analysis and contribution to the manuscript preparation, K.K.-T. All authors have read and agreed to the published version of the manuscript.

Funding: No funding to declare.

Institutional Review Board Statement: Not applicable.

Informed Consent Statement: Not applicable.

Conflicts of Interest: The authors declare no conflict of interest.

References

1. Wainwright, M. Acridine-a neglected antibacterial chromophore. *J. Antimicrob. Chemother.* **2001**, *47*, 1–13. [CrossRef]
2. Prasher, P.; Sharma, M. Medicinal chemistry of acridine and its analogues. *MedChemComm* **2018**, *9*, 1589–1618. [CrossRef] [PubMed]
3. Gabriel, I. "Acridines" as new horizons in antifungal treatment. *Molecules* **2020**, *25*, 1480. [CrossRef] [PubMed]
4. Champoux, J.J. DNA topoisomerases: Structure, function, and mechanism. *Annu. Rev. Biochem.* **2001**, *70*, 369–413. [CrossRef]
5. Li, T.-K.; Houghton, P.J.; Desai, S.D.; Daroui, P.; Liu, A.A.; Hars, E.S.; Ruchelman, A.L.; Lavoie, E.J.; Liu, L.F. Characterization of ARC-111 as a novel topoisomerase I-targeting anticancer drug. *Cancer Res.* **2003**, *63*, 8400–8407.
6. Pommier, Y.; Leo, E.; Zhang, H.; Marchand, C. DNA topoisomerases and their poisoning by anticancer and antibacterial drugs. *Chem. Biol.* **2010**, *17*, 421–433. [CrossRef]

7. Dinardo, S.; Voelkel, K.; Sternglanz, R. DNA topoisomerase II mutant of *Saccharomyces cerevisiae*: *Topoisomerase* II is required for segregation of daughter molecules at the termination of DNA replication. *Proc. Natl. Acad. Sci. USA* **1984**, *81*, 2616–2620. [CrossRef] [PubMed]

8. Holm, C.; Stearns, T.; Botstein, D. DNA topoisomerase II must act at mitosis to prevent nondisjunction and chromosome breakage. *Mol. Cell. Biol.* **1989**, *9*, 159–168. [CrossRef]

9. Kwok, S.C.; Schelenz, S.; Wang, X.; Steverding, D. In Vitro effect of DNA topoisomerase inhibitors on *Candida albicans*. *Med. Mycol.* **2010**, *48*, 155–160. [CrossRef]

10. Khan, S.I.; Nimrod, A.C.; Mehrpooya, M.; Nitiss, J.L.; Walker, L.A.; Clark, A.M. Antifungal activity of eupolauridine and its action on DNA topoisomerases. *Antimicrob. Agents Chemother.* **2002**, *46*, 1845–1850. [CrossRef]

11. Shen, L.L.; Baranowski, J.; Fostel, J.; Montgomery, D.A.; Lartey, P.A. DNA topoisomerases from pathogenic fungi: Targets for the discovery of antifungal drugs. *Antimicrob. Agents Chemother.* **1992**, *36*, 2778–2784. [CrossRef]

12. Del Poeta, M.; Toffaletti, D.L.; Rude, T.H.; Dykstra, C.C.; Heitman, J.; Perfect, J.R. Topoisomerase I is essential in *Cryp-tococcus neoformans*: Role in pathobiology and as an antifungal target. *Genetics* **1999**, *152*, 167–178. [PubMed]

13. Jiang, W.; Gerhold, D.; Kmiec, E.B.; Hauser, M.; Becker, J.M.; Koltin, Y. The topoisomerase I gene from *Candida albicans*. *Microbiology* **1997**, *143*, 377–386. [CrossRef] [PubMed]

14. Augustin, E.; Moś-Rompa, A.; Nowak-Ziatyk, D.; Konopa, J. Antitumor 1-nitroacridine derivative C-1748, induces apoptosis, necrosis or senescence in human colon carcinoma HCT8 and HT29 cells. *Biochem. Pharm.* **2010**, *79*, 1231–1241. [CrossRef]

15. Wiśniewska, A.; Niemira, M.; Jagiełło, K.; Potęga, A.; Świst, M.; Henderson, C.; Skwarska, A.; Augustin, E.; Konopa, J.; Mazerska, Z.; et al. Diminished toxicity of C-1748, 4-methyl-9-hydroxyethylamino-1-nitroacridine, compared with its demethyl analog, C-857, corresponds to its resistance to metabolism in HepG2 cells. *Biochem. Pharm.* **2012**, *84*, 30–42. [CrossRef] [PubMed]

16. Rząd, K.; Paluszkiewicz, E.; Gabriel, I. A new 1-nitro-9-aminoacridine derivative targeting yeast topoisomerase II able to overcome fluconazole-resistance. *Bioorganic Med. Chem. Lett.* **2021**, *35*, 127815. [CrossRef]

17. Tadi, K.; Ashok, B.T.; Chen, Y.; Banerjee, D.; Wysocka-Skrzela, B.; Konopa, J.; Darzynkiewicz, Z.; Tiwari, R.K. Pre-clinical evaluation of 1-nitroacridine derived chemotherapeutic agent that has preferential cytotoxic activity towards prostate cancer. *Cancer Biol.* **2007**, *6*, 1632–1637. [CrossRef]

18. Ashok, B.; Tadi, K.; Banerjee, D.; Konopa, J.; Iatropoulos, M.; Tiwari, R.K. Pre-clinical toxicology and pathology of 9-(2'-hydroxyethylamino)-4-methyl-1-nitroacridine (C-1748), a novel anti-cancer agent in male Beagle dogs. *Life Sci.* **2006**, *79*, 1334–1342. [CrossRef]

19. Ashok, B.T.; Tadi, K.; Garikapaty, V.P.; Chen, Y.; Huang, Q.; Banerjee, D.; Konopa, J.; Tiwari, R.K. Preclinical toxicological examination of a putative prostate cancer-specific 4-methyl-1-nitroacridine derivative in rodents. *Anti-Cancer Drugs* **2007**, *18*, 87–94. [CrossRef]

20. Augustin, E.; Niemira, M.; Hołownia, A.; Mazerska, Z. CYP3A4-dependent cellular response does not relate to CYP3A4-catalysed metabolites of C-1748 and C-1305 acridine antitumor agents in HepG2 cells. *Cell Biol. Int.* **2014**, *38*, 1291–1303. [CrossRef]

21. Tan, X.; Fuchs, B.B.; Wang, Y.; Chen, W.; Yuen, G.J.; Chen, R.B.; Jayamani, E.; Anastassopoulou, C.; Pukkila-Worley, R.; Coleman, J.J.; et al. The Role of *Candida albicans* SPT20 in filamentation, biofilm formation and pathogenesis. *PLoS ONE* **2014**, *9*, e94468. [CrossRef] [PubMed]

22. Bartoszek, A.; Konopa, J. 32P-post-labeling analysis of DNA adduct formation by antitumor drug nitracrine (*Ledakrin*) and other nitroacridines in different biological systems. *Biochem. Pharm.* **1989**, *38*, 1301–1312. [CrossRef]

23. Sirajuddin, M.; Ali, S.; Badshah, A. Drug–DNA interactions and their study by UV–Visible, fluorescence spectroscopies and cyclic voltametry. *J. Photochem. Photobiol. B Biol.* **2013**, *124*, 1–19. [CrossRef] [PubMed]

24. Taraszkiewicz, A.; Grinholc, M.; Bielawski, K.P.; Kawiak, A.; Nakonieczna, J. Imidazoacridinone derivatives as efficient sensitizers in photoantimicrobial chemotherapy. *Appl. Environ. Microbiol.* **2013**, *79*, 3692–3702. [CrossRef] [PubMed]

25. Bailly, C. Contemporary challenges in the design of topoisomerase II inhibitors for cancer chemotherapy. *Chem. Rev.* **2012**, *112*, 3611–3640. [CrossRef] [PubMed]

26. Wu, C.-C.; Li, Y.-C.; Wang, Y.-R.; Li, T.-K.; Chan, N.-L. On the structural basis and design guidelines for type II topoisomerase-targeting anticancer drugs. *Nucleic Acids Res.* **2013**, *41*, 10630–10640. [CrossRef]

27. Lemke, K.; Wojciechowski, M.; Laine, W.; Bailly, C.; Colson, P.; Baginski, M.; Larsen, A.K.; Skladanowski, A. Induction of unique structural changes in guanine-rich DNA regions by the triazoloacridone C-1305, a topoisomerase II inhibitor with antitumor activities. *Nucleic Acids Res.* **2005**, *33*, 6034–6047. [CrossRef]

28. Mazerska, Z.; Sowiński, P.; Konopa, J. Molecular mechanism of the enzymatic oxidation investigated for imidazoacridinone antitumor drug, C-1311. *Biochem. Pharm.* **2003**, *66*, 1727–1736. [CrossRef]

29. Konopa, J.; Wysocka-Skrzela, B.; Tiwari, R.K. *9-Alkilamino-1-Nitroacridine Derivatives*; U.S. Patent 6.589.961 B2; Patent and Trademark Office: Alexandria, VA, USA, 2003.

30. Wysocka-Skrzela, B. Research on tumor-inhibiting compounds: Reactions of 1-nitro-9-aminoacridine derivatives, new antitumor agents, with nucleophiles. *Pol. J. Chem.* **1986**, *60*, 317–318.

31. CLSI. *Reference Method for Broth Dilution Antifungal Susceptibility Testing of Yeasts*; Clinical and Laboratory Standard Institute: Annapolis Junction, MD, USA, 2008.

Article

Sphingolipid Inhibitors as an Alternative to Treat Candidiasis Caused by Fluconazole-Resistant Strains

Rodrigo Rollin-Pinheiro [1,†], Brayan Bayona-Pacheco [2,3,†], Levy Tenorio Sousa Domingos [2,†], Jose Alexandre da Rocha Curvelo [2], Gabriellen Menezes Migliani de Castro [2], Eliana Barreto-Bergter [1] and Antonio Ferreira-Pereira [2,*]

[1] Laboratório de Química Biológica de Microrganismos, Departamento de Microbiologia Geral, Instituto de Microbiologia Paulo de Góes, Universidade Federal do Rio de Janeiro (UFRJ), Rio de Janeiro 21941-902, Brazil; rodrigorollin@gmail.com (R.R.-P.); eliana.bergter@micro.ufrj.br (E.B.-B.)

[2] Laboratório de Bioquímica Microbiana, Departamento de Microbiologia Geral, Instituto de Microbiologia Paulo de Góes, Universidade Federal do Rio de Janeiro (UFRJ), Rio de Janeiro 21941-902, Brazil; bbayona@uninorte.edu.co (B.B.-P.); levydomingos@yahoo.com.br (L.T.S.D.); alexandrecurvelo@hotmail.com (J.A.d.R.C.); g.migliani@gmail.com (G.M.M.d.C.)

[3] Departamento de Medicina, División Ciencias de la Salud, Universidad del Norte, Km 5, vía Puerto Colombia, Área Metropolitana de Barranquilla 081007, Colombia

* Correspondence: apereira@micro.ufrj.br

† These authors contributed equally to the study.

Citation: Rollin-Pinheiro, R.; Bayona-Pacheco, B.; Domingos, L.T.S.; da Rocha Curvelo, J.A.; de Castro, G.M.M.; Barreto-Bergter, E.; Ferreira-Pereira, A. Sphingolipid Inhibitors as an Alternative to Treat Candidiasis Caused by Fluconazole-Resistant Strains. *Pathogens* **2021**, *10*, 856. https://doi.org/10.3390/pathogens10070856

Academic Editor: Jonathan Richardson

Received: 9 June 2021
Accepted: 3 July 2021
Published: 7 July 2021

Publisher's Note: MDPI stays neutral with regard to jurisdictional claims in published maps and institutional affiliations.

Abstract: *Candida* species are fungal pathogens known to cause a wide spectrum of diseases, and *Candida albicans* and *Candida glabrata* are the most common associated with invasive infections. A concerning aspect of invasive candidiasis is the emergence of resistant isolates, especially those highly resistant to fluconazole, the first choice of treatment for these infections. Fungal sphingolipids have been considered a potential target for new therapeutic approaches and some inhibitors have already been tested against pathogenic fungi. The present study therefore aimed to evaluate the action of two sphingolipid synthesis inhibitors, aureobasidin A and myriocin, against different *C. albicans* and *C. glabrata* strains, including clinical isolates resistant to fluconazole. Susceptibility tests of aureobasidin A and myriocin were performed using CLSI protocols, and their interaction with fluconazole was evaluated by a checkerboard protocol. All *Candida* strains tested were sensitive to both inhibitors. Regarding the evaluation of drug interaction, both aureobasidin A and myriocin were synergic with fluconazole, demonstrating that sphingolipid synthesis inhibition could enhance the effect of fluconazole. Thus, these results suggest that sphingolipid inhibitors in conjunction with fluconazole could be useful for treating candidiasis cases, especially those caused by fluconazole resistant isolates.

Keywords: *Candida*; sphingolipids; myriocin; fungal infections

1. Introduction

Candida species cause a wide spectrum of infections in humans, ranging from superficial mycosis, especially associated to skin and vaginal mucosae, to life-threatening disseminated candidiasis [1]. *Candida albicans* and *Candida glabrata* are the most frequent species associated to invasive infections, being responsible for about 46% and 24%, respectively [2]. Candidiasis is commonly associated with different pathologies, such as HIV infection, organ transplantation, cancer and diabetes, which contribute to a mortality rate of up to 60% [3]. In addition, it has been considered the most frequent fungal disease associated to healthcare units and the fourth most prevalent nosocomial infection [4].

The current antifungal drugs available to be used against fungal infections are limited to only four classes: azoles, which block the enzyme lanosterol 14-α demethylase and, thus, disrupt ergosterol synthesis; polyenes, which directly bind to ergosterol found in the plasma membrane and cause the release of cytoplasmic content; echinocandins, which inhibit the enzyme β(1,3)-glucan synthase and affect cell wall synthesis; and fluoropyrimidine analogs, which block DNA synthesis [5]. The first choice of drug to treat candidiasis is fluconazole, an azole antifungal drug, but resistant strains have been emerging over the last decades, causing a significant impact in public health [6,7]. Multi-resistant strains have been isolated, which express different types of transporters or display over-expressed azole targets [8]. In addition, alternative therapeutic options are limited due to low diversity of antifungal classes and high level of toxicity and side effects [9].

For these reasons, there is an urgent need to find alternative therapeutic approaches to obtain better results in treating patients who carry resistant strains. Different fungal cell components have been studied as potent new targets for the development of antifungal drugs. In this context, sphingolipids have been considered interesting candidates. Several studies have shown that sphingolipids, mainly glucosylceramide, play crucial roles in fungal growth, cellular signaling and virulence [10–12]. In *C. albicans*, *Cryptococcus neoformans*, *Penicillium digitatum*, and *Aspergillus fumigatus*, mutants which do not express the glucosylceramide synthase gene displayed alterations in plasma membrane, growth, and virulence in infection models [13–17]. In addition, some compounds that inhibit sphingolipid biosynthesis, such as aureobasidin A and myriocin, have been shown to present antifungal activity with low minimum inhibitory concentration (MIC) in a variety of fungal pathogens, including *Candida* and *Aspergillus* species [18–22]. More recently, a class of drugs called acylhydrazones was described which affect the synthesis of glucosylceramides of *C. neoformans*, *C. albicans*, *A. fumigatus*, and *Pneumocystis murina*, but not those from mammalian cells [23].

Due to the increasing resistance of *Candida* species associated with invasive infections and the potential of targeting sphingolipids, the present study aimed to test two inhibitors of sphingolipid synthesis, aureobasidin A and myriocin. Both drugs were evaluated against different *Candida* clinical isolates, including some that were described as strains that overexpress transmembrane transporters (ABC and MFS) related to multidrug resistance phenotype [24,25]. Aureobasidin A and myriocin inhibit inositolphosphorylceramide (IPC) synthase and glucosylceramide (GlcCer) synthase, respectively, which are two key enzymes for the synthesis of the most important sphingolipids found in fungi such as IPC and GlcCer.

2. Results

2.1. Antifungal Effect against Candida Strains

Aureobasidin A and myriocin were initially tested against all *Candida* strains used in the study (Table 1). ATCC strains were used as a control because they do not to express resistance mechanisms. On the other hand, clinical isolates are highly resistant to fluconazole (MIC > 256 µg/mL) and present different resistance mechanisms as pointed out in Materials and Methods section.

Table 1. *Candida* strains used in the study.

Strains	Resistance Pattern	Reference
C. albicans (ATCC 10231D-5)	No resistance described	American Type Culture Collection *
C. glabrata (ATCC 2001D-5)	No resistance described	American Type Culture Collection *
C. glabrata (109)	*CDR1* gene overexpressed (ABC transporter)	[24]
C. albicans (1114)	*MDR1* gene overexpressed (MFS transporter)	[25]
C. albicans (12-99)	*ERG11*, *CDR1*, *CDR2* and *MDR1* genes overexpressed (ABC and MFS transporters)	[26]

* https://www.atcc.org/, (accessed on 30 June 2021).

Aureobasidin A inhibits fungal growth at 0.5 μg/mL for ATCC strains (*C. albicans* and *C. glabrata*) and at 0.25 μg/mL for clinical isolates (109, 1114, and 1299) (Table 2). Myriocin presents antifungal activity at 2.0 and 1.0 μg/mL for ATCC *C. albicans* and *C. glabrata*, respectively, and at 0.5, 1.0, and 0.25 μg/mL for clinical isolates 109, 1114, and 1299, respectively (Table 2).

Table 2. *Candida* strains susceptibility to Aureobasidin A, Myriocin, and Fluconazole.

	* MIC$_{90}$ (μg/mL)		
	Aureobasidin A	**Myriocin**	**Fluconazole**
C. albicans (ATCC 10231D-5)	0.5	2.0	<8
C. glabrata (ATCC 2001D-5)	0.5	1.0	<8
C. glabrata (109)	0.25	0.5	>256
C. albicans (1114)	0.25	1.0	>256
C. albicans (12-99)	0.25	0.25	>256

* MIC: minimal inhibitory concentration.

Cell viability decreased at MIC values, as evaluated by XTT-reduction assay, indicating that both inhibitors present fungicidal effect (Figure 1).

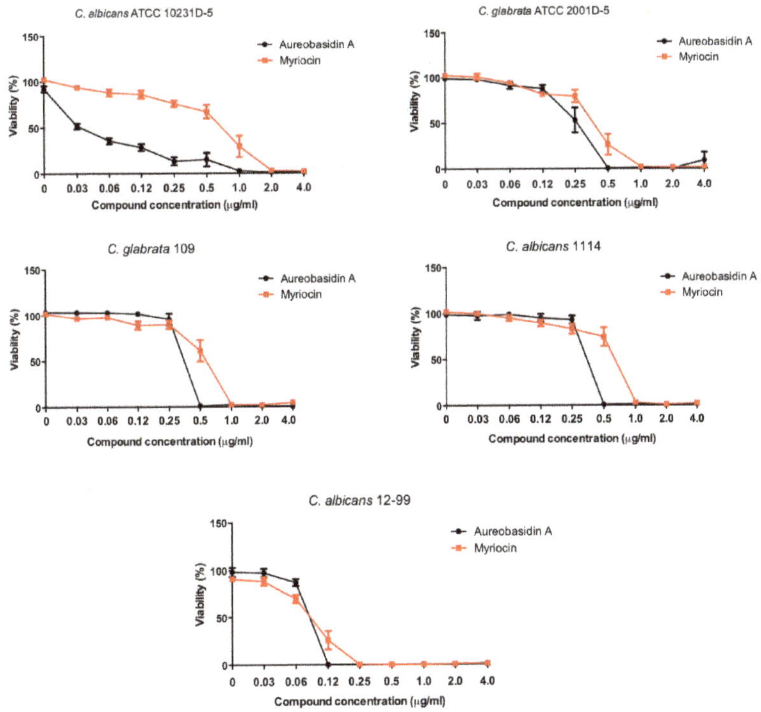

Figure 1. Viability of *Candida* ATCC (*C. albicans* ATCC 10231D-5 and *C. glabrata* ATCC 2001D-5) and clinical strains (*C. albicans* 1114 and 12-99, *C. glabrata* 109) in the presence of aureobasidin A and myriocin. Cells were grown in microplates containing RPMI at 37 °C for 48 h in the absence (0 μg/mL) or in the presence of aureobasidin A and myriocin. After the incubation time, viability was evaluated using the XTT-reduction assay. Cell viability was quantified using a microplate reader (Bio-Rad, Hercules, CA, USA) at 490 nm. Percentage of fungal growth was calculated considering the control (absence of all drugs) as 100%. Errors bars represent standard errors of means of different experiments in different days (*n* = 3).

2.2. Interaction between Aureobasidin A, Myriocin, and Fluconazole

As pointed out previously, fluconazole is the first choice in treating candidiasis and the emergence of fluconazole-resistant strains is a concern in healthcare settings, because it is related to high mortality of infected patients. For this reason, a synergic effect of sphingolipid inhibitors and fluconazole could be useful in order to improve treatment of patients infected with resistant *Candida* species.

Both sphingolipid inhibitors display a synergic effect with fluconazole (Table 3), except aureobasidin A in strain 12-99. Aureobasidin A reduced fluconazole MIC from more than 256 µg/mL (strains 109 and 12-99) or 128 µg/mL (strain 1114) to 32, 16, or 128 µg/mL for strains 109, 1114, and 12-99, respectively. FICI values were 0.1874, 0.25, and 0.56 for strains 109, 1114, and 12-99, respectively, indicating that a synergic effect occurs between aureobasidin A and fluconazole for most of the strains used.

Table 3. Interaction of aureobasidin A or myriocin with fluconazole tested in Candida clinical isolates presenting resistance to fluconazole.

	Candida **Strains**		
	109	**1114**	**12-99**
	MIC_{90} alone (µg/mL)		
Fluconazole	>256	128	>256
Aureobasidin A	0.25	0.25	0.25
Myriocin	0.5	1.0	0.25
	MIC_{90} combined (µg/mL)		
Aureo/Fluco	0.0156/32	0.03125/16	0.015/128
Myr/Fluco	0.0625/64	0.25/32	0.0625/16
	FICI		
Aureo/Fluco	0.1874 (synergic)	0.25 (synergic)	0.56 (no effect)
Myr/Fluco	0.375 (synergic)	0.5 (synergic)	0.31 (synergic)

Aureo: aureobasidin A; Fluco: fluconazole; Myr: myriocin; MIC: minimal inhibitory concentration; FICI: fractional inhibitory index.

On the other hand, myriocin decreased fluconazole MIC to 64 (strain 109), 32 (strain 1114) and 16 µg/mL (strain 12-99) (Table 3). Corresponding FICI values were 0.375, 0.5 and 0.31 for strains 109, 1114, and 1299, respectively, confirming a synergic effect between myriocin and fluconazole.

A graphical representation of synergism data is shown in Figure 2.

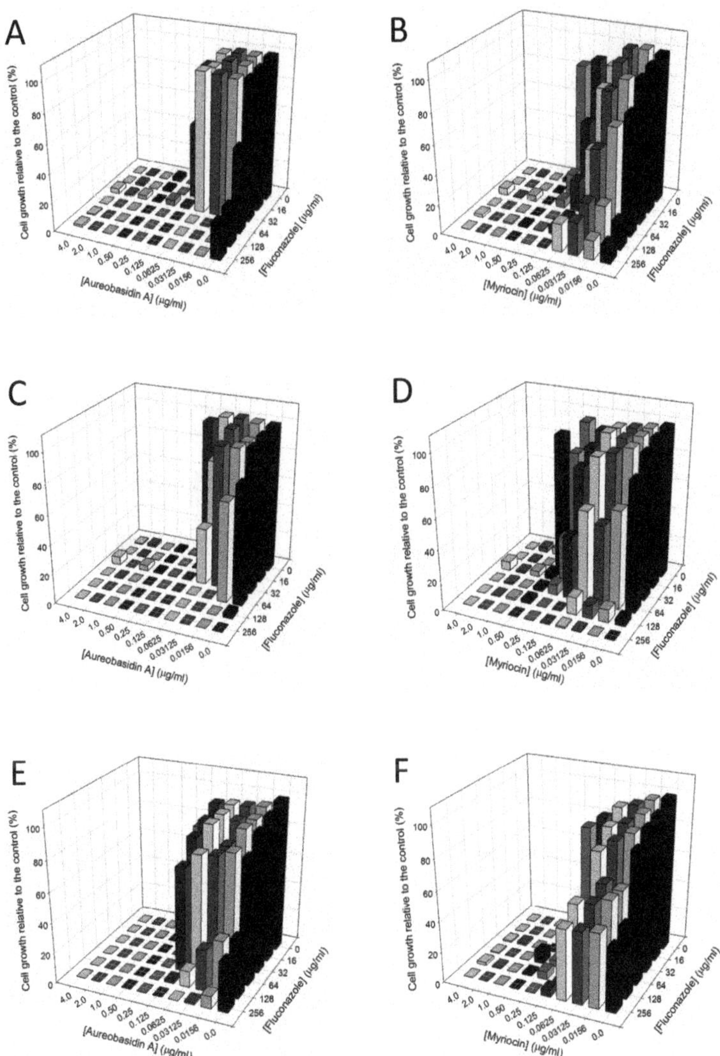

Figure 2. Evaluation of the interaction between aureobasidin A (**A,C,E**) and myriocin (**B,D,F**) with fluconazole. Analysis was performed with the isolates 109 (**A,B**), 1114 (**C,D**), and 1299 (**E,F**) and the results are shown using the software Sigma Plot 12.0. Cells were grown in microplates containing RPMI at 37 °C for 48 h in the absence (0 µg/mL) or in the presence of different combinations of aureobasidin A or myriocin with fluconazole. After the incubation time, cell growth was evaluated using a microplate reader (Bio-Rad, Hercules, CA, USA) at 600 nm. Percentage of fungal growth was calculated considering the control (absence of all drugs) as 100%.

2.3. Cytotoxicity of Aureobasidin A and Myriocin

In order to test if aureobasidin A and myriocin are toxic to mammalian cells at the same concentrations found effective in previous experiments, a cytotoxicity assay was performed on murine macrophages (RAW 264.7). Compared to untreated cells, the control of cells treated with 1% dimethylsulfoxide (DMSO) presented 80% viability whereas myriocin treatment led to 60% viability, demonstrating a decrease of approximately 20% compared to DMSO-treated cells (Figure 3). Regarding aureobasidin A treatment, RAW cell viability was not affected up to 2.5 µg/mL, which is more than 10-fold higher than the concentration presenting synergism with fluconazole.

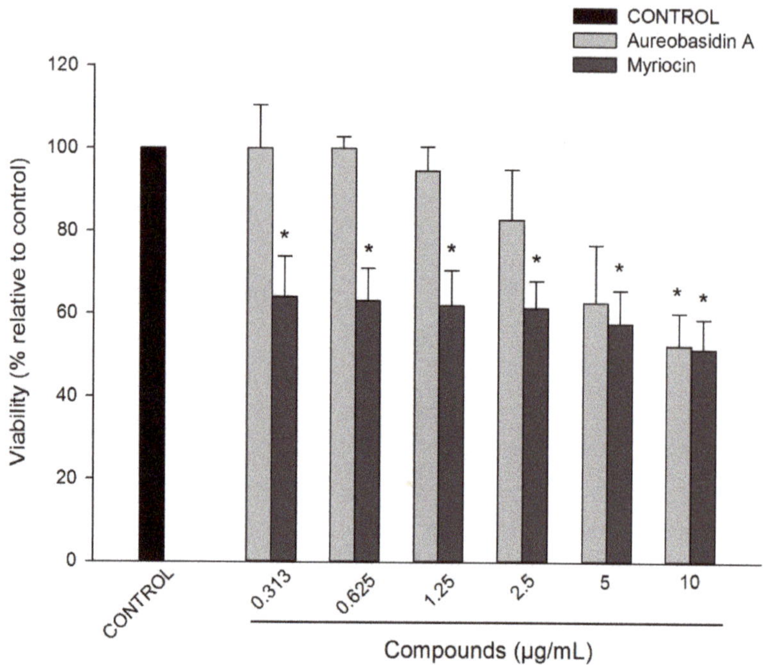

Figure 3. Cytotoxic assay of aureobasidin A and myriocin against murine macrophage (RAW 264.7) cell lineage for 48 h. The data represent the means of three independent experiments and the error bars represent the standard error. Solid black bar represents the control in the absence of compounds. * $p < 0.05$.

These data suggest that aureobasidin A is not toxic at concentrations used in this work, whereas myriocin is partially toxic.

2.4. Effect of Aureobasidin A and Myriocin on the Lifespan of Wild Type Caenorhabditis elegans

To check the toxicity of compounds now against living organisms, we did a survival test using a wild-type of *C. elegans* worm. The worms were tested in the presence of 0.5 µg/mL of both compounds and in the presence of 0.1% DMSO as a control. After 4 days of analysis regarding the survival of the worms in the presence of the compounds, it can be observed that only myrocin at 0.5 µg/mL was toxic to the worms since in the case of aureobasidin A, at the same concentration, the survival was practically the same (approximately 98%) when compared to the control (Figure 4). These results corroborate what was observed in the cytotoxicity assay using macrophages (Figure 3), where only myrocin seemed to be more toxic than aureobasidin A.

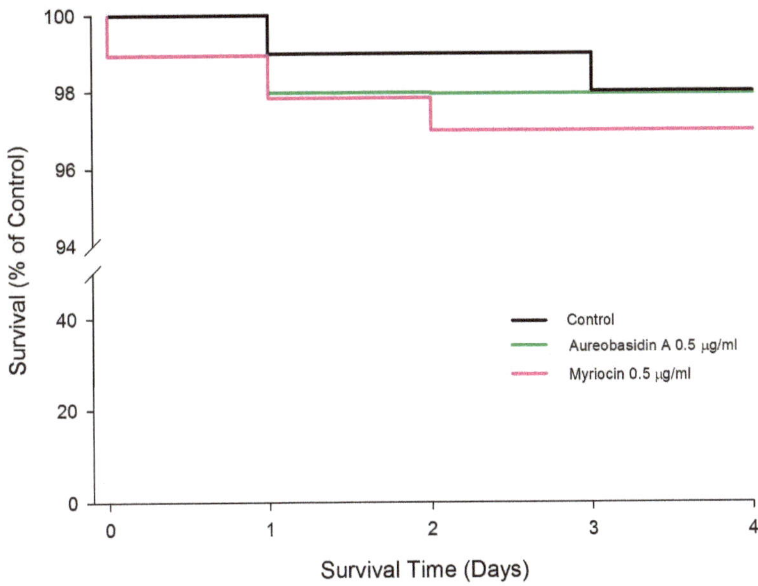

Figure 4. Effect of aureobasidin A and myriocin exposure on lifespan of *C. elegans* wild-type (N2) during five days. The data represent the means of two independent experiments that were collected daily. The control was performed in the absence of compounds but in the presence of same amount of DMSO (0.1%) present at tested compounds.

3. Discussion

Infections caused by *Candida* species represent a significant concern in clinical settings due to their high morbidity and mortality, as well as the emergence of resistant isolates [27,28]. Thus, the study of new alternatives to treat candidiasis, especially those caused by resistant strains, is an urgent need.

Sphingolipids are a potential new target for drug development. They are considered a good candidate because fungal sphingolipids are structurally different from the human counterparts and due to their crucial roles in fungal growth, cellular signaling and pathogenesis also. Several studies demonstrated that different compounds are able to block different steps of sphingolipid biosynthesis and therefore present antifungal activity. These compounds—such as myriocin, fumonisin B1, aureobasidin A, and D-*threo*-1-phenyl-2-decanoylamino-3-morpholino-1-propanol (D-threo-PDMP)—act by blocking serine palmitoyltransferase, ceramide synthases, IPC synthase, and GlcCer synthase, respectively [12].

The present study aimed to use two of these compounds, aureobasidin A and myriocin, in order to evaluate their activity against *Candida* species. *C. albicans* ATCC (10231D-5) and *C. glabrata* ATCC (2001D-5) were used, as well as three clinical isolates highly resistant to fluconazole, *C. glabrata* strain 109 (over-expression of the CDR1 gene), *C. albicans* strain 1114 (over-expression of the MDR1 gene) and *C. albicans* strain 12-99 (over-expression of the genes ERG11, CDR1, CDR2, MDR1) (Table 1).

Aureobasidin A displayed MICs of 0.5 μg/mL for ATCC strains and 0.25 μg/mL for clinical isolates, whereas myriocin presented MICs ranging from 0.25 μg/mL to 2.0 μg/mL (Table 2, Figure 1). These data are in accordance with the literature, since it has also been demonstrated that aureobasidin A at concentrations of 2.0–3.5 μg/mL inhibited ATCC strains of *C. albicans*, *C. glabrata*, *C. tropicalis*, *C. parapsilosis*, and *C. krusei* [29]. Clinical isolates were also evaluated by Tan and Tay [21], who showed MICs for aureobasidin A of 4 μg/mL for *C. albicans* and 1 μg/mL for non-*albicans* isolates. This inhibitory effect has already been described to occur due to ceramide intoxication and deprivation of essential IPCs [19]. In addition, aureobasidin A has already been demonstrated to inhibit IPC synthase activity even at nanomolar levels, suggesting that its antifungal action might be a result of alterations on the biosynthesis of sphingolipids [29]. Kumar and colleagues have also shown that in vitro treatment of *Candida auris* with aureobasidin A leads to a deregulation of many intermediates of sphingolipid biosynthetic pathway [30].

Regarding myriocin, it presented MICs ranging from 0.25 µg/mL to 2.0 µg/mL (Table 2, Figure 1). De Melo and colleagues reported a similar MIC value of 0.12 µg/mL for *C. albicans* SC5314 [31]. Recently, myriocin was tested against some *Candida* strains, including isolates resistant to voriconazole, and MICs were found varying between 0.125–4.0 µg/mL [32]. Aureobasidin A and myriocin also affect *Candida* biofilms, and it was due to modification in lipid composition and to the alteration in lipid raft organization and plasma membrane [20,32]. The effect of myriocin on *Candida* cells has been recently evaluated by Yang and colleagues, who demonstrated that a disruption of plasma membrane is observed when different species are treated with myriocin [32]. Similar data have also been shown in other pathogenic fungi, such as *Scedosporium boydii*, in which myriocin led to alterations on plasma membrane resulting in higher susceptibility to membrane stressors such as SDS [33]. In *Aspergillus fumigatus*, myriocin treatment led to a decrease in phytoceramide content, suggesting that this inhibitor also alter the regulation of sphingolipid production [22]. Thus, the effect of myriocin and aureobasidin A against different pathogenic fungi suggests that the disruption of sphingolipid biosynthesis seems to display a conserved antifungal activity, although more studies are needed using other samples and compounds [18,20,22,34,35].

Synergistic effect of two different drugs is a promising alternative to enhance efficacy of the current antifungals. This approach uses two known drugs combined, which are already approved to be used in clinical settings, and their toxicity was already determined. This is a great advantage when compared to the costly and time-consuming development of new drugs [36]. The best-known example of synergism is the combination of fluorocytosine and amphotericin B, which is the gold standard treatment for cryptococcosis [36,37]. However, very few studies describe synergistic effects of sphingolipid inhibitors and the current antifungal drugs. De Melo and colleagues demonstrated that myriocin is synergistic to amphotericin B but not to fluconazole [31]. However, only one susceptible strain and no fluconazole-resistant isolate was used, so more studies are needed in this field. Since fluconazole is the first choice to treat *Candida* invasive infections with high mortality and resistant strains have been emerging in recent years, it is crucial to develop treatment alternatives. Our data showed that both aureobasidin A and myriocin present synergistic effects with fluconazole on almost all clinical isolates tested in this work (Table 3, Figure 2). As mentioned, the clinical isolates used are highly resistant to fluconazole, which suggests that synergy is a promising option to be used in patients carrying fluconazole-resistant yeast strains. Myriocin also presents synergism with fluconazole in *Scedosporium boydii*, a pathogenic filamentous fungus, suggesting that this effect could be conserved among other pathogens [33], suggesting that targeting fungal sphingolipids in combination with azoles is promising in order to treat fungal infections, especially in cases where resistance to azoles lead to a failure in treatment success.

A key point and concern of using sphingolipid inhibitors to treat fungal infections is their cytotoxic effect in humans. For instance, fumonisin B is a mycotoxin produced by *Fusarium* species that also display toxicity to mammalian cells [38,39]. In the present study, cytotoxic assays showed that myriocin is partially toxic in RAW cells, whereas aureobasidin A is not toxic (Figure 3), and the same profile of toxicity was observed when both compounds were tested against live *C. elegans* (Figure 4), suggesting that both drugs (specially aureobasidin A) could be considered in addition to fluconazole. Considering that fluconazole resistance in *Candida* isolates is a serious problem in clinical healthcare settings, sphingolipid inhibitors were shown to be potential therapeutic options and more studies are needed to explore their use as an alternative approach when administered in combination with fluconazole.

4. Conclusions

The present work showed the two sphingolipid inhibitors, myriocin and aureobasidin A, display antifungal activity against *C. albicans* and *C. glabrata*, not only against ATCC strains but also clinical isolates, suggesting that these compounds are active against strains presenting resistance mechanisms to the current antifungal drugs used in clinical settings. Myriocin and aureobasidin A also presented synergistic interaction with fluconazole, indicating that they could be a promising approach as a combined therapy especially to treat infections caused by resistant strains.

Toxicity analyses revealed that both drugs do not display significant toxic effect, especially in the *C. elegans* model, which reinforces their potential as an alternative therapy when combined with fluconazole. Further studies are needed to evaluate in vivo activity of this approach and to clarify the promising use of sphingolipid inhibitors as alternatives to treat *Candida* infections.

5. Materials and Methods

5.1. Cell Lineages and Reagents

A total of five strains were used in this study (Table 1). *C. albicans* ATCC 10231D-5 and *C. glabrata* ATCC 2001D-5 were used as standard. Three clinical isolates displaying resistance patterns to fluconazole were also evaluated, *C. glabrata* 109 strain (which displays overexpression of *CDR1* gene that encode a ABC transporter), *C. albicans* 1114 strain (which displays overexpression of *MDR1* gene that encode a MFS transporter) and *C. albicans* 12-99 strain (which displays *ERG11*, that confers resistance by mutation or overexpression of 14-α demethylase involved in ergosterol synthesis; *CDR1*, *CDR2* and *MDR1* genes that encode ABC and MFS transporters), kindly provided by Theodore White from University of Missouri, USA. For all experiments, the strains were grown on Yeast Extract Peptone Dextrose (YPD) agar and transferred to YPD broth and incubated at 37 °C for 18 h under agitation.

Cytotoxicity assays were carried out using the murine macrophage-derived cell lines RAW 264.7.

Aureobasidin A (Sigma–Aldrich, St. Louis, MO, USA), myriocin (Sigma–Aldrich, St. Louis, MO, USA) and fluconazole (University pharmacy, UFJF, Juiz de Fora-MG, Brazil) were used in susceptibility and synergism tests.

5.2. Susceptibility Tests with Aureobasidin A and Myriocin and Interaction with Fluconazole

The susceptibility assay was performed to determine the minimal inhibitory concentration (MIC) of aureobasidin A and myriocin, according to Clinical Laboratory Standards Institute (CLSI) M60 protocol. Both compounds were used in a concentration range of 0.031–4.0 µg/mL and MIC_{90} was determined when fungal growth presented 90% of inhibition compared to a positive control of untreated cells. Briefly, yeasts were inoculated in sterile 96-well plates in 200 µL of RPMI medium (Roswell Park Memorial Institute), so that they reached the concentration of 5×10^3 cells/mL in the presence of 1:2 dilutions of each compound. The 96-well plates were incubated at 37 °C for 48 h with shaking (100 rpm). Cell growth was evaluated using a microplate reader (iMark, Bio-Rad, Hercules, CA, USA) at 600 nm.

Candida cell viability was evaluated after MIC determination by using the XTT-reduction technique, according to Rollin-Pinheiro and colleagues [33]. Briefly, after fungal growth as mentioned above, a 0.5 mg/mL XTT solution in PBS was added to the 96-well plates and cells were incubated at 37 °C for 2 h protected from light. Further, optical density was measured using a spectrophotometer (SpectraMax® i3x, Molecular Devices®, San José, CA, EUA) at 490 nm to evaluate cell viability.

Interaction analysis of aureobasidin A and myriocin with fluconazole was performed using the checkerboard method according to Reis de Sá and colleagues [40]. Aureobasidin A and myriocin concentrations ranged from 0.0078–2.0 µg/mL and fluconazole concentration from 16–256 µg/mL. After 48 h of growth at 37 °C under agitation, the fractional inhibitory index (FICI) was calculated according to the formula (MIC combined/MIC drug A alone) + (MIC combined/MIC drug B alone), where A is aureobasidin A or myriocin and drug B is fluconazole. Interaction was classified according to the following parameter: ≤ 0.5, synergistic interaction; >0.5 to ≤ 4, no interaction; >4, antagonistic effect [41].

5.3. Cytotoxicity Assay

Cytotoxicity was analyzed by neutral red (NR) assay with modifications [42]. RAW 264.7 cell monolayer was harvested with a cell scraper and viable cells were counted using the Trypan blue exclusion method. 2×10^5 macrophages per well were seeded in 96-well plates containing Dulbecco's modified Eagle medium (DMEM) with 10% FBS and incubated in a controlled atmosphere of 5% CO_2 at 37 °C for adhesion. Compounds were serial diluted in DMEM and cells were incubated at concentrations of 0.313, 0.625, 1.25, 2.50, 5, and 10 µg/mL at 37 °C, 5% CO_2 for 48h. Cells without compounds were used as control. Absorbance was determined in a spectrophotometer at 595 nm (SpectraMax® i3x, Molecular Devices®, San José, CA, EUA). Each test was performed in triplicate.

5.4. Caenorhabditis elegans Lifespan Assay

C. elegans strain N2 (wild isolate) was obtained from the Caenorhabditis Genetics Center at University of Minnesota (USA) and handled according to standard method [43]. Worms were maintained at 15 °C on nematode growth medium (NGM) and routinely maintained on *Escherichia coli* OP50 strain used as a normal diet for nematodes. Lifespan worm assay was performed as previously described [44,45] with small modifications. Briefly, synchronization of worms was achieved by preparing eggs from gravid adults using a solution containing NaOCl 6% and NaOH 5M; released eggs were washed with M9 buffer and allowed to hatch overnight in NGM agar plates. Synchronized young worms were collected by washing with M9 buffer. Approximately 20 worms were added to each well of 96-well plates containing 100 mL MB medium in the absence or presence of 0.5 µg/mL aureobasidin or 0.5 µg/mL myriocin. Then, the plates were incubated at 25 °C during 4 days without shaking and scored as live and dead-on daily basis. The survival ratio was calculated from the percentage of living worms out of total number of worms including living and dead animals. This experiment was independently conducted in two different days with a twofold analysis in each one.

5.5. Statistical Analyses

All experiments were performed in triplicate, in three independent experimental sets. Statistical analyses were performed using GraphPad Prism version 5.00 for Windows (GraphPad Software, San Diego, CA, USA). One-way analysis of variance using a Kruskal–Wallis nonparametric test was used to compare the differences between groups, and individual comparisons of groups were performed using a Bonferroni posttest. The 90–95% confidence interval was determined in all experiments.

Author Contributions: Conceptualization, R.R.-P., B.B.-P. and L.T.S.D.; Methodology, R.R.-P., B.B.-P., L.T.S.D. and J.A.d.R.C.; Software, R.R.-P., B.B.-P., L.T.S.D. and G.M.M.d.C.; Validation, R.R.-P., B.B.-P. and L.T.S.D.; Formal analysis, R.R.-P., B.B.-P., L.T.S.D. and G.M.M.d.C.; Investigation, R.R.-P., B.B.-P. and L.T.S.D.; Resources, E.B.-B. and A.F.-P.; Data curation, R.R.-P., B.B.-P. and L.T.S.D.; Writing—original draft preparation, R.R.-P.; Writing—review and editing, B.B.-P., L.T.S.D., E.B.-B. and A.F.-P.; Supervision, E.B.-B. and A.F.-P.; Project administration, E.B.-B. and A.F.-P.; Funding acquisition, E.B.-B. and A.F.-P. All authors have read and agreed to the published version of the manuscript.

Funding: This study was financed in part by the Coordenação de Aperfeiçoamento de Pessoal de Nível Superior—Brasil (CAPES)—Finance Code 001; Conselho Nacional de Desenvolvimento Científico e TecnolÓgico (CNPq) Universal-Processo # 408981/2018-0 and Fundação de Amparo à Pesquisa do estado do Rio de Janeiro-CNE (Faperj) Processo # E-26/202974/2017.

Institutional Review Board Statement: Not applicable.

Informed Consent Statement: Not applicable.

Data Availability Statement: Not applicable.

Acknowledgments: The authors thank Walter Oelemann for English revision.

Conflicts of Interest: The authors declare no conflict of interest.

References

1. Poulain, D. Candida albicans, plasticity and pathogenesis. *Crit. Rev. Microbiol.* **2015**, *41*, 208–217. [CrossRef]
2. Dadar, M.; Tiwari, R.; Karthik, K.; Chakraborty, S.; Shahali, Y.; Dhama, K. Candida albicans-Biology, molecular characterization, pathogenicity, and advances in diagnosis and control-An update. *Microb. Pathog.* **2018**, *117*, 128–138. [CrossRef]
3. Singh, A.; Rella, A.; Schwacke, J.; Vacchi-Suzzi, C.; Luberto, C.; Del Poeta, M. Transmembrane transporter expression regulated by the glucosylceramide pathway in Cryptococcus neoformans. *BMC Res. Notes* **2015**, *8*, 681. [CrossRef]
4. Bongomin, F.; Gago, S.; Oladele, R.O.; Denning, D.W. Global and Multi-National Prevalence of Fungal Diseases-Estimate Precision. *J. Fungi* **2017**, *3*, 57. [CrossRef] [PubMed]
5. Kathiravan, M.K.; Salake, A.B.; Chothe, A.S.; Dudhe, P.B.; Watode, R.P.; Mukta, M.S.; Gadhwe, S. The biology and chemistry of antifungal agents: A review. *Bioorg. Med. Chem.* **2012**, *20*, 5678–5698. [CrossRef] [PubMed]
6. Rajendran, R.; Sherry, L.; Nile, C.J.; Sherriff, A.; Johnson, E.M.; Hanson, M.F.; Williams, C.; Munro, C.A.; Jones, B.J.; Ramage, G. Biofilm formation is a risk factor for mortality in patients with Candida albicans bloodstream infection-Scotland, 2012-2013. *Clin. Microbiol. Infect.* **2016**, *22*, 87–93. [CrossRef]
7. Pristov, K.E.; Ghannoum, M.A. Resistance of Candida to azoles and echinocandins worldwide. *Clin. Microbiol. Infect.* **2019**, *25*, 792–798. [CrossRef]
8. Arendrup, M.C.; Patterson, T.F. Multidrug-Resistant Candida: Epidemiology, Molecular Mechanisms, and Treatment. *J. Infect. Dis.* **2017**, *216*, S445–s451. [CrossRef] [PubMed]
9. Roemer, T.; Krysan, D.J. Antifungal drug development: Challenges, unmet clinical needs, and new approaches. *Cold Spring Harb. Perspect. Med.* **2014**, *4*. [CrossRef]
10. Dickson, R.C.; Lester, R.L. Sphingolipid functions in Saccharomyces cerevisiae. *Biochim. Biophys. Acta* **2002**, *1583*, 13–25. [CrossRef]
11. Heung, L.J.; Luberto, C.; Del Poeta, M. Role of sphingolipids in microbial pathogenesis. *Infect. Immun.* **2006**, *74*, 28–39. [CrossRef] [PubMed]
12. Rollin-Pinheiro, R.; Singh, A.; Barreto-Bergter, E.; Del Poeta, M. Sphingolipids as targets for treatment of fungal infections. *Future Med. Chem.* **2016**, *8*, 1469–1484. [CrossRef]
13. Rittershaus, P.C.; Kechichian, T.B.; Allegood, J.C.; Merrill, A.H., Jr.; Hennig, M.; Luberto, C.; Del Poeta, M. Glucosylceramide synthase is an essential regulator of pathogenicity of Cryptococcus neoformans. *J. Clin. Investig.* **2006**, *116*, 1651–1659. [CrossRef]
14. Zhu, C.; Wang, M.; Wang, W.; Ruan, R.; Ma, H.; Mao, C.; Li, H. Glucosylceramides are required for mycelial growth and full virulence in Penicillium digitatum. *Biochem. Biophys. Res. Commun.* **2014**, *455*, 165–171. [CrossRef]
15. Fernandes, C.M.; de Castro, P.A.; Singh, A.; Fonseca, F.L.; Pereira, M.D.; Vila, T.V.; Atella, G.C.; Rozental, S.; Savoldi, M.; Del Poeta, M.; et al. Functional characterization of the Aspergillus nidulans glucosylceramide pathway reveals that LCB Delta8-desaturation and C9-methylation are relevant to filamentous growth, lipid raft localization and Psd1 defensin activity. *Mol. Microbiol.* **2016**, *102*, 488–505. [CrossRef] [PubMed]
16. Oura, T.; Kajiwara, S. Candida albicans sphingolipid C9-methyltransferase is involved in hyphal elongation. *Microbiology* **2010**, *156*, 1234–1243. [CrossRef]
17. Oura, T.; Kajiwara, S. Disruption of the sphingolipid Delta8-desaturase gene causes a delay in morphological changes in Candida albicans. *Microbiology* **2008**, *154*, 3795–3803. [CrossRef]
18. Aeed, P.A.; Young, C.L.; Nagiec, M.M.; Elhammer, A.P. Inhibition of inositol phosphorylceramide synthase by the cyclic peptide aureobasidin A. *Antimicrob. Agents Chemother.* **2009**, *53*, 496–504. [CrossRef] [PubMed]
19. Cerantola, V.; Guillas, I.; Roubaty, C.; Vionnet, C.; Uldry, D.; Knudsen, J.; Conzelmann, A. Aureobasidin A arrests growth of yeast cells through both ceramide intoxication and deprivation of essential inositolphosphorylceramides. *Mol. Microbiol.* **2009**, *71*, 1523–1537. [CrossRef]
20. Lattif, A.A.; Mukherjee, P.K.; Chandra, J.; Roth, M.R.; Welti, R.; Rouabhia, M.; Ghannoum, M.A. Lipidomics of Candida albicans biofilms reveals phase-dependent production of phospholipid molecular classes and role for lipid rafts in biofilm formation. *Microbiology* **2011**, *157*, 3232–3242. [CrossRef]
21. Tan, H.W.; Tay, S.T. The inhibitory effects of aureobasidin A on Candida planktonic and biofilm cells. *Mycoses* **2013**, *56*, 150–156. [CrossRef] [PubMed]
22. Perdoni, F.; Signorelli, P.; Cirasola, D.; Caretti, A.; Galimberti, V.; Biggiogera, M.; Gasco, P.; Musicanti, C.; Morace, G.; Borghi, E. Antifungal activity of Myriocin on clinically relevant Aspergillus fumigatus strains producing biofilm. *BMC Microbiol.* **2015**, *15*, 248. [CrossRef] [PubMed]
23. Lazzarini, C.; Haranahalli, K.; Rieger, R.; Ananthula, H.K.; Desai, P.B.; Ashbaugh, A.; Linke, M.J.; Cushion, M.T.; Ruzsicska, B.; Haley, J.; et al. Acylhydrazones as Antifungal Agents Targeting the Synthesis of Fungal Sphingolipids. *Antimicrob. Agents Chemother.* **2018**, *62*. [CrossRef]
24. Rocha, D.A.S.; Sa, L.F.R.; Pinto, A.C.C.; Junqueira, M.L.; Silva, E.M.D.; Borges, R.M.; Ferreira-Pereira, A. Characterisation of an ABC transporter of a resistant Candida glabrata clinical isolate. *Mem. Inst. Oswaldo Cruz* **2018**, *113*, e170484. [CrossRef] [PubMed]

25. Pinto, A.C.C.; Rocha, D.A.S.; Moraes, D.C.; Junqueira, M.L.; Ferreira-Pereira, A. Candida albicans Clinical Isolates from a Southwest Brazilian Tertiary Hospital Exhibit MFS-mediated Azole Resistance Profile. *An. Acad. Bras. Cienc.* **2019**, *91*, e20180654. [CrossRef]

26. White, T.C.; Holleman, S.; Dy, F.; Mirels, L.F.; Stevens, D.A. Resistance mechanisms in clinical isolates of Candida albicans. *Antimicrob. Agents Chemother.* **2002**, *46*, 1704–1713. [CrossRef]

27. Almirante, B.; Rodríguez, D.; Park, B.J.; Cuenca-Estrella, M.; Planes, A.M.; Almela, M.; Mensa, J.; Sanchez, F.; Ayats, J.; Gimenez, M.; et al. Epidemiology and predictors of mortality in cases of Candida bloodstream infection: Results from population-based surveillance, barcelona, Spain, from 2002 to 2003. *J. Clin. Microbiol.* **2005**, *43*, 1829–1835. [CrossRef]

28. Lackner, M.; Tscherner, M.; Schaller, M.; Kuchler, K.; Mair, C.; Sartori, B.; Istel, F.; Arendrup, M.C.; Lass-Flörl, C. Positions and numbers of FKS mutations in Candida albicans selectively influence in vitro and in vivo susceptibilities to echinocandin treatment. *Antimicrob. Agents Chemother.* **2014**, *58*, 3626–3635. [CrossRef]

29. Zhong, W.; Jeffries, M.W.; Georgopapadakou, N.H. Inhibition of inositol phosphorylceramide synthase by aureobasidin A in Candida and Aspergillus species. *Antimicrob. Agents Chemother.* **2000**, *44*, 651–653. [CrossRef]

30. Kumar, M.; Singh, A.; Kumari, S.; Kumar, P.; Wasi, M.; Mondal, A.K.; Rudramurthy, S.M.; Chakrabarti, A.; Gaur, N.A.; Gow, N.A.R.; et al. Sphingolipidomics of drug resistant Candida auris clinical isolates reveal distinct sphingolipid species signatures. *Biochim. Biophys. Acta Mol. Cell Biol. Lipids* **2021**, *1866*, 158815. [CrossRef]

31. de Melo, N.R.; Abdrahman, A.; Greig, C.; Mukherjee, K.; Thornton, C.; Ratcliffe, N.A.; Vilcinskas, A.; Butt, T.M. Myriocin significantly increases the mortality of a non-mammalian model host during Candida pathogenesis. *PLoS ONE* **2013**, *8*, e78905. [CrossRef]

32. Yang, X.; Pei, Z.; Hu, R.; Zhang, Z.; Lou, Z.; Sun, X. Study on the Inhibitory Activity and Possible Mechanism of Myriocin on Clinically Relevant Drug-Resistant Candida albicans and Its Biofilms. *Biol. Pharm. Bull.* **2021**, *44*, 305–315. [CrossRef]

33. Rollin-Pinheiro, R.; Rochetti, V.P.; Xisto, M.; Liporagi-Lopes, L.C.; Bastos, B.; Rella, A.; Singh, A.; Rozental, S.; Del Poeta, M.; Barreto-Bergter, E. Sphingolipid biosynthetic pathway is crucial for growth, biofilm formation and membrane integrity of Scedosporium boydii. *Future Med. Chem.* **2019**, *11*, 2905–2917. [CrossRef]

34. Heidler, S.A.; Radding, J.A. The AUR1 gene in Saccharomyces cerevisiae encodes dominant resistance to the antifungal agent aureobasidin A (LY295337). *Antimicrob. Agents Chemother.* **1995**, *39*, 2765–2769. [CrossRef] [PubMed]

35. Wuts, P.G.; Simons, L.J.; Metzger, B.P.; Sterling, R.C.; Slightom, J.L.; Elhammer, A.P. Generation of Broad-Spectrum Antifungal Drug Candidates from the Natural Product Compound Aureobasidin A. *ACS Med. Chem. Lett.* **2015**, *6*, 645–649. [CrossRef] [PubMed]

36. Vandeputte, P.; Ferrari, S.; Coste, A.T. Antifungal resistance and new strategies to control fungal infections. *Int. J. Microbiol.* **2012**, *2012*, 713687. [CrossRef] [PubMed]

37. Bennett, J.E.; Dismukes, W.E.; Duma, R.J.; Medoff, G.; Sande, M.A.; Gallis, H.; Leonard, J.; Fields, B.T.; Bradshaw, M.; Haywood, H.; et al. A comparison of amphotericin B alone and combined with flucytosine in the treatment of cryptoccal meningitis. *N. Engl. J. Med.* **1979**, *301*, 126–131. [CrossRef] [PubMed]

38. Delgado, A.; Casas, J.; Llebaria, A.; Abad, J.L.; Fabrias, G. Inhibitors of sphingolipid metabolism enzymes. *Biochim. Biophys. Acta* **2006**, *1758*, 1957–1977. [CrossRef]

39. Streit, E.; Naehrer, K.; Rodrigues, I.; Schatzmayr, G. Mycotoxin occurrence in feed and feed raw materials worldwide: Long-term analysis with special focus on Europe and Asia. *J. Sci. Food Agric.* **2013**, *93*, 2892–2899. [CrossRef]

40. de Sá, L.F.R.; Toledo, F.T.; de Sousa, B.A.; Gonçalves, A.C.; Tessis, A.C.; Wendler, E.P.; Comasseto, J.V.; Dos Santos, A.A.; Ferreira-Pereira, A. Synthetic organotelluride compounds induce the reversal of Pdr5p mediated fluconazole resistance in Saccharomyces cerevisiae. *BMC Microbiol.* **2014**, *14*, 201. [CrossRef]

41. Odds, F.C. Synergy, antagonism, and what the chequerboard puts between them. *J. Antimicrob. Chemother.* **2003**, *52*, 1. [CrossRef]

42. Borenfreund, E.; Puerner, J.A. Toxicity determined in vitro by morphological alterations and neutral red absorption. *Toxicol. Lett.* **1985**, *24*, 119–124. [CrossRef]

43. Breger, J.; Fuchs, B.B.; Aperis, G.; Moy, T.I.; Ausubel, F.M.; Mylonakis, E. Antifungal chemical compounds identified using a C. elegans pathogenicity assay. *PLoS Pathog.* **2007**, *3*, e18. [CrossRef] [PubMed]

44. Tampakakis, E.; Okoli, I.; Mylonakis, E.A.C. Elegans-based, whole animal, in vivo screen for the identification of antifungal compounds. *Nat. Protoc.* **2008**, *3*, 1925–1931. [CrossRef] [PubMed]

45. Singh, S.; Fatima, Z.; Ahmad, K.; Hameed, S. Fungicidal action of geraniol against Candida albicans is potentiated by abrogated CaCdr1p drug efflux and fluconazole synergism. *PLoS ONE* **2018**, *13*, e0203079. [CrossRef] [PubMed]

Article

The Synthesis and Evaluation of Multivalent Glycopeptoids as Inhibitors of the Adhesion of *Candida albicans*

Harlei Martin [1,†], Hannah Masterson [2], Kevin Kavanagh [2,3,*] and Trinidad Velasco-Torrijos [1,3,*]

1   Department of Chemistry, Maynooth University, Maynooth, W23VP22 Co. Kildare, Ireland; harlei.martin.2012@mumail.ie

2   Department of Biology, Maynooth University, Maynooth, W23VP22 Co. Kildare, Ireland; hannah.masterson.2017@mumail.ie

3   The Kathleen Lonsdale Institute for Human Health Research, Maynooth University, Maynooth, W23VP22 Co. Kildare, Ireland

*   Correspondence: Kevin.Kavanagh@mu.ie (K.K.); trinidad.velascotorrijos@mu.ie (T.V.-T.)

†   Present address: Centre de Biophysique Moléculaire, Université d'Orléans, Rue Charles Sadron, F-45071 Orléans 2, France.

Abstract: Multivalency is a strategy commonly used by medicinal carbohydrate chemists to increase the affinity of carbohydrate-based small molecules for their protein targets. Although this approach has been very successful in enhancing binding to isolated carbohydrate-binding proteins, anticipating the multivalent presentations that will improve biological activity in cellular assays remains challenging. In this work we investigate linear molecular scaffolds for the synthesis of a low valency presentation of a divalent galactoside **1**, previously identified by us as an inhibitor of the adhesion of opportunistic fungal pathogen *Candida albicans* to buccal epithelial cells (BECs). Adhesion inhibition assays revealed that multivalent glycoconjugate **3** is more effective at blocking *C. albicans* adherence to BECs upon initial exposure to epithelial cells. Interestingly, **3** did not seem to have any effect when it was pre-incubated with yeast cells, in contrast to the original lead compound **1**, which caused a 25% reduction of adhesion. In competition assays, where yeast cells and BECs were co-incubated, multivalent glycoconjugate **3** inhibited up to 49% *C. albicans* adherence in a dose-dependent manner. The combined effect of compound **1** towards both yeast cells and BECs allowed it to achieve over 60% inhibition of the adhesion of *C. albicans* to BECs in competition assays.

Keywords: multivalency; anti-adhesion glycoconjugates; antifungal agents; *Candida albicans*; glycomimetics

Citation: Martin, H.; Masterson, H.; Kavanagh, K.; Velasco-Torrijos, T. The Synthesis and Evaluation of Multivalent Glycopeptoids as Inhibitors of the Adhesion of *Candida albicans*. *Pathogens* **2021**, *10*, 572. https://doi.org/10.3390/pathogens10050572

Academic Editor: Jonathan Richardson

Received: 1 April 2021
Accepted: 5 May 2021
Published: 8 May 2021

Publisher's Note: MDPI stays neutral with regard to jurisdictional claims in published maps and institutional affiliations.

1. Introduction

Protein-carbohydrate recognition is the first step in the initiation of many human diseases. In many cases, for microorganisms to infect their hosts, microbe proteins initially adhere to carbohydrate epitopes displayed on the host cell surface [1]. Small-molecule inhibitors of the adhesion process have been successfully developed and studied for many years [2]. The use of glycoconjugates as anti-adhesion ligands is desirable since they can mimic the host cell surface glycans and block pathogen attachment, thus preventing infection. However, carbohydrates interact with their protein receptors (lectins) with low affinity (generally millimolar to micromolar dissociation constants) [3]. Consequently, the development of strategies for increasing the lectin–ligand binding affinities to levels required for therapeutic use has received much attention in glycoscience research [4]. In nature, carbohydrate epitopes are expressed multiple times on the cell surface: many lectins have more than one binding site to counteract the low-affinity problem, leading to stronger interactions. This phenomenon has become known as the "multivalent effect" or "cluster glycoside effect", which was first reported by Lee and co-workers in 1995 [5]. Numerous multivalent glycoconjugates with various valencies and spatial arrangement of carbohydrate ligands have been developed to enhance carbohydrate–lectin interactions.

The multivalent glycoconjugates can have well-defined molecular structures and display a specific number of carbohydrate ligands when they are built around scaffolds such as calixarenes [6], dendrimers [7], cyclodextrins [8], cyclopeptides [9] and fullerenes [7]; they can also have higher valencies such as those provided by polymers [10], nanoparticles [11] and quantum dots [12]. There are many reports of multivalent glycoconjugates with increased affinity for isolated lectins, in comparison to their monovalent counterparts [13]. Most of these studies involve lectins whose structures have been well characterized by X-ray crystallography, and the strength of lectin–ligand binding interactions is measured using techniques such as haemagglutination inhibition assays, enzyme-linked lectin assays (ELLA) and isothermal titration calorimetry [14,15]. The multivalent effect in ligand–protein interactions has also been investigated with ligands other than carbohydrates, to potentiate different types of biological effects [16,17]. Despite the many achievements in this field, the design of high-affinity multivalent presentations is still a challenging task, as it strongly depends on multiple factors: these include structural parameters of the glycoconjugates such as linker length, density of the ligands in the glycoconstruct and heterogeneity of the carbohydrate epitopes presented [18]. Moreover, the impact of multivalency on biological activity is particularly difficult to anticipate in cellular assays, given the structural complexity encountered by the ligands at the cell surface [19,20].

C. albicans is an opportunistic pathogenic yeast and the most prevalent cause of hospital-acquired fungal infections worldwide. It is therefore hugely important that new treatments are developed for these infections, which are now becoming resistant to conventional antifungal medicines [21]. The molecular mechanisms of the adherence processes of *C. albicans* to host cells and abiotic surfaces are complex, albeit essential for infection and biofilm formation [22]. Hence, targeting adhesion in *C. albicans* offers an interesting approach towards antifungal therapies as alternatives to conventional drugs [23]. Early reports indicate that *C. albicans* adhesins recognise and bind to many cell surface glycans and carbohydrates, including mono- and disaccharides (e.g., L-fucose, α-D-methyl mannoside and *N*-acetyl-D-glucosamine) [24,25]. Glycosphingolipids have also been shown to act as adhesion receptors for yeasts. *C. albicans* bound specifically to lactosylceramide (Galβ(1-4)Glcβ(1-1)Cer), with the terminal galactose essential for binding [26], and to asialo-GM$_1$ (gangliotetraosylceramide: βGal(1-3)βGalNAc(1-4)βGal(1-4)βGlc(1-1)Cer), with the minimal carbohydrate sequence required for binding being βGalNAc(1-4)βGal [27]. Previous research in our group hence screened a small library of synthetic glycoconjugates with various sugars and valencies, and tested their ability to inhibit the adhesion of *C. albicans* to human BEC. This study showed that the divalent galactoside **1** (Figure 1) was a successful inhibitor of the adhesion of *C. albicans* to BECs [28]. Ongoing research in the group is aiming to identify the adhesin on the cell surface of the *C. albicans* with which this galactosylated compound **1** is interacting. In this work, we developed a low valency-multivalent derivative of the original lead compound **1** to assess the effect of multivalent displays in the inhibition of the adhesion of *C. albicans* to BECs. Recently, we reported the multivalent presentation of lead compound **1** on RAFT cyclopeptide- and polylysine-based scaffolds [29]. Tetra-, hexa- and hexadecavalent displays of compound **1** were thus evaluated as inhibitors of *C. albicans* adhesion, revealing a better performance for the lower valency glycoconjugates. In our investigations towards an optimal multivalent architecture to enhance the activity of lead compound **1**, in the current study we opted for linear peptoid scaffold **2** [30] to prepare glycoconjugate **3** (Figure 1). Since the structure of the target lectin in *C. albicans* is not known, a tetravalent display of compound **1** in a flexible scaffold such as **2** could provide a very useful comparison with the results observed when more rigid and cyclic scaffolds were used. Peptoid scaffolds have been successfully investigated by Taillefumier and co-workers in the design of defined valency glycoconjugates [31,32]. Thus, lead compound **1** was modified with a linker to facilitate connection with this scaffold using Copper(I)-catalyzed Azide–Alkyne Cycloaddition (CuAAC) methodology. The tetravalent display of compound **1** as presented in glycoconjugate **3** may significantly affect how it interacts with carbohydrate-binding proteins in *C. albicans* and BECs surface, which, in turn, should allow for identification of the most suitable structural features to inhibit pathogen–cell interaction.

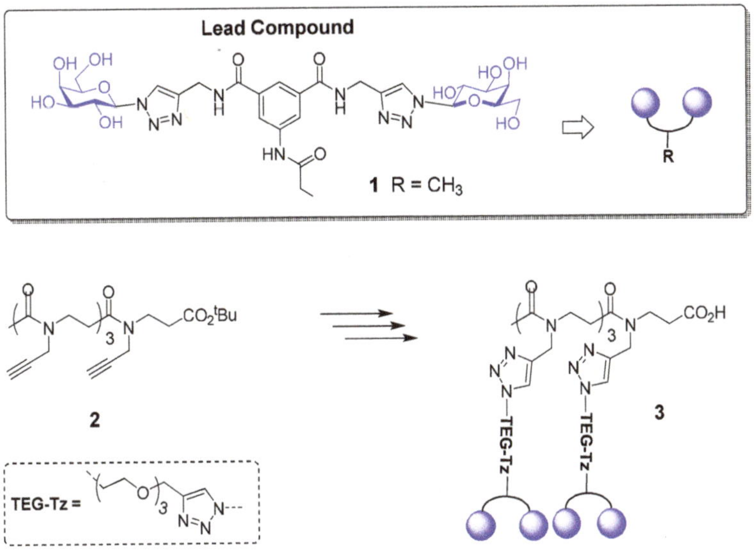

Figure 1. Shows the structure of lead compound **1**, peptoid scaffold **2** and multivalent glycoconjugate **3**.

2. Results and Discussion

2.1. Chemical Synthesis

In order to install lead compound **1** onto the alkyne-featuring scaffold **2** via the CuAAC reaction, it had to be functionalized with an azido group in a manner that would not compromise its interaction with target proteins in the *C. albicans* cell wall. Our previous work showed that functionalization of 5-aminobenzene position in compound **1** with a fluorescent label still allowed for the recognition of the digalactoside motifs and localization of the compound at the cell wall in *C. albicans* [28]. Thus, 5-amino-isophthalic acid was protected using *tert*-butoxycabonyl (Boc) anhydride, followed by reaction with propargylamine and TBTU to give compound **4** (Scheme 1). CuAAC chemistry was used to conjugate the per-acetylated-1-β-azido galactose [33] moieties to dialkyne **4** to give compound **5**, which was then treated with TFA to remove the *N*-Boc-protecting group yielding compound **6** [28]. Reaction with bromoacetyl bromide to give compound **7** and subsequent bromide displacement by treatment with sodium azide gave key azido-functionalized compound **8** [29]. Attempts of direct reaction of this compound with alkynated scaffolds using CuAAC reaction conditions failed to yield any of the desired product, likely due to steric effects. Oligo(ethyleneglycol)s are commonly used as linkers between the carbohydrate moieties and the scaffold in multivalent glycoconjugates, because of their flexibility, water-solubility, the availability of various lengths and the presence of functionalizable hydroxy groups [34]. Therefore, it was decided to use a triethylene glycol (TEG) derivative as a linker to join the divalent galactoside **8** to the alkyne scaffolds **2**.

Using well-documented procedures, TEG was reacted with propargyl bromide to give the mono-propargylated linker **9** [35]. To ensure mono-propargylation, the reaction was carried out in excess of TEG. Compound **9** was then tosylated, yielding compound **10** [36]. This was then reacted with the azido di-galactoside **8** using microwave (MW) assisted CuAAC methodology to give compound **11**. The tosyl group was then replaced by an azide upon treatment with sodium azide to give compound **12**. This key synthetic intermediate presents the acetylated divalent galactoside with a TEG linker functionalized with a terminal azide, available for CuAAC conjugation to various propargylated scaffolds.

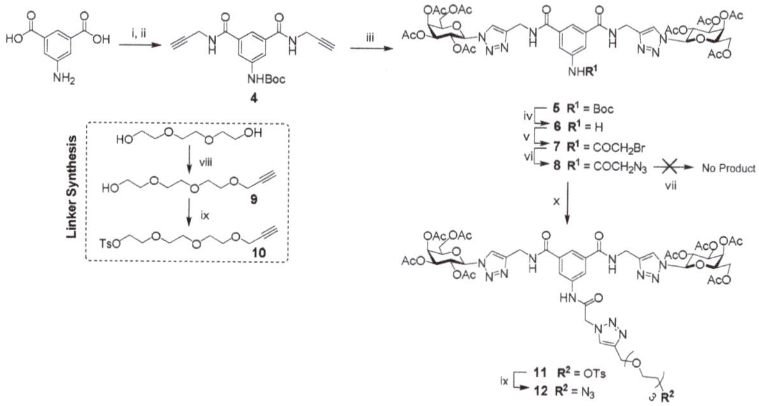

Scheme 1. Synthesis of azido-TEG divalent galactoside **12** (inset shows the synthesis of TEG linker): *Reagents and conditions:* (i) Di-*tert*-butyl dicarbonate, NaOH, 1,4-dioxane, 0 °C to rt, 3 h, 86%; (ii) DMTMM, propargylamine, THF, 48 h, 95%; (iii) 2,3,4,6-tetra-*O*-acetyl-1-β-azido-galactoside, $CuSO_4 \cdot 5H_2O$/Na Asc, CH_3COCH_3/H_2O, rt, 16 h, 71%; (iv) TFA, DCM, 2 h, rt, 99%; (v) bromoacetyl bromide, NEt_3, anhydrous DCM, 16 h, 83%; (vi) NaN_3, anhydrous DMF, N_2, 80 °C, 16 h, quant%; (vii) CuAAC methodology, propargylated scaffolds, no product isolated; (viii) propargyl bromide, NaH, anhydrous THF, N_2, 16 h, 75%; (ix) TsCl, KOH, DCM, 0 °C, 2 h, 87%; (x) **10**, $CuSO_4 \cdot 5H_2O$/Na Asc, CH_3CN/H_2O, MW (microwave), 100 °C, 30 min, 74%; (xi) NaN_3, CH_3CN, DMF, 80 °C, 24 h, 95%.

β-peptoid scaffolds designed by Faure, Taillefumier and coworkers have been used to create linear and cyclic displays of various recognition motifs, including carbohydrates [30–32]. Linear scaffold **2** [30] was reacted with compound **12** using, once again, microwave-assisted CuAAC conditions (Scheme 2); acetylated glycocluster **13** was thus obtained and subjected to mild base hydrolysis that resulted in the removal of the tert-butyl ester and acetyl protecting groups, yielding glycoconjugate **3**, which presents a tetravalent display of the lead divalent galactoside **1**. In addition, CuAAC reaction of linear scaffold **2** with per-acetylated-1-β-azido galactose [33] gave tetragalacoside **14**, which was deprotected under mild basic conditions to give **15** (Scheme 2). This compound, lacking structural features of lead compound **1** other than terminal β-triazolyl-galactosides, would serve as a control in subsequent biological evaluation of anti-adhesion activity against *C. albicans*.

2.2. Biological Evaluation

In our previous work [28,29] we established that compound **1** and derivatives were not toxic to *C. albicans* yeast cells at the concentrations used in the current study. Glycoconjugates **3**, featuring multivalent displays of the original lead compound **1**, along with tetra-β-triazolyl-galactoside **15**, were then evaluated for their anti-adhesive properties against *C. albicans* using several assays.

2.2.1. Exclusion Assay

The initial adherence assay was performed by treating the *C. albicans* with compounds **3** and **15**. After an incubation period of 90 min, the treated yeast cells were exposed to exfoliated BECs. The average number of yeast cells attached to each BEC was calculated (Figure 2a), and the percentage increase or decrease of the number of *C. albicans* cells adhering to the BECs was compared to a control (untreated yeast) and the lead compound **1** (Figure 2b). In this assay, compound **15** only slightly reduced the adhesion of the yeast to the BECs, whereas compound **3** did not have any appreciable effect. Compound **1** showed moderate inhibition of adhesion at this concentration (up to 25% reduction at 10 mg/mL). On the other hand, when the BECs were treated with the above compounds, allowing the incubation period and followed by exposure to *C. albicans* cells, the average number of yeasts attached per BEC was reduced (Figure 2c). The tetra-galactoside **15** caused similar reduction of yeast adhesion as in the previous assay, while multivalent glycoconjugate **3** inhibited *C. albicans* adhesion by 35%, similarly to the inhibition produced by lead compound **1** (Figure 2d). These results suggest that the multivalent presentation in compound **3** disfavours significantly the interactions with structural components of the cell wall in *C. albicans*, which had been previously proposed for compound **1** [28]. However, this effect was not observed upon preliminary exposure to BECs, since both compounds **1** and **3** caused a similar decrease in yeast adhesion.

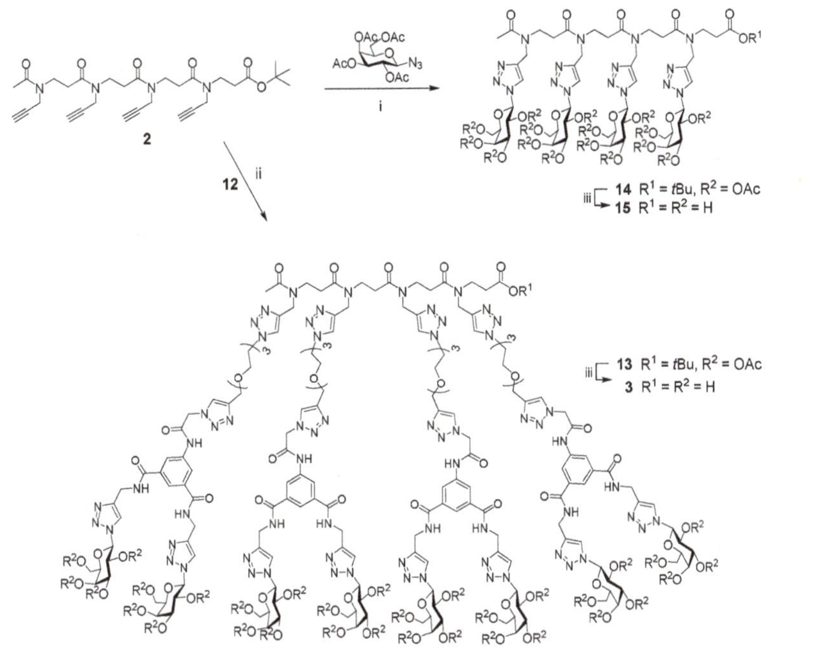

Scheme 2. Synthesis of glycoconjugate **3**. *Reagents and conditions:* (i) CuSO$_4$·5H$_2$O/Na Asc, CH$_3$CN/H$_2$O, MW, 100 °C, 30 min, 58–68%; (ii) MeOH, NEt$_3$, H$_2$O, 45 °C, 6 h, 82–92%.

2.2.2. Competition Assay

The glycoconjugates were then evaluated in a competition assay, in which their anti-adhesion abilities were tested in the presence of both *C. albicans* and BECs. Co-incubation with the compounds resulted in a reduction in yeast adhesion in all cases (Figure 3a), with compound **3** causing up to a 49% decrease. Interestingly, lead compound **1**, which showed activity in both exclusion assays discussed above (i.e., when pre-incubated with *C. albicans* and BECs, independently), was able to inhibit up to 64% the adhesion of *C. albicans* to BECs in the competition assay.

The compounds were then tested at lower concentrations. Glycoconjugate **3** exhibited a typical dose–response pattern, with ca. 50% inhibition of adhesion at the highest concentration (Figure 3b). Compound **15** (Figure 3c) did not show a linear relationship between concentration and activity, where the best inhibition of adhesion was observed at 1 mg/mL.

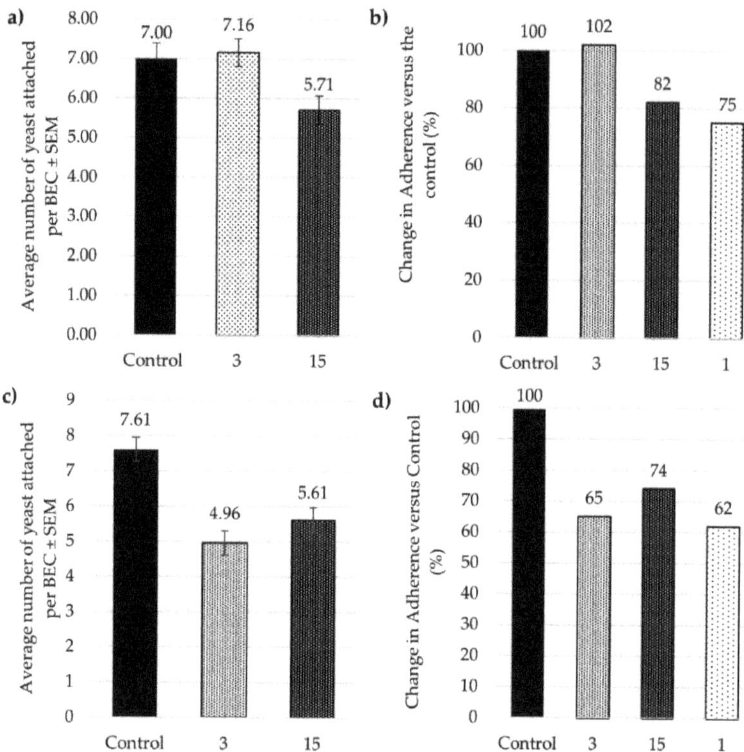

Figure 2. Exclusion assays of glycoconjugates **3** and **15**: (**a**) Shows the average number of yeast attached per BEC (buccal epithelial cells), after the yeast cells were pre-treated with glycoconjugates; (**b**) Shows the percentage change in the adherence of yeast to the BEC versus the control and the lead compound **1** after the yeast were pre-treated; (**c**) Shows the average number of yeast attached per BEC, after the BEC were pre-treated with compounds; (**d**) Shows the percentage change in the adherence of yeast to the BEC versus the control and the lead compound **1** after the BEC were pre-treated. All assays were performed at a glycoconjugate concentration of 10 mg/mL.

Overall, this study shows that the tetravalent presentation of lead compound **1** generated in glycoconjugate **3** significantly impaired its ability to interact with *C. albicans* cells, although it maintained anti-adhesion activity when pre-incubated with BECs. While no significant reduction of adhesion was observed when *C. albicans* cells were pre-treated with glycoconjugate **3**, this compound was able to inhibit adhesion of the yeast when BECs were pre-treated or in competition assays. This clearly indicates that the multivalent compounds in this study do not interact effectively with *C. albicans* cell wall components; the observed reduction in adhesion appears to be due to preferential binding of the compounds to BECs over the yeast cells. Tetra-galactoside **15**, which does not display the structural features of lead compound **1** nor a TEG linker, generally showed the lowest inhibition of adhesion, although at lower concentration it showed better activity; however, this compound did not follow a dose–response pattern. The original lead compound **1** was the most effective inhibitor of yeast adhesion. Interestingly, the exclusion assays show that compound **1** produces a decrease in adhesion not only when pre-exposed to *C. albicans* cells, as our previous work has shown, but also there is a significant contribution to the activity of compound **1** arising from interaction with BECs. These results highlight that the optimization of activity of derivatives of compound **1** should consider also interactions with BECs as a potential approach to broad-spectrum anti-adhesive glycoconjugates.

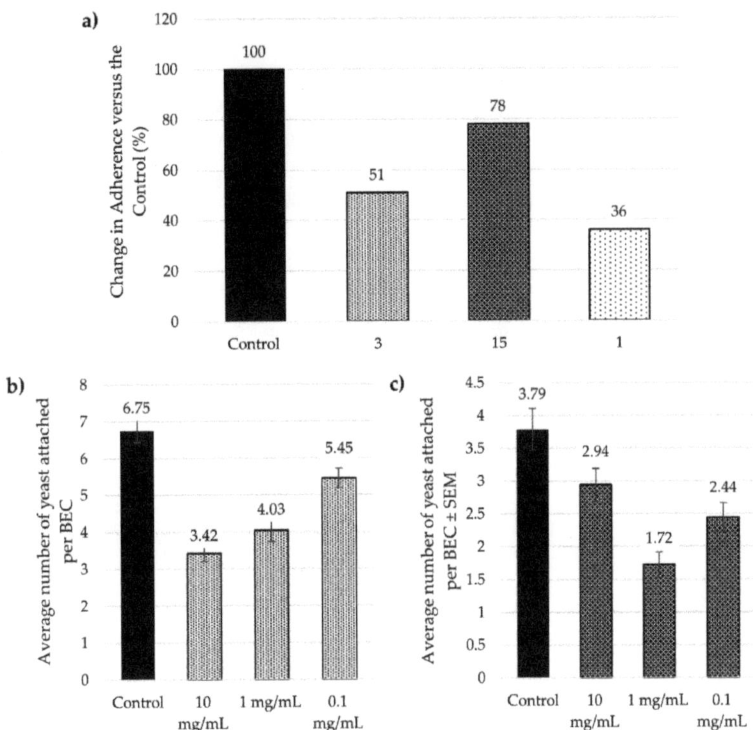

Figure 3. Competition assays of glycoconjugates **1**, **3** and **15**: (**a**) Shows the change in adherence compared to the control and the lead compound **1** at 10 mg/mL; (**b**) Shows the average number of yeast attached to BEC in the presence of compound **3** (10, 1, 0.1 mg/mL); (**c**) Shows the average number of yeast attached to BEC in the presence of compound **15** (10, 1, 0.1 mg/mL).

3. Materials and Methods

3.1. Chemistry

3.1.1. General Methods

All reagents for synthesis were bought commercially and used without further purification. Tetrahydrofuran (THF) was freshly distilled over sodium wire and benzophenone. Dichloromethane (DCM) was freshly distilled over CaH_2 prior to use. Reactions were monitored with thin layer chromatography (TLC) on Merck Silica Gel F_{254} plates. Detection was effected by UV ($\lambda = 254$ nm) or charring in a mixture of 5% sulfuric acid-ethanol. NMR spectra were recorded using Bruker Ascend 500 spectrometer at 293K. All chemical shifts were referenced relative to the relevant deuterated solvent residual peaks. Assignments of the NMR spectra were deduced using ^{1}H NMR and ^{13}C NMR, along with 2D experiments (COSY, HSQC and HMBC). Chemical shifts are reported in ppm. Flash chromatography was performed with Merck Silica Gel 60. Microwave reactions were carried out using a CEM Discover Microwave Synthesizer. Optical rotations were obtained from an AA-100 polarimeter, and $[\alpha]_D$ values are given in 10^{-1} cm^2·g^{-1}. High-performance liquid chromatography analysis (HPLC, Waters Alliance 2695) was performed in final compounds and indicated purity of 95% based on integrations without the use of an internal standard. High-resolution mass spectrometry (HRMS) was performed on an Agilent-LC 1200 Series coupled to a 6210 Agilent Time-Of-Flight (TOF) mass spectrometer equipped with an electrospray source in both positive and negative (ESI+/−) modes. Infrared spectra were obtained as a film on NaCl plates or as KBr disks in the region 4000–400 cm^{-1} on a Perkin

Elmer Spectrum 100 FT-IR spectrophotometer. Spectroscopic data for all compounds are provided in the Supplementary Materials.

3.1.2. Synthetic Procedures

Synthesis of *N*,*N'*-di-(2,3,4,6-tetra-*O*-acetyl-*β*-D-galactopyranosyl-1,2,3-triazol-4-ylmethylamide)-*N''*-(2-bromoacetamido)-5-aminobenzene-1,3-dicarboxamide (**7**)

6 [21] (1.128 g, 1.13 mmol) was dissolved in dry DCM (20 mL). NEt$_3$ (0.19 mL, 1.35 mmol) was added to this solution. Bromoacetyl bromide (0.12 mL, 1.35 mmol) was dissolved in dry DCM (5 mL) in a separate round-bottom flask. The first solution was added to the second dropwise via a cannula and the resulting reaction mixture was allowed to stir for 16 h. The reaction mixture was washed with water (20 mL), HCl (1 N, 20 mL), sat. NaHCO$_3$ solution (20 mL), followed by brine (20 mL). The organic phase was dried (MgSO$_4$) and the solvent was removed in vacuo to obtain the pure product **7** without further purification as a brown, sticky solid (1.056 g, 83%). R$_f$ = 0.65 (DCM, 5% MeOH). $[\alpha]_D^{24}$ −4.0 (c 1.0, DCM). ^{1}H NMR (500 MHz, CDCl$_3$) δ 9.10 (s, 1H, N*H*COCH$_2$Br), 8.09–7.90 (m, 6H, triaz-H, CON*H*CH$_2$-triaz and Ar-H), 7.75 (s, 1H, Ar-H), 5.93 (d, *J* = 9.2 Hz, 2H, H-1), 5.60 (t, *J* = 9.7 Hz, 2H, H-2), 5.54 (d, *J* = 2.9 Hz, 2H, H-4), 5.32–5.26 (m, 2H, H-3), 4.67 (ddd, *J* = 39.3, 15.1, 5.4 Hz, 4H, CH$_2$-triaz), 4.31 (t, *J* = 6.4 Hz, 2H, H-5), 4.22–4.11 (m, 4H, H-6 and H-6′), 3.98 (s, 2H, C*H$_2$*-Br), 2.21 (s, 6H, OAc), 2.00 (app d, *J* = 2.7 Hz, 12H, OAc × 2), 1.82 (s, 6H, OAc). ^{13}C NMR (125 MHz, CDCl$_3$) δ 170.4 (CO of OAc), 170.1 (CO of OAc), 169.8 (CO of OAc), 169.4 (CO of OAc), 166.5 (CONHCH$_2$-triaz), 165.0 (COCH$_2$Br), 145.6 (C-triaz), 138.3 (Ar-C), 135.0 (Ar-C), 121.6 (CH-triaz), 121.4 (Ar-CH), 121.2 (Ar-CH), 86.2 (C-1), 74.0 (C-5), 70.8 (C-3), 68.1 (C-2), 66.8 (C-4), 61.2 (C-6), 35.5 (CH$_2$-triaz), 29.6 (NHCOCH$_2$Br), 20.7 (CH$_3$ of OAc), 20.6 (CH$_3$ of OAc), 20.5 (CH$_3$ of OAc), 20.3 (CH$_3$ of OAc). IR (film on NaCl): 3345, 3087, 2975, 1752, 1651, 1536, 1446, 1371, 1227, 1063, 924 732 cm^{-1}. HRMS (ESI+): *m/z* calculated for C$_{44}$H$_{52}$BrN$_{12}$O$_{21}$ + H$^+$ [M+H$^+$]: 1122.2539, found 1122.2545.

Synthesis of *N*,*N'*-di-(2,3,4,6-tetra-*O*-acetyl-*β*-D-galactopyranosyl-1,2,3-triazol-4-ylmethyl amide)-*N''*-(2-azidoacetamido)-5-aminobenzene-1,3-dicarboxamide (**8**)

7 (231 mg, 0.206 mmol) and NaN$_3$ (30 mg, 0.412 mmol) were dissolved in anhydrous DMF (10 mL) and heated to 80 °C. The reaction mixture was allowed to stir for 16 h. The solvent was removed in vacuo, and the resulting residue was re-dissolved in DCM (20 mL) and was washed with brine (20 mL × 3). The organic phase was dried over MgSO$_4$ and the solvent was removed in vacuo to obtain the pure product **8** without further purification as a yellow solid (1.056 g, 83%). R$_f$ = 0.41 (DCM:MeOH 9:1). $[\alpha]_D^{22}$ -5.6 (c 0.9, DCM). ^{1}H NMR (500 MHz, CDCl$_3$) δ 9.10 (s, 1H, N*H*COCH$_2$N$_3$), 8.18 (s, 2H, N*H*CH$_2$CCH), 8.02 (s, 2H, Ar-H), 7.97 (s, 2H, C*H*-triaz), 7.82 (s, 1H, Ar-H), 5.95 (d, *J* = 9.2 Hz, 2H, H-1), 5.61 (t, *J* = 9.7 Hz, 2H, H-2), 5.56 (d, *J* = 3.1 Hz, 2H, H-4), 5.32 (dd, *J* = 10.1, 3.5 Hz, 2H, H-3), 4.67 (ddd, *J* = 20.4, 15.4, 5.5 Hz, 4H, CH$_2$-triaz), 4.34 (t, *J* = 6.6 Hz, 2H, H-5), 4.23–4.13 (m, 4H, H-6 and H-6′), 4.06 (s, 2H, C*H$_2$*-N$_3$), 2.21 (s, 6H, OAc), 2.01 (s, 12H, OAc × 2), 1.82 (s, 6H, OAc). ^{13}C NMR (125 MHz, CDCl$_3$) δ 170.4 (CO of OAc), 170.1 (CO of OAc), 169.9 (CO of OAc), 169.3 (CO of OAc), 166.5 (CONHCH$_2$CCH), 166.3 (COCH$_2$N$_3$), 145.5 (C-triaz), 138.0 (Ar-C), 134.9 (Ar-C), 121.6 (Ar-CH and CH-triaz), 121.4 (Ar-CH), 86.1 (C-1), 73.9 (C-5), 70.8 (C-3), 68.1 (C-2), 66.9 (C-4), 61.2 (C-6), 52.5 (CH$_2$N$_3$), 35.4 (CH$_2$-triaz), 20.7 (CH$_3$ of OAc), 20.6 (CH$_3$ of OAc), 20.5 (CH$_3$ of OAc), 20.2 (CH$_3$ of OAc). IR (film on NaCl): 3342, 2942, 2110, 1747, 1655, 1528, 1427, 1368, 1211, 1046, 923, 733 cm^{-1}. HRMS (ESI+): *m/z* calculated for C$_{44}$H$_{52}$N$_{12}$O$_{21}$ + Na$^+$ [M+Na$^+$]: 1107.3268, found 1107.3303.

Synthesis of 2-[2-(2-Propargyloxyethoxy)ethoxy]ethanol (**9**)

Procedure adapted from Weil et al. [30]. Triethylene glycol (1 mL, 7.48 mmol, 3 equiv) was diluted with dry THF (10 mL) under N$_2$. The solution was cooled to 0 °C and NaH (60% oil dispersion) (0.1 g, 2.49 mmol) was added portion-wise. The reaction was allowed to warm up to rt and was stirred for 20 min. Propargyl bromide (0.27 mL, 2.49 mmol) was added dropwise. The reaction mixture was allowed to stir overnight. Column chromatography (100% EtOAc) eluted the pure product **9** as a clear oil (0.292 g, 75%). (R_f = 0.42: EtOAc) ^{1}H NMR (500 MHz, CDCl$_3$) δ 4.09–4.08 (m, 2H, CH_2CCH), 3.63–3.53 (m, 10H, CH$_2$ × 5), 3.50–3.46 (m, 2H, CH_2-OH), 3.11 (s, 1H, OH), 2.39 (t, *J* = 2.8 Hz, 1H, CH$_2$CCH). The NMR data is in agreement with the data reported in the literature [30].

Synthesis of 2-(2-(2-Propargyloxyethoxy)ethoxy)ethyl-4-methylbenzenesulfonate (**10**)

This procedure was adapted from Ramström et al. [31]. **9** (0.288 g, 1.53 mmol) was dissolved in DCM (5 mL). TsCl (0.321 g, 1.68 mmol, 1.1 equiv) was added and the mixture was cooled to 0 °C on ice. KOH (0.343 g, 6.12 mmol, 4 equiv) was added slowly after grinding. The mixture was vigorously stirred for 2 h. The mixture was poured onto ice-water and extracted with DCM (3 × 20 mL). The combined organic layers were dried over MgSO$_4$, filtered and concentrated in vacuo to give the pure product **10** as a clear oil (0.457 g, 87%). ^{1}H NMR (500 MHz, CDCl$_3$) δ 7.79 (d, *J* = 8.3 Hz, 2H, Ar-H), 7.33 (d, *J* = 8.2 Hz, 2H, Ar-H), 4.18 (d, *J* = 2.4 Hz, 2H, CH_2CCH), 4.16–4.13 (m, 2H, CH_2OTs), 3.69–3.65 (m, 4H, CH$_2$ × 2), 3.65–3.61 (m, 2H, CH$_2$), 3.58 (s, 4H, CH$_2$ × 2), 2.43 (s, 3H, CH$_3$-Ar), 2.42 (t, *J* = 2.4 Hz, 1H, CH$_2$CCH). The NMR data are in agreement with the data reported in the literature [31].

Synthesis of *N*, *N′*-di-(2,3,4,6-tetra-*O*-acetyl-β-D-galactopyranosyl-1,2,3-triazol-4-ylmethyl amide)-*N″*-(2-4-((2-(2-(2-(4-methylbenzenesulfonate)ethoxy)ethoxy)ethoxy) methyl)-1*H*-1,2,3-triazol-1-yl)acetamido)-5-aminobenzene-1,3-dicarboxamide (**11**)

Copper sulphate pentahydrate (20 mg) and sodium ascorbate (40 mg) were added to a solution **8** (0.516 g, 0.476 mmol) and **10** (0.163 g, 0.476 mmol) in CH$_3$CN/H$_2$O (4 mL/2 mL). The reaction was allowed to stir in the MW at 100 °C until deemed complete by TLC analysis (15 min). The solvent was removed in vacuo. The residue was dissolved in DCM (30 mL), washed with brine (20 mL × 3), and dried (MgSO$_4$). The mixture was filtered and the solvent was removed in vacuo to yield the crude product, which was purified by silica gel column chromatography (DCM:MeOH 98:2–95:5) to give the pure product **11** as a yellow solid (0.307 g, 74%). R_f = 0.45 (DCM:MeOH 9:1). $[\alpha]_D^{22}$-5 (c 1, DCM). ^{1}H NMR (500 MHz, CDCl$_3$) δ 9.82 (s, 1H, NHCH$_2$N$_3$), 8.16 (s, 2H, NHCH$_2$CCH), 7.97 (s, 2H, CH-triaz), 7.84 (appd, *J* = 2.8 Hz, 3H, Ar-H × 2 and CH-triaz), 7.76 (s, 1H, Ar-H), 7.69 (d, *J* = 8.3 Hz, 2H, Ar-H of OTs), 7.26 (d, *J* = 8.1 Hz, 2H, Ar-H of OTs), 5.89 (d, *J* = 9.2 Hz, 2H, H-1), 5.60 (t, *J* = 9.7 Hz, 2H, H-2) 5.50 (d, *J* = 2.9 Hz, 2H, H-4), 5.26 (dd, *J* = 10.3, 3.3 Hz, 2H, H-3), 5.21 (s, 2H, CH$_2$), 4.67–4.55 (m, 6H, CH_2-triaz and CH_2CN$_3$), 4.28 (t, *J* = 6.6 Hz, 2H, H-5), 4.11 (qd, *J* = 11.5, 6.6 Hz, 2H, H-6 and H-6′), 4.06–4.04 (m, 2H, CH$_2$), 3.65–3.59 (m, 2H, CH$_2$), 3.58–3.54 (m, 4H, CH$_2$ × 2), 3.48 (s, 2H, CH$_2$), 2.37 (s, 3H, CH$_3$ of OTs), 2.14 (s, 6H, CH$_3$ of OAc), 1.96 (s, 6H, CH$_3$ of OAc), 1.95 (s, 6H, CH$_3$ of OAc), 1.75 (s, 6H, CH$_3$ of OAc). ^{13}C NMR (125 MHz, CDCl$_3$) δ 170.4 (CO of OAc), 170.1 (CO of OAc), 169.9 (CO of OAc), 169.3 (CO of OAc), 166.6 (CONHCH$_2$CCH), 164.5 (COCH$_2$N$_3$), 145.4 (*C*-triaz), 144.9 (Ar-*C* of OTs), 144.8 (*C*HCN$_3$), 138.1 (Ar-*C*), 134.8 (Ar-*C*), 132.7 (Ar-*C* of OTs), 129.9 (Ar-*C*H of OTs), 127.9 (Ar-*C*H of OTs), 125.3 (*C*HCN$_3$), 121.9 (*C*H-triaz), 121.6 (Ar-*C*H), 121.3 (Ar-*C*H), 86.0 (C-1), 73.8 (C-5), 70.9 (C-3), 70.5 (CH$_2$), 70.4 (CH$_2$), 70.3 (CH$_2$), 69.7 (CH$_2$), 69.4 (CH$_2$), 68.6 (CH$_2$), 68.0 (C-2), 66.9 (C-4), 64.3 (NHCOCH$_2$N$_3$), 61.1 (C-6), 52.8 (CH$_2$), 35.4 (CH$_2$-triaz), 21.6 (CH$_3$ of OAc), 20.6 (CH$_3$ of OAc), 20.5 (CH$_3$ of OAc), 20.2 (CH$_3$ of OAc). IR (film on NaCl): 3344, 3091, 2939, 1754, 1657, 1599, 1535, 1448, 1370, 1223, 1176, 1095, 1054, 924 cm^{-1}. HRMS (ESI+): *m/z* calculated for C$_{60}$H$_{74}$N$_{12}$O$_{27}$S + Na$^+$ [M+Na$^+$]: 1449.4405, found 1449.4332.

Synthesis of *N,N′*-di-(2,3,4,6-tetra-*O*-acetyl-β-D-galactopyranosyl-1,2,3-triazol-4-ylmethylamide)-*N″*-(2-4-((2-(2-(2-azidoethoxy)ethoxy)ethoxy)methyl)-1*H*-1,2,3-triazol-1-yl)acetamido)-5-aminobenzene-1,3-dicarboxamide (**12**)

Compound **11** (183 mg, 0.128 mmol) and NaN$_3$ (17 mg, 0.256 mmol) were dissolved in anhydrous DMF (10 mL) and heated to 80 °C. The reaction mixture was allowed to stir for 16 h. The solvent was removed in vacuo, and the resulting residue was dissolved in DCM (20 mL) and was washed with brine (20 mL × 3). The organic phase was dried (MgSO$_4$) and the solvent was removed in vacuo to obtain the pure product **12** without further purification as a yellow solid (167 g, 100%). R_f = 0.42 (DCM:MeOH 9:1). $[\alpha]_D^{22}$-3 (c 1, DCM). ^{1}H NMR (500 MHz, CDCl$_3$) δ 9.79 (s, 1H, NHCH$_2$N$_3$), 8.13 (s, 2H, NHCH$_2$-triaz), 7.94 (s, 2H, CH-triaz), 7.82 (s, 1H, CH-triaz), 7.77 (s, 2H, Ar-H), 7.70 (s, 1H, Ar-H), 5.85 (d, J = 9.2 Hz, 2H, H-1), 5.57 (t, J = 9.8 Hz, 2H, H-2), 5.47 (d, J = 2.7 Hz, 2H, H-4), 5.25–5.15 (m, 4H, H-3 and CH$_2$), 4.66–4.48 (m, 6H, CH_2-triaz × 3), 4.24 (t, J = 6.3 Hz, 2H, H-5), 4.13–4.04 (m, 4H, H-6 and H-6′), 3.67–3.47 (m, 8H, CH$_2$ × 4), 3.27–3.24 (m, 2H, CH$_2$), 2.12 (s, 6H, CH$_3$ of OAc), 1.94 (s, 6H, CH$_3$ of OAc), 1.93 (s, 6H, CH$_3$ of OAc), 1.73 (s, 6H, CH$_3$ of OAc). ^{13}C NMR (125 MHz, CDCl$_3$) δ 170.5 (CO of OAc), 170.2 (CO of OAc), 170.0 (CO of OAc), 169.4 (CO of OAc), 166.7 (CONHCH$_2$-triaz), 164.5 (COCH$_2$N$_3$), 145.6 (C-triaz), 145.0 (CHCN$_3$), 134.9 (Ar-C), 125.4 (CHCN$_3$), 121.9 (CH-triaz), 121.7 (Ar-CH), 121.3 (Ar-CH), 86.2 (C-1), 74.0 (C-5), 71.0 (C-3), 70.6 (CH$_2$ × 2), 70.0 (CH$_2$), 69.9 (CH$_2$), 68.1 (C-2), 67.0 (C-4), 64.5 (NHCOCH$_2$N$_3$), 61.2 (C-6), 50.7 (CH$_2$), 35.5 (CH$_2$-triaz), 20.7 (CH$_3$ of OAc × 2), 20.6 (CH$_3$ of OAc), 20.4 (CH$_3$ of OAc). IR (film on NaCl): 3335, 3088, 2924, 2109, 1754, 1658, 1600, 1534, 1447, 1370, 1222, 1054, 924 cm^{-1}. HRMS (ESI+): *m/z* calculated for C$_{53}$H$_{67}$N$_{15}$O$_{24}$ + Na$^+$ [M+Na$^+$]: 1320.4381, found 1320.4375.

Synthesis of tert-butyl-(4,8,12,16-tetra-aza)(5,9,13,17-tetra-oxo)(4,8,12,16-tetra-*N*-propargyl) octadeconoate (**2**)

The method outlined by Faure, Taillefumier and coworkers was adapted to synthesise the tetrameric propargylated scaffold [25]. The secondary amine was then acetylated using the following procedure:

Tetrameric propargylated scaffold (0.490 g, 0.96 mmol) was dissolved in DCM (20 mL). Acetic anhydride (0.91 mL, 9.6 mmol) was added to the solution. The reaction mixture was allowed to stir for 6 h at rt and was concentrated under reduced pressure. The crude mixture was dissolved in ethyl acetate (20 mL) and was washed with sat NaHCO$_3$ (20 mL × 2) and brine (20 mL × 2). The organic layer was dried (MgSO$_4$), filtered and then concentrated in vacuo. The crude product was purified by silica gel column chromatography (DCM:MeOH 9:1) to give the pure product as a pale-yellow oil (515 mg, 97%). ^{1}H NMR (500 MHz, CDCl$_3$) δ 4.24–4.06 (m, 8H), 3.83–3.70 (m, 3H), 3.72–3.60 (m, 5H), 2.91–2.68 (m, 6H), 2.55 (m, 2H), 2.37–2.23 (m, 2H), 2.24–2.10 (m, 4H), 1.43 (s, 9H, C(CH$_3$)$_3$). ^{13}C NMR (125 MHz, CDCl$_3$) δ 171.5, 171.2, 170.9, 170.1, 81.7, 81.6, 81.4, 80.9, 78.9, 78.9, 78.6, 72.9, 72.8, 72.6, 72.5, 72.5, 72.1, 71.9, 71.7, 44.4, 44.2, 43.9, 43.7, 43.6, 43.5, 43.0, 42.8, 39.8, 39.1, 39.0, 38.9, 38.6, 38.4, 34.8, 34.5, 34.4, 34.3, 34.1, 32.2, 31.8, 29.7, 28.1, 21.8, 21.5. HRMS (ESI+): m/z calculated for C$_{30}$H$_{40}$N$_4$O$_6$ + H$^+$ [M+H$^+$]:553.3026, found 553.3015. ^{1}H NMR and ^{13}C NMR spectroscopic data corresponded to that found in the literature [25].

Synthesis of Acetylated Tetravalent β-Peptoid Glycocluster (**13**)

Copper sulphate pentahydrate (40 mg) and sodium ascorbate (80 mg) were added to a solution of **10** (30 mg, 0.0231 mmol) and **14** (13 mg, 0.0231 mmol) in CH$_3$CN/H$_2$O (4 mL/ 2 mL). The reaction was allowed to stir in the MW at 100 °C for 10 min × 2. The solvent was removed in vacuo. The residue was dissolved in DCM (30 mL), washed with brine (20 mL × 3), and dried (MgSO$_4$). The mixture was filtered and the solvent was removed in vacuo to yield the crude product, which was purified by silica gel column chromatography (DCM:MeOH 98:2–95:5) to give the pure product **13** as a yellow sticky solid (77 mg, 58%). R_f = 0.48 (DCM:MeOH 9:1). $[\alpha]_D^{24}$-7 (c 1, DCM). ^{1}H NMR (500 MHz, CDCl$_3$) δ 10.08 (s, 4H, NHCOCH$_2$N$_3$), 8.56–7.59 (m, 36H, NHCH$_2$CCH × 8, CH$_2$CCH × 16 and Ar-H × 12), 5.91 (s, 8H, H-1), 5.63 (t, J = 10.2 Hz, 8H, H-2), 5.53 (s, 8H, H-4), 5.28 (d, J = 10.2 Hz, 16H, H-3 and CH$_2$ × 4), 4.80–4.39 (m, 6H, CH_2-triaz, CH$_{2'}$s), 4.38–4.05 (m, H, H-5 and H-6, CH$_{2'}$s), 3.86–3.31 (m, 13H, CH$_{2'}$s), 3.11–2.31 (m, 4H, CH$_{2'}$s), 2.18 (s, 7H, OAc and NAc), 1.99 (s, 15H, OAc), 1.79 (s, 6H, OAc), 1.43 (s, 3H, *tert*-butyl). ^{13}C NMR (125 MHz, CDCl$_3$) δ 170.59, 170.35, 170.09, 122.1, 121.6, 120.7, 86.3 (C-1), 74.1 (C-5), 71.1 (C-3), 71.0 (CH$_2$), 70.7 (CH$_2$), 70.6 (CH$_2$), 70.5 (CH$_2$), 69.8 (CH$_2$), 69.3 (CH$_2$), 68.2 (C-2), 67.1 (C-4), 64.7 (CH$_2$), 61.3 (C-6), 52.9 (CH$_2$), 50.3 (CH$_2$), 44.2 (CH$_2$), 43.5 (CH$_2$), 43.2 (CH$_2$), 42.8 (CH$_2$), 40.0 (CH$_2$), 39.8 (CH$_2$), 38.7 (CH$_2$), 35.5 (CH$_2$-triaz), 34.3 (CH$_2$), 33.8 (CH$_2$), 32.1 (CH$_2$), 29.89, 28.3

(C(CH$_3$)), 21.7 (CH$_3$ of NHAc), 20.84 (CH$_3$ of OAc), 20.72 (CH$_3$ of OAc), 20.45 (CH$_3$ of OAc), 1.21. IR (film on NaCl): 3392, 2927, 1753, 1647, 1536, 1448, 1370, 1223, 1092, 1060, 923, 732 cm^{-1}. MALDI-TOF-MS [M+H]$^+$: *m/z* calculated for C$_{242}$H$_{309}$N$_{64}$O$_{102}$ +H$^+$: 5744.104, found 5744.346.

Synthesis of Tetravalent β-Peptoid Glycocluster (**3**)

Compound **13** (70 mg, 0.0122 mmol) was dissolved in methanol/H$_2$O (4 mL, 2 mL). NEt$_3$ (0.1 mL) was added, and the reaction mixture was allowed to stir at 45 °C for 6 h. The solution was cooled, Amberlite H$^+$ was added and the mixture was allowed to stir for 30 min. The solution was filtered, and the solvent was removed in vacuo. Excess NEt$_3$ was removed using the Schlenk line. The product was freeze-dried overnight to yield the pure product **3** as a white fluffy solid (44 mg, 82%). ^{1}H NMR (500 MHz, D$_2$O) δ 8.62–7.72 (m, 28H, Ar-H and triaz-H), 5.69 (s, 10H, H-1 and CH$_2$s), 5.47 (s, 6H, CH$_2$s), 4.74–4.44 (m, 24H, CH$_2$-triaz and CH$_2$s), 4.22 (s, 14H, H-2 and CH$_2$s), 4.11 (d, *J* = 20.3 Hz, 14H, H-4 and CH$_2$s), 3.99 (s, 8H, H-5), 3.88 (d, *J* = 9.8 Hz, 14H, H-3 and CH$_2$s), 3.84–3.36 (m, 60H, H-6, H-6$'$ and CH$_2$s), 2.99–2.31 (m, 24H, CH$_2$s), 1.99–1.86 (m, 3H CH$_3$). ^{13}C NMR (125 MHz, DMSO) δ 171.1, 170.9, 166.1, 166.0, 165.1, 162.8, 145.5, 144.3, 139.1, 135.6, 126.2, 124.0, 122.3, 121.7, 88.5, 78.9, 74.2, 70.2, 70.1, 70.0, 69.8, 69.5, 69.4, 69.1, 68.9, 63.9, 60.9, 60.7, 52.6, 49.9, 49.8; 45.9, 40.5, 40.4, 40.2, 40.0, 39.9, 39.7, 39.5, 36.3, 35.4, 31.3, 28.2, 9.1. IR (ATR): 3301, 2925, 1650, 1540, 1443, 1388, 1253, 1092, 1055, 891 cm^{-1}. MALDI-TOF-MS [M+H]$^+$: *m/z* calculated for C$_{242}$H$_{309}$N$_{64}$O$_{102}$ +Na$^+$: 4422.7465, found 4422.807.

Synthesis of 2,6,10,14-tetraoxo-3,7,11,15-tetrakis((1-(2,3,4,6-tetra-*O*-acetyl-β-D-galactopyranosyl -1*H*-1,2,3-triazol-4-yl)methyl)-3,7,11,15-tetraazaoctadecan-18-oic acid (**14**)

Copper sulphate pentahydrate (40 mg) and sodium ascorbate (80 mg) were added to a solution of per-acetylated-1-β-azido galactose [28] (567 mg, 1.520 mmol) and **2** [25] (200 mg, 0.362 mmol) in CH$_3$CN/H$_2$O (4 mL/2 mL). The reaction was stirred in the MW at 100 °C for 20 min (10 min × 2). The solvent was removed in vacuo. The residue was dissolved in DCM (30 mL), washed with brine (20 mL × 3), and dried (MgSO$_4$). The mixture was filtered and the solvent was removed in vacuo to yield the crude product, which was purified by silica gel column chromatography (DCM:MeOH 98:2–95:5) to give the pure product **14** as a yellow sticky solid (503 mg, 68%). R$_f$ = 0.47 (DCM:MeOH 9:1). [α]$_D^{27}$-4.71 (c 0.85, DCM). ^{1}H NMR (500 MHz, CDCl$_3$) δ 7.98–7.77 (m, 4H, triaz-H), 5.98–5.74 (m, 4H, H-1), 5.59–5.42 (m, 8H, H-2 and H-4), 5.37–5.18 (m, 4H, H-3), 4.81–4.49 (m, 8H, CH$_2$ × 4), 4.35–4.08 (m, 12H, H-5, H-6 and H-6$'$), 3.83–3.53 (m, 8H, CH$_2$ × 4), 3.02–2.68 (m, 6H, CH$_2$ × 3), 2.65–2.47 (m, 2H, CH$_2$), 2.27–2.21 (m, 12H, OAc), 2.20–2.15 (m, 3H, NAc), 2.06–1.98 (m, 24H, OAc), 1.92–1.80 (m, 12H, OAc), 1.47–1.40 (m, 9H, C(CH$_3$)$_3$). ^{13}C NMR (125 MHz, CDCl$_3$) δ 171.3, 170.3, 170.1, 169.8, 168.8, 144.8, 144.6, 122.3, 122.2, 86.3, 77.3, 77.0, 76.8, 74.0, 70.8, 68.0, 66.8, 61.1, 45.2, 44.2, 40.8, 34.6, 31.9, 31.8, 28.1, 22.0, 21.5, 20.7, 20.6, 20.5, 20.2. IR (ATR): 2938, 1746, 1638, 1423, 1368, 1211, 1157, 1091, 1044, 952, 922, 841, 733 cm^{-1}. HRMS (ESI+): *m/z* calculated for C$_{86}$H$_{116}$N$_{16}$O$_{42}$ + Na$^+$ [M+Na$^+$]: 2067.7331, found 2067.7223.

Synthesis of 2,6,10,14-tetraoxo-3,7,11,15-tetrakis((1-(β-D-galactopyranosyl-1*H*-1,2,3-triazol-4-yl)methyl)-3,7,11,15-tetraazaoctadecan-18-oic acid (**15**)

Compound **14** (452 mg, 0.221 mmol) was dissolved in methanol/H$_2$O (4 mL, 2 mL). NEt$_3$ (0.1 mL) was added, and the reaction mixture was allowed to stir at 45 °C for 6 h. The solution was cooled, Amberlite H$^+$ was added and the mixture was allowed to stir for 30 min. The solution was filtered, and the solvent was removed in vacuo. Excess NEt$_3$ was removed using the Schlenk line. The product was freeze-dried overnight to yield the pure product **15** as a white fluffy solid (267 mg, 92%). [α]$_D^{24}$+14.0 (c 0.5, H$_2$O). ^{1}H NMR (500 MHz, D$_2$O) δ 8.23–8.16 (m, 2H), 8.08 (s, 2H), 5.64–5.54 (m, 4H, H-1), 4.65–4.42 (m, 8H, CH$_2$-triaz and CH$_2$), 4.19–4.06 (m, 4H, H-2), 4.01 (appdd, *J* = 5.2, 3.2 Hz, 4H, H-4), 3.97–3.86 (m, 4H, H-5), 3.85–3.76 (m, 4H, H-3), 3.75–3.61 (m, 12H, H-6, H-6$'$ and CH$_2$ × 2), 3.58–3.51 (m, 4H), 2.80–2.46 (m, 8H, CH$_2$ × 4), 2.13–2.04 (m, 1H), 1.16 (s, 3H, CH$_3$). ^{13}C

NMR (126 MHz, DMSO) δ 173.1, 170.5, 170.3, 144.6, 122.5, 122.3, 88.6, 88.5, 78.9, 75.6, 74.2, 69.9, 68.9, 60.9, 58.6, 45.7, 44.7, 43.8, 43.0, 42.9, 39.9, 22.1, 21.7, 9.0. IR (film on NaCl): 3299, 2916, 1719, 1620, 1451, 1421, 1365, 1224, 1087, 1053, 885 cm^{-1}. HRMS (ESI+): *m/z* calculated for $C_{50}H_{76}N_{16}O_{26} + Na^+$ [M+Na$^+$]: 1339.5014, found 1339.5032.

3.2. Biological Evaluation

Fungal Strain: C. albicans (clinical isolate from a corneal infection) was maintained on sabouraud dextrose agar and cultures were grow to the stationary phase (1–2×10^8 cells/mL) overnight in YEPD broth (1% (*w/v*) yeast extract, 2% (*w/v*) bacteriological peptone, 2% (*w/v*) glucose) at 30 °C and 200 rpm. Stationary phase yeast cells were harvested, washed with PBS and resuspended at a density of 1×10^8 cells/mL in PBS.

Buccal epithelial cells: Buccal epithelial cells (BECs) were harvested from healthy volunteers by gently scraping the inside of the cheek with a sterile tongue depressor. Cells were washed in PBS and resuspended at a density of 5×10^5 cells/mL.

Adherence assays: Yeast cells were mixed with BECs in a ratio of 50:1 in a final volume of 2 mL and incubated at 30 °C and 200 rpm for 90 min. The BEC/yeast cell mixture was harvested by passing through a polycarbonate membrane containing 30 µm pores which trapped the BECs but allowed unattached yeast cells to pass through. This was washed ×2 with 10 mL PBS and cells remaining on the membrane were collected and placed on glass slides which were left to air-dry overnight. The cells were heat fixed and stained using 0.5% (*w/v*) crystal violet, rinsed using cold water to remove any surplus stain and left to air-dry for 30 min. The number of *C. albicans* cells adhering to a sample of 200 BECs per treatment was assessed microscopically. In the *exclusion assay*, the yeast cells or BECs were incubated for 90 min in the presence of each compound at the given concentration. After this time, the cells were harvested and washed twice with PBS before being resuspended in 1 mL PBS before being mixed with BECs or yeast cells (as described). In the *competition assay* format yeast cells, BECs and compound (at the given concentration) were co-incubated for 90 min prior to harvesting.

Statistics: All experiments were performed on three independent occasions. In each assay, the number of yeast cells adhering to 200 randomly chosen BECs was determined. Results are mean ± SEM.

4. Conclusions

In summary, we report the synthesis of a multivalent display of divalent galactoside **1**, a potent inhibitor of the adhesion of *C. albicans* to BECSs. Glycoconjugate **3** was built upon a linear peptoid scaffold, to access distinct spatial presentations of the recognition motifs provided by lead compound **1**. This compound, together with control compound **15**, were successfully synthesised using a series of sequential CuAAC reactions. The ability of these compounds to inhibit adhesion of *C. albicans* to BECs was then evaluated. Although multivalent presentations of carbohydrate epitopes are commonly used to increase their biological activity, we did not observe this effect for the compounds studied herein. In fact, the exclusion assays indicated that the multivalent glycoconjugates did not bind effectively to *C. albicans* cell wall, in contrast to their parent compound **1**. Since the fungal target for lead compound **1** is not known, the results highlight the importance of detailed structural knowledge of the target proteins when considering the design of multivalent presentation. Interestingly, the inhibition of *C. albicans* adhesion observed for the compounds in this study is then likely due to interactions with BECs. Further studies are currently underway in our laboratory to establish if these are specific interactions and to optimize structural parameters, as this could extend the scope of activity of these compounds to other pathogens in the oral cavity.

Supplementary Materials: The following are available online at https://www.mdpi.com/article/10.3390/pathogens10050572/s1; it includes spectroscopic data for the synthetic compounds described herein.

Author Contributions: Conceptualization, K.K. and T.V.-T.; methodology, H.M. (Harlei Martin) and H.M. (Hannah Masterson); formal analysis, H.M. (Harlei Martin); investigation, H.M. (Harlei Martin); resources, K.K. and T.V.-T.; data curation, all authors; writing—original draft preparation, H.M. (Harlei Martin); writing—review and editing, H.M. (Harlei Martin), K.K. and T.V.-T.; supervision, K.K. and T.V.-T. All authors have read and agreed to the published version of the manuscript.

Funding: We would like to acknowledge Maynooth University for the provision of the John and Pat Hume Scholarship to H. Martin.

Institutional Review Board Statement: All experiments were performed in accordance with the Guidelines stipulated in Directive 2004/23/EC of the European Parliament and of the Council (31 March 2004). Experiments were conducted according to the guidelines by the Ethics Committee at Maynooth University.

Informed Consent Statement: Informed consent was not necessary for this study.

Data Availability Statement: The data presented in this study are available within the article and in the supplementary material.

Acknowledgments: We would like to acknowledge Maynooth University for the provision of the John and Pat Hume Scholarship to H. Martin.

Conflicts of Interest: The authors declare no conflict of interest.

References

1.   Varki, A. Biological roles of glycans. *Glycobiology* **2012**, *27*, 3–49. [CrossRef]
2.   Cozens, D.; Read, R.C. Anti-adhesion methods as novel therapeutics for bacterial infections. *Expert Rev. Anti-Infect. Ther.* **2012**, *10*, 1457–1468. [CrossRef]
3.   Linman, M.J.; Taylor, J.D.; Yu, H.; Chen, X.; Cheng, Q. Surface Plasmon Resonance Study of Protein−Carbohydrate Interactions Using Biotinylated Sialosides. *Anal. Chem.* **2008**, *80*, 4007–4013. [CrossRef] [PubMed]
4.   Dimick, S.M.; Powell, S.C.; McMahon, S.A.; Moothoo, D.N.; Naismith, J.H.; Toone, E.J. On the Meaning of Affinity: Cluster Glycoside Effects and Concanavalin, A.J. *Am. Chem. Soc.* **1999**, *121*, 10286–10296. [CrossRef]
5.   Lee, Y.C.; Lee, R.T. Carbohydrate-Protein Interactions: Basis of Glycobiology. *Acc. Chem. Res.* **1995**, *28*, 321–327. [CrossRef]
6.   Boukerb, A.M.; Rousset, A.; Galanos, N.; Méar, J.-B.; Thépaut, M.; Grandjean, T.; Gillon, E.; Cecioni, S.; Abderrahmen, C.; Faure, K.; et al. Antiadhesive Properties of Glycoclusters against *Pseudomonas aeruginosa* Lung Infection. *J. Med. Chem.* **2014**, *57*, 10275–10289. [CrossRef] [PubMed]
7.   Chabre, Y.M.; Roy, R. Multivalent glycoconjugate syntheses and applications using aromatic scaffolds. *Chem. Soc. Rev.* **2013**, *42*, 4657–4708. [CrossRef]
8.   Martínez, Á.; Ortiz Mellet, C.; García Fernández, J.M. Cyclodextrin-based multivalent glycodisplays: Covalent and supramolecular conjugates to assess carbohydrate–protein interactions. *Chem. Soc. Rev.* **2013**, *42*, 4746–4773. [CrossRef]
9.   Galan, M.C.; Dumy, P.; Renaudet, O. Multivalent glyco(cyclo)peptides. *Chem. Soc. Rev.* **2013**, *42*, 4599–4612. [CrossRef]
10. Miura, Y. Design and synthesis of well-defined glycopolymers for the control of biological functionalities. *Polym. J.* **2012**, *44*, 679–689. [CrossRef]
11. Yilmaz, G.; Becer, C.R. Glyconanoparticles and their interactions with lectins. *Polym. Chem.* **2015**, *6*, 5503–5514. [CrossRef]
12. Hill, S.; Galan, M.C. Fluorescent carbon dots from mono- and polysaccharides: Synthesis, properties and applications. *Beilstein J. Org. Chem.* **2017**, *13*, 675–693. [CrossRef] [PubMed]
13. Bernardi, A.; Jimenez-Barbero, J.; Casnati, A.; De Castro, C.; Darbre, T.; Fieschi, F.; Finne, J.; Funken, H.; Jaeger, K.-E.; Lahmann, M.; et al. Multivalent glycoconjugates as anti-pathogenic agents. *Chem. Soc. Rev.* **2013**, *42*, 4709–4727. [CrossRef]
14. Pieters, R.J. Maximising multivalency effects in protein–carbohydrate interactions. *Org. Biomol. Chem.* **2009**, *7*, 2013–2025. [CrossRef]
15. Pifferi, C.; Goyard, D.; Gillon, E.; Imberty, A.; Renaudet, O. Synthesis of Mannosylated Glycodendrimers and Evaluation against BC2L-A Lectin from *Burkholderia cenocepacia*. *ChemPlusChem* **2017**, *82*, 390–398. [CrossRef] [PubMed]
16. Arsiwala, A.; Castro, A.; Frey, S.; Stathos, M.; Kane, R.S. Designing Multivalent Ligands to Control Biological Interactions: From Vaccines and Cellular Effectors to Targeted Drug Delivery. *Chem. Asian J.* **2019**, *14*, 244–255. [CrossRef]
17. Brissonnet, Y.; Araoz, R.; Sousa, R.; Percevault, L.; Brument, S.; Deniaud, D.; Servant, D.; Le Questel, J.Y.; Lebreton, J.; Gouin, S.G. Di- and heptavalent nicotinic analogues to interfere with α7 nicotinic acetylcholine receptors. *Bioorg. Med. Chem.* **2019**, *27*, 700–707. [CrossRef]
18. González-Cuesta, M.; Ortiz-Mellet, C.; García Fernández, J.M. Carbohydrate supramolecular chemistry: Beyond the multivalent effect. *Chem. Commun.* **2020**, *56*, 5207–5222. [CrossRef] [PubMed]
19. Chalopin, T.; Brissonnet, Y.; Sivignon, A.; Deniaud, D.; Cremet, L.; Barnich, N.; Bouckaert, J.; Gouin, S. Inhibition profiles of mono- and polyvalent FimH antagonists against 10 different *Escherichia coli* strains. *Org. Biomol. Chem.* **2015**, *13*, 11369–11375. [CrossRef]

20. Lehot, V.; Brissonnet, Y.; Dussouy, C.; Brument, S.; Cabanettes, A.; Gillon, E.; Deniaud, D.; Varrot, A.; Pape, P.L. Multivalent fucosides with nanomolar affinity for the *Aspergillus fumigatus* lectin FleA prevent spore adhesion to Pneumocytes. *Chem. Eur. J.* **2018**, *24*, 19243–19249. [CrossRef]

21. Revie, N.M.; Iyer, K.R.; Robbins, R.; Cowen, L.E. Antifungal drug resistance: Evolution, mechanisms and impact. *Curr. Opin. Microbiol.* **2018**, *45*, 70–76. [CrossRef]

22. Ciurea, C.N.; Kosovski, I.B.; Mare, A.D.; Toma, F.; Pintea-Simon, I.A.; Man, A. *Candida* and Candidiasis-Opportunism Versus Pathogenicity: A Review of the Virulence Traits. *Microorganisms* **2020**, *8*, 857. [CrossRef] [PubMed]

23. Martin, H.; Govern, M.M.; Abbey, L.; Gilroy, A.; Mullins, S.; Howell, S.; Kavanagh, K.; Velasco-Torrijos, T. Inhibition of adherence of the yeast *Candida albicans* to buccal epithelial cells by synthetic aromatic glycoconjugates. *Eur. J. Med. Chem.* **2018**, *160*, 82–93. [CrossRef] [PubMed]

24. Critchley, I.A.; Douglas, L.J. Role of glycosides as epithelial cell receptors for Candida albicans. *J. Gen. Microbiol.* **1987**, *133*, 637–643. [CrossRef]

25. Tosh, F.D.; Douglas, L.J. Characterization of a fucoside-binding adhesin of Candida albicans. *Infect. Immun.* **1992**, *60*, 4734–4739. [CrossRef]

26. Jimenez-Lucho, V.; Ginsburg, V.; Krivan, H.C. *Cryptococcus neoformans, Candida albicans,* and other fungi bind specifically to the glycosphingolipid lactosylceramide (Galβ1-4Glcβ3-lCer), a possible adhesion receptor for yeasts. *Infect. Immun.* **1990**, *58*, 2085–2090. [CrossRef]

27. Yu, L.; Lee, K.K.; Hodges, R.S.; Paranchych, W.; Irvin, R.T. Adherence of *Pseudomonas aeruginosa* and *Candida albicans* to Glycosphingolipid (Asialo-GM1) Receptors is Achieved by a Conserved Receptor-Binding Domain Present on Their Adhesins. *Infect. Immun.* **1994**, *62*, 5213–5219. [CrossRef] [PubMed]

28. Martin, H.; Kavanagh, K.; Velasco-Torrijos, T. Targeting adhesion in fungal pathogen *Candida albicans. Future Med. Chem.* **2021**, *13*, 313–334. [CrossRef]

29. Martin, H.; Goyard, D.; Margalit, A.; Doherty, K.; Renaudet, O.; Kavanagh, K.; Velasco-Torrijos, T. Multivalent Presentations of Glycomimetic Inhibitor of the Adhesion of Fungal Pathogen Candida albicans to Human Buccal Epithelial Cells. *Bioconjug. Chem.* **2021**, ahead of print. [CrossRef]

30. Roy, O.; Faure, S.; Thery, V.; Didierjean, C.; Taillefumier, C. Cyclic β-Peptoids. *Org. Lett.* **2008**, *10*, 921–924. [CrossRef]

31. Cecioni, S.; Faure, S.; Darbost, U.; Bonnamour, I.; Parrot-Lopez, H.; Roy, O.; Taillefumier, C.; Wimmerová, M.; Praly, J.P.; Imberty, A.; et al. Selectivity among Two Lectins: Probing the Effect of Topology, Multivalency and Flexibility of "Clicked" Multivalent Glycoclusters. *Chem. Eur. J.* **2011**, *17*, 2146–2159. [CrossRef]

32. Szekely, T.; Roy, O.; Dériaud, E.; Job, A.; Lo-Man, R.; Leclerc, C.; Taillefumier, C. Design, Synthesis, and Immunological Evaluation of a Multicomponent Construct Based on a Glycotripeptoid Core Comprising B and T Cell Epitopes and a Toll-like Receptor 7 Agonist That Elicits Potent Immune Responses. *J. Med. Chem.* **2018**, *61*, 9568–9582. [CrossRef] [PubMed]

33. Tropper, F.D.; Andersson, F.O.; Braun, S.; Roy, R. Phase Transfer Catalysis as a General and Stereoselective Entry into Glycosyl Azides from Glycosyl Halides. *Synthesis* **1992**, *1992*, 618–620. [CrossRef]

34. Lu, G.; Lam, S.; Burgess, K. An iterative route to "decorated" ethylene glycol-based linkers. *Chem. Comm.* **2006**, *15*, 1652–1654. [CrossRef] [PubMed]

35. Ng, D.Y.W.; Fahrer, J.; Wu, Y.; Eisele, K.; Kuan, S.L.; Barth, H.; Weil, T. Efficient Delivery of p53 and Cytochrome C by Supramolecular Assembly of a Dendritic Multi-Domain Delivery System. *Adv. Healthc. Mater.* **2013**, *2*, 1620–1629. [CrossRef] [PubMed]

36. Norberg, O.; Deng, L.; Yan, M.; Ramström, O. Photo-Click Immobilization of Carbohydrates on Polymeric Surfaces—A Quick Method to Functionalize Surfaces for Biomolecular Recognition Studies. *Bioconjug. Chem.* **2009**, *20*, 2364–2370. [CrossRef] [PubMed]

MDPI

St. Alban-Anlage 66

4052 Basel

Switzerland

Tel. +41 61 683 77 34

Fax +41 61 302 89 18

www.mdpi.com

Pathogens Editorial Office

E-mail: pathogens@mdpi.com

www.mdpi.com/journal/pathogens